T0254211

Economic Policy of the People's Republic of China

Barbara Darimont
Editor

Economic Policy of the People's Republic of China

Springer Gabler

Editor
Barbara Darimont
East Asia Institute
Ludwigshafen University of Business and Society
Ludwigshafen am Rhein, Germany

ISBN 978-3-658-38466-1 ISBN 978-3-658-38467-8 (eBook)
https://doi.org/10.1007/978-3-658-38467-8

This book is a translation of the original German edition „Wirtschaftspolitik der Volksrepublik China“ by Darimont, Barbara, published by Springer Fachmedien Wiesbaden GmbH in 2020. The translation was done with the help of artificial intelligence (machine translation by the service DeepL.com). A subsequent human revision was done primarily in terms of content, so that the book will read stylistically differently from a conventional translation. Springer Nature works continuously to further the development of tools for the production of books and on the related technologies to support the authors.

This Springer Gabler imprint is published by the registered company Springer Fachmedien Wiesbaden GmbH, part of Springer Nature.
The registered company address is: Abraham-Lincoln-Str. 46, 65189 Wiesbaden, Germany

The original version of the book has been revised. A correction to this book can be found at https://doi.org/10.1007/978-3-658-38467-8_17

Foreword

The economic policy of the People's Republic of China is an omnipresent topic. The German economy is closely intertwined with the Chinese economy so that it is necessary to deal with this matter. In the context of my lectures at the East Asia Institute of the Ludwigshafen University of Applied Sciences, it turns out time and again that the topicality of issues is usually only reflected in newspaper articles and the information content is limited due to the focus on current events. There is a very good journalism in Germany concerning the PRC, but many topics need a certain distance and above all the question arose in the lectures on how individual aspects are to be classified in the overall context. On the other hand, there is very good academic literature on individual topics, but this needs to be researched. As a single academic, it is hardly possible for me to cover all of China's economic policy issues. The academic mid-level staff ground in a university is limited or non-existent, so the idea came up to write a book in the final semester of 2019 of the China Department of the East Asia Institute.

The book provides a comprehensive overview of Chinese economic policy and highlights current issues in their entirety. In this way, the high diversity of China should become clear. The PRC is and remains a country of extremes. The book is aimed at a broad audience, such as companies and laypersons, but also those interested in economic policy.

The production of the book would not have been possible without the contributions of the students of the East Asia Institute of the Ludwigshafen University of Applied Sciences. I am especially grateful to the students who are not listed as authors; many of their thoughts or ideas have been incorporated into the book in one way or another. Research is an exploration of the unknown, and therefore, it is not always possible to predict whether a topic will yield a gain in knowledge. Students have always motivated and stimulated me, so I consider my work as a lecturer at a university a privilege. Therefore, I would at this point like to thank the East Asia Institute and the University of Applied Sciences as a whole for making such a project possible.

I would especially like to thank Louis Margraf, who helped me as a research assistant throughout the summer of 2019. Countless times he went through the manuscripts and made corrections as well as suggestions. Without his collaboration, the book would not have been written.

Finally, I would like to thank all the other people involved who provided suggestions without me being able to list them all individually here. I would like to explicitly mention the collaboration of Patricia Fuchs, Jan-Sören Garnost, Franziska Kunter, Noah Kurz, Mike Laub, Benedict Pham, Fabian Schmidt, Leonardo Tolve, and Konstantin Krone.

Munich, Germany
Fall 2019

Contents

Abbreviations

ACFIC	National All-China Federation for Industry and Commerce
BAT	Baidu, Alibaba and Tencent
GDP	Gross domestic product
BNSC campaign	Building a New Socialist Countryside
B2B	Business-to-Business
B2C	Business-to-Customer
C2C	Customer-to-Customer
CGN	China General Nuclear Power Group
CNNC	China National Nuclear Corporation
ha	Hectare
IAEA	International Atomic Energy Agency
INDGs	Intended Nationally Determined Contributions
kg	Kilogram
SME	Small and medium-sized enterprises
CCP	Communist Party of China
MEP	Ministry of the Environment
MEE	Ministry of Ecology and Environment
billion	Billion
m.	Million
MOFCOM	Ministry of Foreign Commerce
MOHURD	Ministry of Housing and Urban Development
MURCEP	Ministry of Rural–Urban Construction and Environmental Protection
NABU	Nature and Biodiversity Conservation Union Germany
NEA	National Energy Administration
PRC	People's Republic of China
RMB	Renminbi (Chinese currency unit)
SAC	Securities Association of China
SAIC	State Administration for Industry and Commerce
SASAC	State-Owned Assets Supervision and Administration Commission
SAUA	Sino-Australia Uranium Agreement

SEPA	State Environmental Protection Agency
SNERDI	Shanghai Nuclear Engineering Research and Design Institute
SNWDP	South-North Water Transfer Project
SOE	State-owned enterprise
t	Tons
TPOP	Third-Party Online Payment
VR	People's Republic of
WHO	World Health Organization
WTO	World Trade Organization

1 Introduction

Barbara Darimont

The People's Republic of China (PRC) is trying to replace the United States of America as a world power. This is a potential power shift that is taking place on a global, not a national, scale. It is a process that will take several years – if not decades – and will lead to eruptions in the world economy. In democracies, this transfer of power takes place through elections, and thus usually peacefully. This is a great achievement of democracy. The outcome of this potential transition of power is unclear; it is to be hoped that it will take place peacefully. In the context of economic policy discussions, democracy and its liberal economic concept are sometimes seen as inferior to the "China model" with an autocratic system. In Germany, voices are being raised calling for state intervention to protect against China as an economic power. If it is now assumed that the PR China could become the future world power, then it is indispensable to deal with the Chinese economy and economic policy. In some cases, the accusation is made that there is too little interest in the PRC in Germany (Deuber 2019; Yang 2019). Although the German Ministry of Education and Research has already launched a China Competence Initiative to impart broader knowledge and skills about China among the population (Bundesministerium für Bildung und Forschung 2018), the reasons for the possible lack of interest have not been clarified, so that no measures can yet be taken.

Why is Germany so little interested in the PRC? The PRC is one of the most important – if not the most important – trading partners of Germany. Germany's prosperity depends not least on China as an economic power. One explanation would be that Chinese culture is a shame culture and German culture is a guilt culture (Benedict 2014/1946). This

B. Darimont (✉)
East Asia Institute of Ludwigshafen University of Business and Society,
Ludwigshafen am Rhein, Germany
e-mail: darimont@oai.de

B. Darimont (ed.), *Economic Policy of the People's Republic of China*,
https://doi.org/10.1007/978-3-658-38467-8_1

cultural difference may lead to misunderstandings. In a shame culture, problems are not directly addressed, many things are glossed over. The backdrop and the facade are important. The "loss of face", the exposure is to be avoided at all costs. Provocation is interpreted as loss of face and is generally seen as the worst possible stylistic device in the PRC. American President Trump engages in this to some extent to excess. In a shame culture, this is about the worst thing that can happen to someone – an entire nation is being paraded here. Government officials could hardly act more antithetically: On the one hand, Trump, who shares everything possible with the outside world, whether it wants to hear it or not. On the other, Xi, who would never show an expression of emotion. Conversely, in a blame culture, self-expression often has a negative connotation. Germany, the country of complainers, which often talks itself down, is a very good example of this (von Borstel 2012). In Germany, the PRC is often portrayed badly, which is referred to as "China bashing" (Wagner 2016). Since Germans like to portray themselves badly, they also look for the downside in other nations and tend to report negatively. Conversely, in the shame culture, the bad portrayal of China in the German media is incomprehensible – an affront. A particularly negative portrayal of Germany in the Chinese media would be an affront to Germans from the Chinese perspective. Curiosities perceived from the Chinese side are excluded here. Under this assumption, mutual misunderstanding is pre-programmed.

Writing about the PRC is a tightrope walk. Writing negatively about the People's Republic could amount to "China bashing". If the situation is portrayed too positively, the accusation of "anticipatory obedience" with regard to the Chinese Communist Party (CCP) is quickly raised. Against this background, an attempt will be made to present an objective account of the Chinese economy. In addition, the Chinese perspective will be taken into account to provide a better understanding. Within China the opinion is not uniform in many areas, this is not always apparent due to the restrictive censorship. Criticism is very often expressed in ironic ways that are difficult for a non-native speaker to access, such as parodies on social media that use puns and whose biting mockery brings a blush to one's face. The book is intended to help people understand the Chinese economy in Germany. Last but not least, it is an attempt to see China's enormous development potential amidst the global political hubbub, because without China, global politics will hardly be possible in the future.

1.1 Aim of the Book

Writing an overview of China's economy is fraught with the risk that the situation will have become outdated by the time it is published. In addition, much is now written about the PRC, so the reason for a large summary is not necessarily obvious. For students dealing with the Chinese economy, however, the many individual pieces of information from newspapers and institutions are difficult to categorize, especially at the beginning of their studies. One or the other businessman, entrepreneur or interested person may feel the same way.

The book's design makes it suitable for both practitioners and students. Many chapters are structured in such a way that, after a general introduction, specific topics are addressed that are particularly topical. One focus is on environmental policy issues, which students in particular repeatedly raise in lectures. This is a legitimate concern, since the PRC is the country with the greatest economic growth and, at the same time, major environmental problems. The limits of growth are being tested in the PRC more than in any other country. How to limit growth will have to be decided by Chinese politicians, this can serve as a model for other countries. With its one-child policy, China already tried to introduce a limit to growth in the late 1970s. The consequences of the demographic factor in China can hardly be foreseen. It is possible that the future of the world will be decided in China, whether in terms of the seemingly limitless economic growth to date, or in terms of the limits of natural resources, or even in terms of the ageing of a society in which there seem to be only limited resources for care and provision. All the more reason to take a closer look at the country; the individual contributions are intended to serve as a basis for discussion. The aim of the book, however, is not to present the latest research, but to provide as broad an overview as possible of the Chinese economy.

The book does not claim to cover all the issues that are virulent in the PRC economy. This will hardly be possible because of the enormous speed of Chinese economic development. Therefore, the contributions are snapshots. The consideration is to develop more snapshots in the future. Does it then make sense at all to publish an entire work for this purpose? The overall impression is what counts. If the many different strands of development are taken together, a partly different picture emerges than if only individual aspects are illuminated. When dealing with individual topics, it became clear time and again that many things did not correspond to what was assumed in advance. The overall picture shows that the People's Republic of China works with a trial-and-error procedure not only in politics, but also in the economy in particular, where any planning is superfluous. Almost exclusively, what works and what doesn't is tested in the here and now. This generates a speed in overall economic development that is currently an advantage. The Chinese master plan is found in the five-year plans, but the bottom line is that they are also based on trial-and-error. Tracing these developments is a challenge that can only be seen as a process. Presenting the various test fields as an overview should give an impression of the diversity of Chinese economic developments.

1.2 Overview of Economic Literature

For those looking for a systematic overview of the Chinese economy, two works are recommended:

- Arthur Kroeber: China's Economy. What everyone needs to know, 2016
- Barry Naughton, The Chinese Economy. Adaptation and Growth, second edition, 2018

Kroeber gives a tight overview of all economically relevant topics. His introduction already hints at the questions that are currently being debated. When a political leader has to choose between enormous economic growth and maximum political control, what does he choose? According to Kroeber (2016, pp. 23–26), this contradiction is not perceived as a contradiction in the PRC. President Xi Jinping stood for reform at the beginning of his presidency. Many place hope in him that he will implement economic reforms that are seen as urgently needed. To date, however, his reign has been characterized by an increase in the CCP's political power and much tighter control. Economic growth is currently estimated to be below 6% of gross national product, and in some cases even lower (Stevenson 2019). Does the choice between economic growth and absolute political control therefore remain a contradiction in terms in the PRC? But this would abrogate the unspoken contract between the population and the state not to demand political participation when prosperity is assured.

Naughton provides a very comprehensive overview, which is characterized by the recording of the historical background and richness of detail. In the new edition of his book, he devotes special attention to the question of whether the Chinese "economic miracle" is really a miracle or whether it follows the general paths of economic development. The various strands of growth and rural development are explained in detail. The classic economic sectors are presented in a very clear and well-founded manner.

Anyone wishing to keep abreast of current developments should consult the website of the Mercator Institute for Chinese Studies (Merics 2019). Here, many of the current topics are presented and discussed in newsletters or blogs. For Germany, the institute is unique in its dedication to a better understanding of China.

1.3 Principles of the Chinese Economy

Among the most important political principles influencing China's economy is that China is a one-party autocratic state that is centrally run but often decentralized in practice. The principle of the autocratic one-party state manifests itself in the fact that all policy is dictated by the party – the CCP. Although concepts such as the rule of law and democracy are in the Chinese Constitution of 1982 and are frequently mentioned in political pronouncements, these concepts are not identical to the Western understanding. This is particularly clear with the term rule of law, since in the PRC it means that the state uses law to enforce its policies and order. In contrast, in Germany the rule of law means that the state – like the citizen – is bound by law and order. Another peculiarity in Chinese politics is the admiration of authority. Under Mao Zedong, the cult of personality and thus the allegiance to authority were taken to extremes. The subsequent leadership tried to introduce a rule of law instead of a rule of person. Therefore, age limits were introduced for the leading politicians. With the Xi Jinping era, the relationship seems to be reversed again (Kroeber 2016, pp. 1–3).

The PRC sees itself as a centralized unitary state. However, due to its size, a crackdown by the national government in Beijing is not always possible. Regional economic differences alone require different treatment, so that uniform policy enforcement rarely exists (Kroeber 2016, p. 4). The central government may be of good will, as in the case of environmental policy, but the implementation of environmental regulations proves difficult. Furthermore, provinces in China engage in local protectionism, for example, by imposing some kind of tariff on the import of goods from other provinces. For this reason, the PRC has sometimes been referred to as an informal federalism or de facto federal state, which makes it difficult to implement a unified economic policy (Heilmann 2016, pp. 75–77).

The People's Republic of China sees itself as a communist state. From the outside, or from a Western perspective, the PRC looks like a turbo-capitalist state. In part, the PRC is also referred to as post-communist (Kroeber 2016, p. 6). There is hardly anything that money cannot buy in China. The distribution of income is a case in point. At the beginning of the 1970s, there was hardly any other country on earth where incomes were distributed so equally. Almost everyone had a similarly high income, with the possible exception of the top leadership elite. Currently, the PRC is one of the countries with the largest income gap in the world. These are extremes that are hard to comprehend from a European perspective. This highlights another special feature of the PRC: Hardly any other country has so many extremes. Bringing these extremes, be it education, be it income, be it climatic or economic conditions, under one leadership is a challenge per se. For this endeavor, the PRC has no overarching concept or strategy. Step by step, the political leadership is trying to address the problems. This was the case at least until the mid or late 2000s. With the inauguration of Xi Jinping, an overarching strategy became visible that is much more aggressive in seeing the PRC's role as a global player and world power. However, whether there is indeed an overarching plan to conquer world power or the trial-and-error strategy is being applied to the world is difficult to see.

1.4 Structure of the Book

Developments in the Chinese economy are extremely fast-moving. Some contemporary topics can be found in the daily press diurnally, while other issues fade into the background. In 2019, in addition to the trade conflict between the USA and the PRC, which will certainly accompany the global community for a few more years, the social credit system is discussed, which from 2020 will evaluate Chinese companies and also foreign companies with regard to their conduct, in addition to Chinese and foreign citizens. This is certainly a major intrusion into the private and corporate sphere, but in principle, the effects could only be speculated at the time of writing this book. For this reason, the subject has only been touched upon in this book. During the research, it became apparent that the e-commerce company Alibaba, for example, is much more intertwined with Chinese politics at the local level than is commonly known. This is why it is discussed in more detail than the social credit system.

The aim was to work as much as possible with Chinese literature, because many topics are dealt with in greater detail and more critically there than when they are published by the state. Many aspects only become apparent with the examination of Chinese reality. However, some topics had to be left out due to a lack of or insufficient sources, such as the topic of infrastructure policy. Another complex of topics that is not dealt with is social policy, which is a problem that can hardly be underestimated due to the aging of Chinese society. The PRC will be the first country to have to solve the problems of an industrialized nation regarding the growing proportion of old people in the population. It is likely to be one of the most pressing problems for the Chinese economy. One approach to the solution will be artificial intelligence and digitization, not so much as caring robots, but rather as replacements for missing workers. Other approaches, such as long-term care insurance, are just beginning to be experimented with. It therefore seems premature to take stock, as many activities are currently being initiated in this area.

The basic structure of the book is as follows: The second chapter first explains the difficulties regarding the reliability of Chinese statistics and data. As a consequence, Chinese data from the Bureau of Statistics will be used, as everything else is speculation. It would be difficult to imagine that foreign institutions have more reliable information than the Chinese state. It is true that foreign institutions make their own calculations, but their basis is often unclear.

The third chapter outlines the historical background, because this is still necessary for understanding the Chinese economy. The transition from a planned economy to a market economy has not yet been fully completed. Many economic policy instruments of central administration and market economy are applied in parallel side by side. This leads to contradictions that are often difficult for Western economists to understand.

The fourth chapter deals with the structure of the state and the current leadership. The power structures of the Chinese ruling elite are still difficult to understand, not only for foreigners but also for Chinese citizens. Communist China is led by a party elite consisting of about 200,000 people. A few hundred are the actual decision-makers and it is to be presumed that they are defending their possessions. This would suggest that a moneyed aristocracy rules China, thus communist China and the capitalist United States may be closer in their political power structures than they appear (Brown 2018).

The fifth chapter shows the current (as of late 2019) economic plans and development. A controversial discussion among Chinese economists is emerging. Does a so-called China model exist in which an authoritarian state guarantees economic growth? Or does

China's economic development take the path of most developing states? If so, the gigantic economic growth would be the result of a catch-up process that should soon be completed. The discussion presented here shows how divergent opinion is in China away from the party line. Conversely, this also applies to foreign observers: there are voices that assume that the China model is superior to the open market economy or at least poses a challenge. Others deny the existence of the model.

In this context, the role of state-owned enterprises is important, as the sixth chapter shows. They continue to serve as the backbone of the Chinese economy. Although until a few years ago the private sector was the driving force behind innovation, economic growth and the creation of new jobs, which is why the private sector was encouraged, there has been a rethink in about 2016. State-owned enterprises are again being subsidized and promoted to a greater extent. We can only speculate about the reasons, but at least the Chinese government can control state-owned enterprises more strictly and thus steer the economy more strongly in times of crisis.

Nevertheless, the private sector remains relevant and is discussed in chapter seven. In recent years, private entrepreneurs have created highly innovative companies that operate worldwide. The Chinese state encourages the creation of companies, making the start-up scene one of the most virulent in the world. Cities such as Shenzhen and Hangzhou have their own start-up scene that is second to none in the world. The establishment of companies has been made easier in many areas, the only question is what happens to the companies whose business ideas then fail to work out in implementation. For this reason, insolvency law, which has received little attention to date, is presented. Although insolvency means a loss of face in the Chinese context, the closure or liquidation of unprofitable companies is advantageous for legal security and thus for economic development, in order to be able to better calculate one's own business results.

In the eighth chapter, we first look at consumption in general and then analyze luxury consumption. The global luxury goods market is dominated by Chinese consumers. The phenomenon that communist China is home to the most luxury consumers in the world can only be explained by Chinese circumstances. Through brands, Chinese consumers try to create their own identity, to stand out from the crowd. Confucianism, which is currently being revived as a Chinese cultural asset in the People's Republic of China, promotes this luxury consumption, since according to the current interpretation of Confucianism it is permissible to flaunt one's own achievements – and thus one's own money. The future of international luxury consumption therefore lies in the People's Republic of China.

After the actors in the market, namely the state, companies and consumers, have been presented, the various markets are analysed. The labour market is examined in the ninth chapter, where various contradictions clash, such as a shortage of skilled workers combined with a high unemployment rate of university graduates. After the development of the labour market, the special features and the situation of migrant workers have been presented, the chapter is devoted to the question of whether China has reached or passed the Lewis Turning Point.

The e-commerce market, which is the subject of the tenth chapter, is one of the largest growth markets in the PRC. The Chinese economy has developed exceptionally well in this area over the last 10–20 years. Globally, the Chinese e-commerce market is likely to lead in terms of innovation, development and revenue. This is unlikely to change in the future as the PRC has a gigantic number of users on these platforms. Users, and therefore data, are the new "petroleum" of the economy because they can be converted into knowledge. The platform economy is ubiquitous in China and is primarily served by the rural population, who gain job opportunities as a result. In the countryside, it is evident that the influence of e-commerce companies is immense as a result. In this context, the state has almost only a controlling function. For the foreign observer the development is exciting, the PR China will generate and determine the progress in this area.

Financial policy is presented in the eleventh chapter, and then the topics of FinTech and crowdfunding are addressed. The Chinese financial system is state-run and therefore there are no private banks or competition between banks. Public debt has been a hot topic for the past decade or so, and few other countries have borrowed from private investors at the pace that the PRC has in recent years. Whether it will be possible to continue on this course remains questionable. Bank loans are only granted through state-owned banks, making it difficult for private companies in particular to obtain credit. In addition, investors receive only very low interest rates at the bank, so that various forms of lending have developed. On the one hand, there is a shadow banking sector where these needs can be met. But crowdfunding is also an attractive option for both borrowers and investors. In this area, government regulations have become increasingly restrictive over the past 3–4 years. Investors in the PRC now have hardly any opportunities to invest their money profitably. Capital flight remains an option, but it is also being rigidly prevented by the state.

In the twelfth chapter, agricultural policy is examined from the perspective of whether the PRC can feed itself. The question is important in the context of the trade conflict with the USA, since American soy production is affected. In the article, it becomes clear that soy is primarily used as animal feed in China. The PRC is one of the largest meat exporters. Pork is one of the staple foods in the PRC. However, due to the rampant African fever, animal stocks are massively reduced, so that the PRC now has to import pork again.

Environmental policy has become one of the most important issues in the People's Republic of China. Especially among the Chinese population, the smog of the big cities has led to a distinct environmental awareness. The Chinese state has taken many measures in the last 5 years to reduce environmental pollution. Environmental policy is a priority. One of the biggest problems facing metropolitan areas is water supply. For years, there were fears that the PRC would secure water resources in the Himalayas. However, the gigantic North-West Project, in which water is to be piped from the Himalayan region to Beijing, has recently enjoyed less media attention. The thirteenth chapter sheds light on the possible reasons. In the context of waste management, China has set new goals. A ban on plastic imports has been enacted in order to avoid having to dispose of other countries' waste in addition to its own. In this area, too, the PRC is forced to act progressively and will thus probably become one of the innovative nations.

Energy policy, which is the subject of the fourteenth chapter, is closely related to environmental policy, since for many years the PRC generated energy mainly by burning coal. This contributed to immense air pollution, which in turn led to rethinking among the political leadership. The PRC's energy policy now emphasizes energy that does not produce visible pollution. On the one hand, this involves nuclear energy, which China intends to expand more in the future. On the other hand, China is investing heavily in solar, wind and hydropower. The People's Republic of China is likely to develop new technologies in this area in the future.

The fifteenth chapter looks at China's foreign trade. The People's Republic of China is the largest trading partner for many countries internationally. In 2001, China joined the World Trade Organization. It is therefore not surprising that the Chinese government is increasingly becoming a global player and wants to play this role actively. This poses a challenge for the previous industrialized nations, as they see the old order in danger and do not necessarily accept the conditions for trade that China sets. Europe has agreed to a dialogue with the PRC in which possible disputes will be discussed. In the future, investment protection agreements and free trade agreements will be negotiated. The US has chosen a different path by going on a confrontational course with the PRC. The trade conflict between the US and China is likely to lead to a global recession. In order not to depend solely on the US and its trade, the PRC started the Silk Road Initiative several years ago. The project envisages building infrastructure first and then increasing the value chain in individual countries. It is a gigantic project that could boost global trade. It is viewed with suspicion by many foreign nations because it spreads China's influence to vast regions of the world.

References

Benedict, Ruth. 2014/1946. *Chrysantheme und Schwert. Formen der japanischen Kultur, Aus dem Englischen von Jobst-Mathias Spannagel*. Berlin: Suhrkamp.

Brown, Kerry. 2018. *China's dream: The culture of Chinese communism and the secret sources of its power*. Cambridge: Polity Press.

Bundesministerium für Bildung und Forschung. 2018. *Gemeinsame Pressemitteilung mit dem Auswärtigen Amt und der Kultusministerkonferenz – Bund und Länder wollen die China-Kompetenz fördern*. https://www.bmbf.de/de/gemeinsame-pressemitteilung-mit-dem-auswaertigen-amt-und-der-kultusministerkonferenz-bund-6160.html. Accessed on 03.10.2019.

Deuber, Lea. 2019. *Schaut auf dieses Land*. Süddeutsche Zeitung. https://www.sueddeutsche.de/leben/china-europa-verhaeltnis-1.4509728?reduced=true. Accessed on 03.10.2019.

Heilmann, Sebastian. 2016. *Das Politische System der Volksrepublik China*, 3. Aufl. Wiesbaden: Springer.

Kroeber, Arthur R. 2016. *China's economy. What everyone needs to know*. Oxford: Oxford University Press.

Merics. 2019. *Home page*. https://www.merics.org/de. Accessed on 03.10.2019.

Naughton, Barry. 2018. *The Chinese economy. Adaptation and growth*, 2. Aufl. Cambridge: The MIT Press.

Stevenson, Alexandra. 2019. China injects $126 billion into its slowing economy. *The New York times*. https://www.nytimes.com/2019/09/06/business/china-economy-reserve.html. Accessed on 03.10.2019.

von Borstel, Stefan. 2012. *Die Deutschen – Nation der Nörgler und Jammerer.* Die Welt. https://www.welt.de/debatte/kommentare/article108305231/Die-Deutschen-Nation-der-Noergler-und-Jammerer.html. Accessed on 07.10.2019.

Wagner, Daniel. 2016. *With its China bashing, America risks breaking a profitable partnership.* South China morning post. https://www.scmp.com/comment/insight-opinion/article/1928704/its-china-bashing-america-risks-breaking-profitable. Accessed on 03.10.2019.

Yang, Xifan. 2019. Ob in China … ein Sack Reis umfällt oder das Land zur Weltmacht aufsteigt – im Westen weiß man wenig über den Konkurrenten. Doch auch die Chinesen selbst haben immer größere Schwierigkeiten, sich zu verstehen, S. 3. *Die Zeit*, 11.04.2019.

2 Data Validity

Alessa Mondorf

In the People's Republic of China (PRC), economic data is collected, evaluated and published by the National Bureau of Statistics. The validity of the official statistics is doubted, especially in the Western media, and accusations are made that these data are at least partially manipulated. These accusations concern statistics in the areas of gross domestic product, retail sales and the labor market. Increasing attention to official Chinese data can be seen in the rising number of publications on the subject in recent years: From 2006 to 2010, about three studies were published annually, and from 2011 to 2015, the annual average was seven (Plekhanov 2017, p. 78).

Data on a country's economy, environment, labor market, and educational situation can have a major impact on how the global public, politicians, and business community perceive the respective nation. While governments adjust their policies based on statistical data, companies consult such data when making investment decisions. Since the human capital factor played an essential role in the economic rise of the PRC, the unemployment rate is examined in more detail. The situation on the labour market influences the development of wages, which is relevant for companies. But the unemployment rate is not only important for companies, it is also important for policy makers. Unemployment has far-reaching consequences for any economy: for individual citizens, a layoff has a very concrete impact on income and thus on purchasing power. Even though the unemployment rate is a so-called lagging indicator, it can confirm trends and thus provides certainty in assessing the situation of an economy (Eurostat 2016).

Since the provision of statistics requires correct implementation and methodology, factors such as regional characteristics, informative value of the data collected, must be taken into account. A certain degree of deviation from the real situation is present in every

A. Mondorf (✉)
Mannheim, Germany

B. Darimont (ed.), *Economic Policy of the People's Republic of China*,
https://doi.org/10.1007/978-3-658-38467-8_2

statistic and can hardly be avoided. Surveys can never depict the "whole truth", because it must always be assumed that a sample is used as a basis for calculating an average value. In small, regulated countries, even a relatively small sample size is sufficient to derive trends. However, the larger and more dynamic a country is, the larger the one sample needs to be in order to be able to create a comparable picture (Von der Lippe 2011). This process is laborious and costly. In the case of the PRC, this is a factor that is often underestimated. One reason for the difficulty of data collection is the size of the People's Republic. The PRC currently has a population of about 1.4 billion people in one area, which is about 19% of the total world population (United Nations Department of Economic and Social Affairs 2017). Accurately capturing the economic data of such a large economy is difficult per se. In the case of China, this is compounded by the fact that the economy is developing rapidly and that regional levels of development vary widely. This means that data collection in China is fundamentally non-trivial and requires a high degree of planning. This planning is done centrally so that one authority always has an overview and sovereignty over the totality of the figures.

2.1 Static Data Collection

In the PRC, various institutions and authorities deal with data collection and the compilation of statistics. The independent collection of data by private companies without the approval of the National Bureau of Statistics of China or a corresponding authority is not permitted. All data not based on those of the National Bureau of Statistics are consequently estimates, or based on a study approved by the Chinese government.

2.1.1 National Statistical Office

In the PRC, the National Bureau of Statistics is responsible for collecting and evaluating data on the economy and society (National Bureau of Statistics of China 2007). In addition to its central headquarters in Beijing, the National Bureau of Statistics maintains a separate office for each province and city with direct access to the government (National Bureau of Statistics of China 2019). It reports directly to the State Council, headed by Premier Li Keqiang, and ranks below that of a ministry. The State Council, which is the highest administrative organ in China, takes over the management of the bureau and provides the managerial staff and financial resources (Holz 2014, p. 321). That is, unlike the Federal Statistical Office in Germany, for example, this office is not an independent institution (Statistisches Bundesamt 2019). Since March 2016, it has been headed by Ning Jizhe, who is also Vice Chairman of the National Development and Reform Commission of the PRC (Bloomberg News 2019). All statistical data are published annually in the Statistical Yearbook and on the Statistics Bureau's website on a monthly, quarterly, or annual basis (National Bureau of Statistics of China 2018a).

In addition to the collection and analysis of data, it is the task of the statistical office to provide the political leadership with the data relevant to them. The public is not the focus as a recipient of the statistics; the priority is politics (Plekhanov 2017, p. 322). The Office not only analyses and uses data collected by its own staff, but also receives further information and statistics from, for example, the Ministry of Human Resources and Social Security. However, it only has direct supervisory authority over its own employees, for example survey teams, but not over the work of employees of other ministries (Holz 2014, p. 321). The statistical office cannot check the completeness and accuracy of data that is passed on from outside.

A frequent criticism of the work of the statistical office is the lack of transparency with regard to the methods used for data collection and their evaluation. It is often unclear which method was used to obtain the data in the first place, or whether this method was modified in any way. Additional guidance on how to read and interpret the published datasets is often not provided (Plekhanov 2017, p. 81). Moreover, again and again, the inconsistency of the data can be observed: For example, in the same yearbook the same data are given in different sections with different heights. For example, when the annual GDP data for each of the country's provinces are added together in 1 year, their sum is significantly higher than the national GDP: the amount is, in fact, so much higher that the New York Times refers to it as a "phantom province" (Tatlow 2013). Inevitably, this means that the underlying datasets from individual provinces must differ from the central data from the Bureau of Statistics. What causes this, however, has not been explained to date.

For these reasons, economists at home and abroad regularly attempt to calculate or estimate various data sets. These alternative figures are then used to make a comparison with the official data. In the best case, the researchers conduct surveys in China on their own as part of their projects. However, such independent surveys and data collection, i.e. all surveys and data collection not initiated by the Bureau of Statistics, are only possible with the explicit permission of the Bureau of Statistics. Many alternative sources within China are also still in the development stage and are not fully developed. They often refer only to individual cities or provinces, but not to the country as a whole, so that hardly any statements about the overall situation can be derived on their basis (Plekhanov 2017, p. 82). The lack of reliable alternative sources makes independent evaluation of the data more difficult, as it is almost impossible to compare them. In addition, the approaches of various scientists are criticized (Plekhanov 2017, pp. 89–90).

Nevertheless, since the economic reforms under Deng Xiaoping in 1978, there is no evidence that data are manipulated at central levels of government (Lau 2018). Rather, it appears that the data received by the Bureau of Statistics from officials or institutions at different levels are inaccurate or manipulated. As a result, the data provided deviates from reality after the individual datasets are merged. It is often argued that the methods used to collect data are outdated or not fit for purpose (Plekhanov 2017, p. 82). At the national level, there can hardly be any interest in inaccurate data, since important decisions have to be made on this basis and incorrect data lead to incorrect decisions.

2.1.2 Civil Servants

In addition to the employees of the statistics office, there are approximately 40 million civil servants at various local and national levels who are engaged in administrative tasks in the PRC (Crabbe 2014, p. 4). The high number of stakeholders and organizational levels means that the scope to change numbers on a small scale is quite large. With 40 million civil servants, it is almost impossible to trace individual records back to a specific person. Moreover, there are various incentives for a local official to manipulate the data in one direction or another, as long as it is not discovered, it can have a positive impact on his career. Ning Jizhe, the director of the National Bureau of Statistics since 2016, claimed in an interview that deliberately altering the data only has negative consequences for everyone involved and promises no career advancement. Therefore, there would be no incentives to manipulate at all (CBNEditor 2018). However, other researchers cite a number of reasons that make manipulation appear profitable (Crabbe 2014, p. 23).

The local statistics are basically controlled by the statistical office. This is to verify whether the growth targets set by the government have been met or whether there have been any deviations. Local officials know that all statistics are monitored and checked against targets. If growth targets have not been met in the region they manage, it means a loss of face for officials and can have negative consequences, such as pay cuts or demotion. Therefore, it is in officials' interest that "bad" results do not become known, creating an incentive for manipulation (Shira and Associates 2017). Chinese authorities are aware of this issue. As part of the anti-corruption campaign, investigative teams have therefore been sent to various provinces to address the problems on the ground and improve the situation (Qi 2016). However, there are incentives for manipulation not only at the professional level, but also those of a private nature, such as the misappropriation of public funds for private purposes. According to an article in the Global Times public spending on education and healthcare for instance amounted to eight trillion RMB in 2011 (Song 2013), which would account for 17% of GDP of the same year. This percentage is not reflected in official statistics. Official purchases made by local officials were found to be well above market price levels. This concerns, for example, furniture for school building equipment or electronics for administrative office equipment. Apparently, local officials entered into these overpriced deals because the companies favored them either with cash, expensive gifts, or other benefits. Public funds were also used to buy luxurious private cars, which were then in turn disguised as company assets. This method of misappropriating funds is not unknown even in the Chinese People's Liberation Army: in 2013, a review of army-owned vehicles in the army's repertoire was carried out there, as there was a surprisingly high stock of luxury cars from manufacturers such as Mercedes-Benz and Porsche, which are not at all suitable for military purposes. In fact, army officers had spent state money to then use said cars privately (Crabbe 2014, p. 30).

Such loopholes sometimes arise due to the hierarchy of staff in provincial administrations. The staff of the statistical offices at the local level report directly to the respective members of the government in the provinces, but not to the National Statistics Office.

Thus, if data are to be "touched up" at the local level, this is illegal in principle, but it is unlikely that an employee will defy the instructions of his or her direct superior (Holz 2014).

2.2 Case Study: Unemployment Statistics

The following section will use the example of the labour market to illustrate which factors influence the data in official statistics. First, the problem areas on the labour market will be briefly discussed in order to put them in relation to the unemployment rate, which has remained consistently high for decades.

2.2.1 Specificities of the Labour Market Situation

China is currently home to about 1.4 billion people, the majority of whom are located in cities due to urbanization policies in recent years. According to the National Bureau of Statistics, more than 776 million people were employed in China in 2017 (National Bureau of Statistics of China 2017). About 280 million of them came as migrant workers to the big cities and factories, where there is still a need for cheap labor to construct buildings or manufacture smartphones (National Bureau of Statistics of China 2018b). Migrant workers make up about 35% of China's total labor force. About 11% of all workers are self-employed and the remaining 54% are employees (National Bureau of Statistics of China 2017, own calculation).

China's development from a country that became known as the workbench of the world due to its cheap labour to an economically strong industrial nation brought with it massive job losses in the secondary sector. The dwindling factory and construction jobs are having a massive impact on the labor market, as the emerging vacancies in the tertiary sector are still insufficient to provide employment for all potential workers. More than 200 million Chinese are employed in the primary sector, i.e. in fishing, forestry and agriculture or mining. The largest share, nearly 350 million people, are employed in the tertiary sector (National Bureau of Statistics of China 2018a). Workers without education find it difficult to find employment due to the decreasing number of jobs in the primary and secondary sectors. In the tertiary sector, there are jobs that do not require a skilled worker, but here they are competing with numerous better educated people who take jobs for which they are actually overqualified in order to at least have an income (Crabbe 2014, p. 182).

Migrant workers are often only employed for a very short time and are then looking for work again. As a result, many of them are usually unemployed and without income for several months. Lin Yanling, a professor at the Chinese University of Industrial Relations in Beijing, describes this group as "particularly vulnerable in the labour market". According to Lin, economic downturns particularly affect workers without education, who already have a weak position in the labour market (Zuo and Xin 2018).

Although, as a result of the one-child policy, there are now fewer young people entering the labour market than retirees leaving it, the current development does not necessarily lead to an easing of the labour market. Today, a large proportion of young people are well educated. In contrast to the generation born in 1950, which attended school for an average of 7–8 years, today the period of schooling and education is about 11 years (Du and Yang 2014). Meanwhile, a large number of young people are acquiring higher education degrees in the hope of better earnings. In 2017, more than 7.3 million Chinese students earned their university degrees and entered the labor market (Statista 2019). University graduates often face great competition in the job market. Although they are highly educated and motivated, surveys by the Organisation for Economic Co-operation and Development (OECD) have found that some struggle to find employment within the first 6 months of obtaining their degree (OECD 2015). Only graduates from highly regarded universities, that is, those from the Chinese government's 985 or 211 projects, manage to find work quickly after graduation. It is difficult with a degree certificate from an "average" university (Yuen 2013).

The current trade conflict with the USA has far-reaching consequences not only for the Chinese economy, but also for the Chinese labour market. 35% of American companies that have a location in the People's Republic have already transferred it to other states in Southeast Asia or are considering the option (American Chamber of Commerce 2018). Among Japanese firms, the figure is as high as 60% that have already transferred or are in the process of doing so, according to Kyodo News. Approximately 45 million jobs are provided by foreign companies, but in the supply chain, many more Chinese companies depend to a greater or lesser extent on orders from foreign companies. The relocation of these companies to other countries means the loss of jobs in China. According to an estimate by the major Swiss bank UBS, the US-China trade conflict could result in a loss of up to 1.5 million jobs in China (Cheng 2019).

There is disagreement among scholars on whether unemployment is an acute problem or not and whether government intervention is necessary. This highlights the importance of accurate statistics to understand the real situation and develop effective solutions (Zuo and Xin 2018).

2.2.2 Special Features of Unemployment Statistics

Data on the number and percentage of registered urban unemployed in the total labor force are released quarterly by the Bureau of Statistics. Until April 2018, the statistics on registered unemployed in urban areas was the only indicator of unemployment in the People's Republic. After many economists criticized the quality and informative value of the statistics, it was reformed and supplemented last year (Plekhanov 2017, p. 92).

The unemployment rate in the PRC hardly changed over many years. Since 2002, the percentage of unemployed has fluctuated minimally between values of four to a maximum of 4.3%. The rate remains seemingly completely unaffected by external influences, such as the 2009 global economic crisis, the declining GDP growth rate since then (National

Bureau of Statistics of China 2018a). As a result of the 2008 global financial crisis, at least 20 million migrant workers lost their jobs, but this did not show up in the figures. Over five quarters, the unemployment rate remained consistently at 4% (Tang 2017). An unemployment rate of 5% can generally be assumed to be full employment, as there will always be some number of people in frictional, voluntary, or seasonal unemployment (Blanchard and Katz 1997). An unemployment rate even below this 5% mark would be a flagship for any economy and shows that the economy as a whole is doing well. The problems of graduates mentioned above do not show up in the statistics in any way.

The registered unemployment rate published by the statistics bureau has been criticized by many scholars. Cai, the vice president of the Chinese Academy of Social Sciences, even called the official unemployment statistics "almost useless." (Cai et al. 2013). In 2017, the Youth Media Association published an investigative report on the true level of unemployment: based on 838.87 million working-age citizens (that is, 16–59, not in schools or universities), the unemployment rate was more than 22% after subtracting retirees over 60, who are still recorded as employed in the official statistics (Timm 2018). This figure appears in stark contrast to the 4% estimated by the statistics office. By other calculation, unemployment in 2013 and 2014 in major Chinese cities was at least 8% but no more than 9% (Yang 2014). However, even these figures, which are conservatively calculated compared to the Youth Media Association, are twice as high as those of the statistics office.

The first thing to note is that the Chinese government does not want to admit unemployment for ideological reasons. According to its own statement, unemployment is a "product of capitalism", and Chinese people without jobs are not unemployed, but only between different jobs (Wang 2018). Until the 1990s, the problem of unemployment did not exist in China for ideological reasons. Today, this could still be the reason why the quarterly urban unemployment rate issued is reported so low. The fear that high unemployment could lead to social unrest and thus criticism of the communist party weighs heavily (Cheng 2019). Furthermore, only registered unemployed people are included in the statistics. Moreover, anyone who does not register as unemployed with the employment office is not included in the figures (Crabbe 2014, p. 181).

This procedure is quite common and makes sense; in Germany an unemployed person must also first register as unemployed in order to be entitled to support. However, the hurdles in China are much higher: registering as unemployed and thus applying for financial support can only be done at the respective place of residence where a person is registered. This residence is called *hukou* in China, it is inherited virtually by birth and can only be changed under strict conditions (Crabbe 2014, p. 179). Therefore, millions of people who work far from home cannot register for unemployment locally (Yang 2014). Furthermore, there is an age and occupation restriction as far as unemployment is concerned. Men over the age of 50 and women over the age of 45 are not included in the quota (Crabbe 2014, p. 181). They are considered to be out of the labor force because of their age. People who are farmers living in the countryside are also not included as they are considered to be employed for life. Workers whose jobs are eliminated, for example in times of economic difficulty, also often fall through the cracks: before impending layoffs, companies offer their employees severance packages. At the same time, there is often another offer: If the affected employee quits himself or herself, he or she receives a higher compensation payment than the legally required severance pay. In this way, companies want to avoid conflicts with their employees who "voluntarily" terminate their employment in this way. At the same time, this approach benefits the unemployment rate, because "voluntarily" unemployed persons are not included in the statistics (Crabbe 2014, p. 182).

Another hurdle to registering as unemployed is unemployment insurance. Only those who have paid into the insurance scheme for more than 12 months are entitled to up to 24 months of state support. However, most social insurance branches in China have been in existence for less than 20 years and many people are not even insured. In 2014, for example, barely more than 10% of migrant workers were insured (National Bureau of Statistics of China 2015). Even if a worker meets all the criteria and is thus eligible to register as unemployed, the financial incentive to do so is relatively low: the amount of unemployment benefits is based on local minimum wages and averages only 17.8% of the average local monthly salary (Plekhanov 2017, p. 92). This is not enough to cover current living expenses, which is why some unemployed people do not even register as such. In addition, applying for the money involves a lot of bureaucracy. Some people do not even know about the state support option, as they were not informed about it by local authorities or their previous employers (Plekhanov 2017, p. 92). The statistics also do not capture what proportion of the working population is in marginal employment but is actually looking for a full-time job (Crabbe 2014, p. 182). Only those who have worked less than 1 h in the last week are considered unemployed. Thus, anyone who has taken a casual job, even if only once for an hour, cannot register as unemployed. This means that the rate of urban registered unemployment is both partly lower stated than it actually is and unsuitable to reflect the labour market situation in the country.

2.2.3 Change in Unemployment Statistics Since 2018

Ongoing criticism of the figures prompted the Chinese government to reform unemployment statistics: In April 2018, a new approach to measuring unemployment was introduced. For the first time, the statistics are no longer based solely on the number of registered unemployed in cities, but capture the rate through household surveys (Wang 2018). It includes all citizens aged 16 and older living in cities, as well as migrant workers who have resided in a city for more than 6 months. This method follows the standards of the International Labour Organization, a specialized agency of the United Nations, which are used by many countries around the world (Beddor 2018). According to Sheng Laiyun, the highest-ranking statistician at the Bureau of Statistics, the data obtained through this method is more comprehensive than before, while having higher comparability with indicators in other countries (Tang 2017). The monthly survey of households is now conducted by the National Bureau of Statistics itself, rather than by staff at the local level. Because the researchers conducting the survey report to the national agency, they are independent of local governments. The data collected are sent directly to the National Bureau of Statistics and are free from influence by other agencies. Therefore, many scholars hope that the data obtained will be more reliable and thus comparable in different fields (Wang 2018). Press reports from the 2000s claim that the National Bureau of Statistics had been collecting data via direct survey for a number of years, but these were only released at irregular intervals, if at all, and were from different cities, making them difficult to compare (Bloomberg News 2018).

In September 2018, the urban registered unemployment rate was 3.82%, while the rate obtained through surveys was 4.9% (Xin 2018). Thus, there is a difference between the two sets of data collected. The new method is apparently an improvement on the previous survey method, but the statistics bureau has not published how the survey of citizens is conducted, nor how many citizens participated in such a survey in the first place, nor what criteria were used to select them. Thus, the data remains opaque after the 2018 reform. In order to determine how representative the figures are, it would have to be known on what basis they are based, that is, for example, what the age structure of the respondents is, whether both sexes are represented in appropriate proportions and where the respondents live.

2.3 Conclusion

In the meantime, the fact that official data in China reflect a distorted picture of reality has been analysed by many scientists around the world. Criticism of the validity and resilience of the published data has increased rather than decreased in recent years. Above all, the lack of transparency with regard to the collection and analysis of the data, as well as the possibility of interfering with the data at the local level, have been criticised. The reform of unemployment statistics in 2018 was a step in the right direction. The Chinese government is making efforts to introduce international standards.

Ultimately, it is hardly possible for an external observer to assess how high unemployment actually is in China. Even on the development of individual sectors, estimates can at best be made, as there is little official information available. Since this problem affects not only unemployment figures, but also those for gross domestic product, average income and, for example, retail sales, the issue of validity will not lose any of its media attention in the coming years, and further publications on this topic can be expected. For the study of the Chinese economy, this means that all data should be viewed with caution.

References

American Chamber of Commerce in Shanghai. 2018. *Impact of U.S. and Chinese tariffs on American companies in China*. https://www.amcham-shanghai.org/sites/default/files/2018-09/2018%20 U.S.-China%20tariff%20report.pdf. Accessed on 20.09.2019.

Beddor, Christopher. 2018. *Breakingviews – China's new unemployment measure has big job ahead*. Reuters. https://www.reuters.com/article/us-china-economy-jobs-breakingviews/breakingviews-chinas-new-unemployment-measure-has-big-job-ahead-idUSKBN1HP0MC. Accessed on 20.09.2019.

Blanchard, Oliver, und Lawrence Katz. 1997. What we know and do not know about the natural rate of unemployment. Journal of Economic Perspectives 11: 51–72. Nashville: American Economic Association.

Bloomberg News. 2018. *China's stats chief defends quality of data. Nation will meet 2018 growth target amid trade threats: Ning statistics bureau has zero tolerance of any data manipulations*. https://www.bloomberg.com/markets/fixed-income. Accessed on 20.09.2019.

———. 2019. *Ning Jizhe*. https://www.bloomberg.com/profiles/people/19687022-ning-jizhe. Accessed on 20.09.2019.

Cai, Fang, Du Yang, and Meiyan Wang. 2013. Demistify the labour statistics in China. *China Economic Journal* 6 (2–3): 123–133. New York: Routledge.

CBNEditor. 2018. *China's statistics Bureau says days of fake data are over. Shedding light on China's monetary system and macroeconomic trends (China Banking News)*. http://www.chinabanking-news.com/2018/08/22/chinas-statistics-bureau-says-days-fake-data-following-crack/. Accessed on 20.09.2019.

Cheng, Evelyn. 2019. *Chinese unemployment worries are growing as Beijing beefs up stimulus*. CNBC. https://www.cnbc.com/2019/01/16/unemployment-worries-in-china-grow-as-beijing-beefs-up-stimulus.html. Accessed on 20.09.2019.

Crabbe, Matthew. 2014. *Myth-busting China's numbers. Understanding and using China's statistics*. Basingstoke: Palgrave Macmillan.

Du, Yang, und Cuifen Yang. 2014. Demographic transition and labour market changes: Implications for economic development in China. *Journal of Economic Surveys* 28(4): 617–635. Chichester: Wiley.

Eurostat. 2016. *Glossary: Lagging indicator*. https://ec.europa.eu/eurostat/statistics-explained/index.php/Glossary:Lagging_indicator. Accessed on 20.09.2019.

Holz, Carsten A. 2014. The quality of China's GDP statistics. *China Economic Review* 30: 309–338. Amsterdam.

Lau, Lawrence. 2018. *Are Chinese economic statistics reliable?* https://www.chinausfocus.com/finance-economy/are-chinese-economic-statistics-reliable. Accessed on 20.09.2019.

National Bureau of Statistics of China. 2007. *About NBS*. http://www.stats.gov.cn/english/nbs/200701/t20070104_59235.html. Accessed on 20.09.2019.

———. 2015. *2014 Nián quánguó nóngmín gōng jiāncè diàochá bàogào*. http://www.stats.gov.cn/tjsj/zxfb/201504/t20150429_797821.html. Accessed on 20.09.2019.

———. 2017. China Statistical Yearbook. Abschn. 4.2–4.9. http://www.stats.gov.cn/tjsj/ndsj/2017/indexeh.htm. Accessed on 18.01.2020.

———. 2018a. *China statistical yearbook 2017 4–2 number of employed persons at year-end in urban and rural areas*. http://www.stats.gov.cn/tjsj/ndsj/2017/html/EN0402.jpg. Accessed on 20.09.2019.

———. 2018b. *2017 Nián nóngmín gōng jiāncè diàochá bàogào*. http://www.stats.gov.cn/tjsj/zxfb/201804/t20180427_1596389.html. Accessed on 20.09.2019.

———. 2019. *Dìfāng tǒngjì wǎngzhàn.* http://www.stats.gov.cn/tjgz/wzlj/dftjwz/. Accessed on 20.09.2019.

OECD. 2015. *China. Paris.* (OECD economic surveys). https://read.oecd-ilibrary.org/economics/oecd-economic-surveys-china-2015_eco_surveys-chn-2015-en#page1. Accessed on 20.09.2019.

Plekhanov, Dmitriy. 2017. Quality of China's official statistics: A brief review of academic perspectives. *Copenhagen Journal of Asian Studies* 35 (1): 76–101. Copenhagen: Copenhagen Business School.

Qi, Liyan. 2016. *Number's game: One province's statistics problem is a drag for China.* The wall street journal. https://blogs.wsj.com/chinarealtime/2016/09/30/numbers-game-one-provinces-statistics-problem-is-a-drag-for-china/. Accessed on 20.09.2019.

Shira, Dezan & Associates. 2017. China's Official Statistics. *China Business Review.* https://www.chinabusinessreview.com/chinas-official-statistics/. Accessed on 18.01.2020.

Song, Shengxia. 2013. *Procurement problem.* Global Times. Accessed on 20.09.2019.

Statista. 2019. *Number of university graduates in China between 2007 and 2017 (in thousands).* https://www.statista.com/statistics/227272/number-of-university-graduates-in-china/. Accessed on 20.09.2019.

Statistisches Bundesamt. 2019. *Über Uns.* https://www.destatis.de/DE/UeberUns/UeberUns.html. Accessed on 20.09.2019.

Tang, Frank. 2017. *China to start releasing proper unemployment figures in 2018 after decades of downplaying the problem.* South China morning post. https://www.scmp.com/news/china/economy/article/2125671/china-start-releasing-proper-unemployment-figures-after-decades. Accessed on 20.09.2019.

Tatlow, Didi Kirsten. 2013. The phantom province in China's economy. *New York Times.* https://rendezvous.blogs.nytimes.com/2013/02/06/the-phantom-province-in-chinas-economy/. Accessed on 20.09.2019.

Timm, Leo. 2018. Unemployment: China's Achilles' heel? *Vision times.* http://www.visiontimes.com/2018/10/24/unemployment-chinas-achilles-heel.html. Accessed on 20.09.2019.

United Nations Department of Economic and Social Affairs. 2017. *World Population Prospects: The 2017 Revision.* https://www.un.org/development/desa/publications/world-population-prospects-the-2017-revision.html. Accessed on 18.01.2020.

Von der Lippe, Peter. 2011. *Wie groß muss meine Stichprobe sein, damit sie „repräsentativ" ist?* http://von-der-lippe.org/dokumente/Wieviele.pdf. Accessed on 20.09.2019.

Wang, Orange. 2018. China is taking a new approach to its jobless rate, but is it enough? *South China morning post* (international). https://www.scmp.com/news/china/economy/article/2142753/china-taking-new-approach-its-jobless-rate-it-enough. Accessed on 20.09.2019.

Xin, Zhou. 2018. *What is China's unemployment rate? State survey says it's falling, private survey disagrees.* South China morning post. https://www.scmp.com/economy/china-economy/article/2171176/what-chinas-unemployment-rate-state-survey-says-its-falling. Accessed on 20.09.2019.

Yang, Liu. 2014. *The Chinese labour market: High unemployment coexisting with a labour shortage.* https://voxeu.org/article/china-s-unemployment-and-labour-shortage. Accessed on 20.09.2019.

Yuen, Lotus. 2013. Why Chinese college graduates aren't getting jobs. *The Atlantic.* https://www.theatlantic.com/china/archive/2013/05/why-chinese-college-graduates-arent-getting-jobs/276187/. Accessed on 20.09.2019.

Zuo, Mandy, und Zhou Xin. 2018. The hidden cracks in China's employment figures. *South China morning post.* https://www.scmp.com/news/china/economy/article/1904879/hidden-cracks-chinas-employment-figures. Accessed on 20.09.2019.

3 Economic Development Since 1949

Barbara Darimont

The economic history of China spans several millennia. Sometimes very different conditions emerged, so China was at times a huge unitary state or fragmented into many small kingdoms. Characteristic is an alternation of opening and isolation in the relations with other nations and peoples. While in times of opening the economy flourished, it stagnated in times of isolation. Coinage, salt monopolies, paper money, inflation, and foreign trade were influential in Chinese economic history (von Glahn 2016). In the nineteenth and twentieth centuries, civil wars and mass campaigns prevailed, leading to economic stagnation. The death of Mao Zedong in 1976 marked a turnaround and since then the PRC has experienced unprecedented economic growth. In the following, the modern Chinese economic history from the founding of the PRC in 1949 will be presented in order to fathom the basis for this phenomenon.

Contemporary Chinese economic history can be divided into the era of the Maoist period and the period of opening since 1978 (Taube 2014). The division into five-year plans seems to make little sense because economic development has been shaped by political interventions, such as the "Great Leap Forward" or the Cultural Revolution. Hereinafter, a differentiation into four phases is made, especially since there were different motives for the respective economic policies: the reconstruction phase from 1949 to 1956, the period of power struggles from 1957 to 1976, the opening period from 1977 to 1990, and the economic growth from 1990 onwards (Mühlhahn 2017, p. 1).

Overall, the development of China's economy does not follow a coherent concept, but rather, especially in recent decades, economic policy has resembled a gigantic

B. Darimont (✉)
East Asia Institute of Ludwigshafen University of Business and Society,
Ludwigshafen am Rhein, Germany
e-mail: darimont@oai.de

B. Darimont (ed.), *Economic Policy of the People's Republic of China*,
https://doi.org/10.1007/978-3-658-38467-8_3

experimentation in which it is unclear how the whole thing will turn out. Partly, the economic policy strategy especially of the last decades is called "Big Push Industrialization" (Naughton 2018, p. 65).

3.1 Start-Up Phase 1949 to 1956

Due to the Second World War and the civil war that followed, there was a lack of industrial facilities, infrastructure and almost all factors of production except labor after the founding of the PRC on October 1, 1949. In agriculture, inefficient methods were used and the population could hardly be fed. Eighty percent of the labor force worked in the agricultural sector. In addition, inflation was rampant. During the Korean War in the 1950s, a trade boycott on Chinese goods further exacerbated the situation (Naughton 2018, p. 75). Key institutions of economic policy were absent, such as offices to collect economic data. These years were primarily characterized by reconstruction. Inflation was eventually contained and crucial infrastructure projects were initiated (Taube 2014, pp. 648–649).

Many institutions did not function at the beginning of the People's Republic due to the previous war events, so that the new government and the respective city governments could not record any revenues. Taxes and fees were levied, which increased the cost of production, causing unemployment. In the early 1950s, the new leadership was able to stabilize the economic situation, introduce a new currency the "people's currency" (renminbi), which stopped inflation towards the end of 1952 (Naughton 2018, p. 76). At the beginning, the new government relied on the proletariat, the peasants, the petty bourgeoisie and the national bourgeoisie to avoid an economic collapse and to gain support in the cities (Mühlhahn 2017, p. 15).

3.1.1 Domestic Political Discourse

In the early 1950s, the political leadership debated whether to abolish private property. Liu Shaoqi believed that agriculture should be developed first, followed by light and heavy industry. He assumed that industry had to be developed first to produce machinery for agriculture. Only then collectivization in the countryside would make sense. In contrast, Mao Zedong wanted to introduce the planned economy as soon as possible, with which heavy industry was to be developed. He was finally able to push through this view, so that in the middle of 1955 the first five-year plan was passed and the private enterprises were nationalized. For economic policy, the Soviet model served as an example (Mühlhahn 2017, pp. 37–41). Although the Soviet model was followed, there was no consensus on the degree of centralization, as well as on the salary policy in the cities and other economic policy issues (Naughton 2018, p. 76).

After the first years, the CCP realized that many reforms did not have the desired effect. For this reason, the "Let Hundred Flowers Blossom Movement" was initiated. The idea was to criticize in order to identify and eventually correct problems. Often the criticism voiced was directed against the cult of personality around Mao and the lack of democracy. The Party put an end to the Hundred Flowers Movement and criticism, which it felt was disproportionate, and ushered in the "Right-Wing Deviation Campaign," in which critics were denounced. In the Right-Wing Deviation Campaign, approximately 800,000 intellectuals were placed in labor camps (Naughton 2018, p. 79). Subsequently, the "Three Red Banners," as well as the "Great Leap Forward," were proclaimed to advance socialist construction and silence criticism (Mühlhahn 2017, pp. 56–57).

3.1.2 Economic Development and Changes in Ownership Structures

In terms of regulatory policy, this period saw the emergence of a centrally planned economy. A land reform expropriated large landowners. In the cities, industrial and commercial enterprises were nationalized, leaving only about 20% of businesses in private hands (Taube 2014, pp. 648–649). Heavy industry was massively expanded, and as a result economic growth increased, from which only a small part of the population benefited. Agriculture hardly developed at all (Mühlhahn 2017, pp. 37–41).

Changing ownership and equitable distribution of land was a priority for the CCP. In the countryside, the CCP carried out radical reforms, redistributing 42% of arable land in the years from 1950 to 1952 alone (Naughton 2018, p. 75). While the land reform followed the theoretical terms of Marxism, it was not compatible with Chinese realities. Most peasants were either tenants or owners of their farmland. The number of large landowners, on the other hand, was relatively small, although they owned about a third of land (Mühlhahn 2017, p. 33).

In 1953, the land reform was finally completed. However, the reform did not lead to a real change in conditions; rather, the original structures of land ownership continued to exist. This was because poor peasants did not have enough capital to invest in implements, etc. Therefore, land inequality remained. However, many fields were declared that had not been registered before in order to save taxes (Mühlhahn 2017, pp. 32–36).

3.1.3 Administrative Measures

A household registration was introduced in 1958, called *hukou* (hùkǒu), which still exists today. At birth, every Chinese citizen is given a residence that they cannot change. This measure serves to be able to plan the supply of the population. The supply is thus linked to the place of residence. In addition, Residents' Committees were formed to control and manage the population and were responsible for all social issues in their respective districts. Work units (dānwèi) were created (Mühlhahn 2017, pp. 23–24). In principle, the work unit was the company or institution for which someone worked. This unit was responsible for allocating housing, social security, and all social issues such as daycare space, etc. Personal files were introduced for each person, which documented the entire life of a citizen. The society could be classified into politically correct or not with the help of these personal files and could be differentiated and evaluated with the help of household registration (Perry 2007, p. 11).

3.1.4 Foreign Policy Development

In 1958, the so-called Taiwan Crisis occurred, in which the People's Republic shelled an island off Taiwan, Jinmen (Vogelsang 2012, p. 547). As a result, the United States moved a large fleet into the Taiwan Strait. Under Soviet and American pressure, Mao was forced to end the Taiwan crisis. This subsequently led to a break with the Soviet Union. Mao's criticism was directed against the Soviet model, thus laying the groundwork for the "Great Leap Forward" and the Cultural Revolution, both of which were a departure from the Soviet model. During this time, China was an opponent of the USA and the Soviet Union and thus isolated in terms of world politics (Mühlhahn 2017, pp. 45–51).

3.2 Period of Power Struggles from 1957 to 1976

In 1958, Mao launched the "Great Leap Forward" campaign, through which Chinese society was to leapfrog several industrial economic stages. The goal was to outdo the Soviet Union in a competition for the best socialist model. Basic economic knowledge was ignored, so this stage ended disastrously for the population. This campaign can be considered an experiment in which the planned economy was reduced to absurdity (Vogelsang 2012, pp. 551–554).

3.2.1 Second Five-Year Plan (1958–1962)

The "Great Leap Forward" was intended to solve China's growing food problems and at the same time drive forward industrialization (Mühlhahn 2017, p. 58). The Second Five-Year Plan finally deviated from the original Soviet line. The population was meant to be enthused about socialist development and replace the experts. State fixed investment was increased by 85% and allocated to heavy industry (Taube 2014, pp. 651–654). Since industry was to be carried out in the countryside, so-called backyard blast furnaces were built to produce steel. However, the projects were poorly organized and led to immense environmental problems. The steel produced during the Great Leap Forward was not recyclable (Naughton 2018, p. 81).

From the mid-1950s, Mao sought collectivization in the countryside. This was contrary to the view of the CCP Central Committee, which first called for mechanization in the countryside in order to subsequently make agriculture more effective. A unified buying and selling system was implemented to control grain prices and optimize food supply. Under this system, farmers had to sell all of their crops to the government at a set price that barely justified production. There were countless protests against this practice. Due to the smaller harvest, food became scarce and was rationed for the urban population from 1955. Distribution took place via the hukou (Mühlhahn 2017, pp. 51–55).

Agricultural cooperatives were established from 1954 to 1956, and by the end of 1956, 98% of the rural population was organized into agricultural cooperatives (Naughton 2018, p. 77). By the mid-1950s, agriculture could no longer meet demand due to population growth and the expansion of heavy industry. Crop failures from 1959 to 1961 led to a devastating famine with approximately 40 million deaths (Mühlhahn 2017, p. 61). Other scholars assume a higher number and cite 55 million dead (Dikötter 2014, p. 431). The famine had various causes, but above all it had been a wrong planning.

In 1960, the break with the USSR and the misallocations by the planned economy led to disaster. According to the plan figures, there was enough food, so resources were diverted from this area. In some cases, arable land was used by crops for industry. Furthermore, Mao wanted to continue payments by grain to the USSR at any cost. The crop failures further aggravated the situation. Warning signals about the impending disaster were dispelled by Mao (Dikötter 2014, pp. 109–115). The famine then led to years of stagnation in the Chinese economy, which did not end until 1978. In the summer of 1961, Mao Zedong's resignation was forced (Taube 2014, pp. 651–654).

From 1961 onwards, Liu Shaoqi, as President of the People's Republic, pursued a realpolitik in which private enterprise was again permitted in order to remedy the grievances. Individual private initiatives were allowed, so that smaller markets emerged. Overall, this phase was characterized by recovery. However, this economic policy strategy could not prevail because the "Great Leap Forward" was considered a disaster initiated by hostile powers. Therefore, military development was pushed from 1960 onwards (Taube 2014, pp. 654–655). The focus on military and ideological disputes also had a negative impact on the economy.

3.2.2 Foreign Policy Development

The "Third Front" strategy, initiated in 1964, was a response to the threat of international disputes with the Soviet Union, the disputes with India and, indirectly, the United States, and the political events on Taiwan. The aim was to build up an industrial capital stock in the Chinese hinterland in order to be prepared for possible belligerent confrontations. Since the coastal strips were generally poorly defensible, there was consideration of waging guerrilla warfare from the hinterland. Enormous financial and human resources were mobilized to build this Third Front. However, from a national economic point of view, the design was neither efficient nor sensible, as the plants were often difficult to access, such as in side valleys. The factories were too difficult to reach. The military value of this strategy was low, as there was no warlike conflict. However, this strategy was counterproductive for the Chinese economy, as many unprofitable state-owned enterprises were established and given grandfather status (Taube 2014, pp. 351–354).

3.2.3 Cultural Revolution

The "Great Leap Forward" was followed by a political readjustment in the early 1960s to contain the consequences of the famine. Zhou Enlai proclaimed the "Four Modernizations" in 1964, but they were not implemented until 1975. The economic policies of this transitional period consisted of balancing the state budget through cuts, using financial indicators to evaluate management, allocating reconstruction aid, eliminating extra-budgetary funds, supporting slow and sustainable growth, controlling inflation, and installing the market to complement planning (Mühlhahn 2017, p. 63). Medical care was transformed as barefoot doctors offered minimal health care and educated about hygiene (Mühlhahn 2017, p. 75). These efforts were partially overturned by the onset of the Cultural Revolution.

The Cultural Revolution lasted from 1966 to 1976 and cannot be explained by concepts of economic policy; it was an ideologically and militarily led power struggle within the party, in which Mao wanted to rise to old greatness. During the Cultural Revolution, many central institutions were non-functional due to personnel purges. Civil war-like conditions prevailed in the countryside, hampering agriculture; the situation did not improve until the early 1970s. Industrial development stagnated due to work stoppages, and it was not until 1978 that reform considerations took hold (Taube 2014, pp. 658–659).

Deng Xiaoping had been stripped of all political posts during the Cultural Revolution; when he returned to the political stage in mid-1973, he was discredited by Jiang Qing, Mao's wife, and her supporters. Unlike Deng, they wanted political mobilization, class struggle, as well as foreign policy isolation; Deng's goals, on the other hand, were economic growth, stability, and a pragmatic foreign policy. Mao sought a balance between the opposing groups. Finally, from 1975 onwards, Zhou Enlai's "Four Modernizations" (agriculture, industry, science and technology) gained influence as a political guideline (Mühlhahn 2017, pp. 79–80).

The closure of educational institutions, especially universities, had drastic consequences for the development of the economy that began in 1978 (Taube 2014, pp. 655–658). The youth, who had acted as Red Guards, were poorly educated or not educated at all, so that an entire generation had an inadequate education. Furthermore, they were disillusioned about power struggles and corruption in the party (Mühlhahn 2017, p. 79).

3.3 Opening Period from 1977 to 1990

In 1978, at the Third Plenum of the Central Committee, under the leadership of Deng Xiaoping, it was decided to open up to foreign countries, as well as market reforms, and thus a completely new economic model was finally introduced. Political campaigns were put on the back burner. Economic development took precedence. For reforms from the year 1978, political leaders had no clear concept (Naughton 2018, p. 97). At the 14th Party Congress in September 1992, the transition to a "socialist market economy" was proclaimed, with the socialist market economy seen as a transitional phase to a socialist society. Accession to the World Trade Organization (WTO) in 2001 represented a further step in opening up and economic development (Taube 2014, pp. 659–660).

3.3.1 Pragmatic Economic Policy

After Mao's death, a growing crisis of legitimacy emerged. Initially, Hua Guofeng was appointed as Mao's successor, but he was already deposed in 1977. Urban unemployment, stagnating food supply, poor housing conditions, as well as falling wages and poverty in the countryside led to the decision of a new economic policy. Deng Xiaoping initially started the reform policy only to make the planned economy more efficient (Mühlhahn 2017, p. 89). Besides that, Deng initialized the limitation of terms of office according to age in order to counteract a cult of personality or personal rule, as under Mao. The slogan "Chinese-style socialism" emerged. From this developed the transition to a market economy (Mühlhahn 2017, p. 94).

In principle, economic policy changes were made, but a change of system was not indicated. In 1978, for example, the "Fifth Modernization" which meant political modernization, i.e. democracy, was demanded, but Wei Jingsheng, a well-known dissident, was imprisoned for decades for this demand (Perry 2007, p. 13; Vogelsang 2012, p. 591). Initially, economic policy was to be strengthened. Various experiments were carried out, such as the introduction of fiscal responsibility or special economic zones.

In 1982, the people's communes were abolished. They were replaced by communes at the administrative level (Vogelsang 2012, p. 584). During decollectivization, many cadres were able to secure the best land and technical equipment for themselves and their relatives. With the granting of land use rights, there was a kind of ownership. A 1984 government directive allowed land to be leased for up to 15 years (Mühlhahn 2017, p. 100). The massive increases in agricultural production in the years from 1979 to 1984 can be attributed to the previous reforms. The reforms were initially limited to the poor regions, but were then extended to the whole country (Lin 1992, pp. 34–37). The so-called household responsibility system promoted self-reliance in the countryside. Farmers were allowed to cultivate fields on their own and sell parts of their harvest in small markets. Local governments were dissolved in the early 1980s. The household responsibility system was a huge success, but it led to an increase in the price of agricultural products. Slowly a market with flexible prices developed and from 1985 market prices were liberalised (Mühlhahn 2017, p. 96).

In addition, township and village enterprises (TVEs) were established from 1979 onwards. Many of the surplus workers in the countryside could be employed here. TVEs were considered part of the collective economy; they gradually became competitors of state enterprises. In state enterprises, managers were given responsibility for production. They were allowed to generate profits and reinvest them or distribute them to employees. From 1984, private enterprises were allowed in cities with a maximum of seven employees and then from 1988 with no limit (Vogelsang 2012, p. 585). The combination of increasing competition, monetary incentives and effective monitoring of company developments led to an improvement in the situation of state-owned enterprises and was an alternative to radical privatisation (Naughton 2018, p. 107).

The special economic zones were intended to attract foreign companies to help build the economy. The risks were low because the foreign companies bore the investment risks. They mostly took over the production costs and infrastructure services. In the process, it can be seen that it was not an imposed process by the central administration, but that many reforms and further developments were demanded by the local people. Further developments were often initially carried out in a grey area and only subsequently approved by the government. Consequently, one cannot speak of a uniform process initiated by the central government, but rather the market economy developed almost "by accident" (Mühlhahn 2017, p. 67). Reforms were also initiated in industry. A dual pricing system was introduced, which led to the spread of corruption and made fair competition among companies difficult. From 1993 onwards, the dual price system consisting of a price set by the state and the price of private demand was dissolved because the direct allocation of raw materials was abolished (Mühlhahn 2017, p. 99).

These developments led to the emergence of an entrepreneurial class that included both private entrepreneurs and managers of state enterprises (Mühlhahn 2017, p. 100). In toto, the standard of living rose, equality as well as collectivism gradually dissolved. From 1988 to 1989, inflation in products of living was in some cases 28% over the previous year (Naughton 2018, p. 109). This inflation was one of several reasons why students and citizens protested in Tiananmen Square in 1989 (Mühlhahn 2017, pp. 82–102). Thus, an authoritarian system emerged until 1990, where no political participation could be successfully introduced. This system is referred to and discussed as the "China model" (Mühlhahn 2017, p. 112).

3.3.2 One-Child Policy

At the end of the 1970s, China comprised one billion people. This meant that 20% of the world's population lived there and had to be fed from 7% of the world's arable land. From that point on, China had problems feeding its own population (Vogelsang 2012, p. 593). In particular, it was feared that the population would be hard to feed and that this would hinder economic growth (Feng et al. 2016, p. 83). In 1980, therefore, the so-called one-child policy was introduced in the PRC. Each married couple was only allowed to have one child, unless they belonged to an ethnic minority, in which case they were allowed to have more than one. The policy was strictly enforced in urban areas, while in rural areas it could often be circumvented because enforcement was more difficult. Sanctions were usually enforced through the employer in the form of wage cuts, etc. Further, gynaecological inspections are more feasible in urban areas than in rural areas. The increasing ageing of Chinese society led to a rethink from the turn of the millennium and in 2016 the one-child policy was abandoned. Since then, Chinese couples have been allowed to have two children. Moreover, they no longer need permission from the authorities (Feng et al. 2016, p. 83).

One consequence of this one-child policy was that many female fetuses were aborted. In 1986, abortions based on sex were banned, and in 1993, they were banned. This ban was carried over into the Population and Family Planning Act (Davis 2014, p. 31). Without massive international migration, there is currently a surplus of 30 million men in China who cannot find a bride (Davis 2014, p. 36).

Furthermore, urbanization influences demographics. The People's Republic of China plans to relocate up to two-thirds of its population to cities in the future. This means that China is currently transforming itself from a village society to an urban society at a very high pace. A comparable experiment has not been undertaken in the history of the world. This raises problems, such as the fact that many old people in the countryside will be without children and therefore without care. In many developed countries, the proportion of old people is also very high, but the per capita income has already reached a certain level, from which the PRC is still far away, so that the balance between generations is difficult (Davis 2014, p. 28). It is questionable whether the demographic factor will actually lead to a slowdown in economic growth (Davis 2014, p. 36).

3.4 Economic Growth Since 1990

The military suppression of the civil protests on 04 June 1989 initially led to an economic standstill. Foreign companies withdrew. With his trip to southern China in 1992, Deng Xiaoping was able to revive the stagnating economy, which was followed by a boom of unprecedented proportions. The new leadership under Jiang Zemin and Zhu Rongji initiated a number of economic reforms. In the process, the private sector was promoted in a special way (Mühlhahn 2017, p. 115). From 1993 onwards, inflation started to rise again and reached 20% in 1995. This prompted Zhu Rongji, as president of the central bank, to aim for a so-called soft landing: inflation was controlled and financial institutions were reformed. Through direct political control, he was able to curb inflation (Naughton 2018, p. 111). Due to the reforms of the state-owned enterprises, many of them went bankrupt. As a result, the actual unemployment rate was over 10% of the urban population (Naughton 2018, p. 117).

Various reforms were initiated for WTO accession, such as strengthening competition between counties and provinces, devaluing the renminbi, which made foreign direct investment more attractive. The hukou system was relaxed. A tax reform was implemented in 1994, raising the national economic tax rate from 11% in 1994 to 21% in 2008. Social safety nets were established in parallel. State-owned enterprises were exposed to open competition. So-called national champions were selected to remain as state-owned enterprises. Foreign companies were increasingly admitted and in 2001 the PRC officially joined the WTO (Mühlhahn 2017, pp. 118–122). The goal in principle was to achieve high economic growth. China had a high investment ratio for years. Foreign capital sometimes contributed to these growth rates, furthermore also an intensive technology transfer. State-owned enterprises were further reformed and the State Assets Control and Administration Commission (SASAC) was established (Mühlhahn 2017, pp. 121–125).

Under the leadership of Hu Jintao and Wen Jiabao, economic growth was further accelerated. Under Presidents Jiang and Hu, China became a world power with an enormous economic volume. In 2009, China first replaced Germany as the world export champion and finally Japan in 2010. China seems to have any kind of resources in abundance, but also the corresponding problems (Vogelsang 2012, p. 603).

The Hu Jintao and Wen Jiabao period of government is characterized by the fact that social security was expanded and especially the rural population was freed from too high taxes. However, other social services, such as education and health, remained underfunded. Wen Jiabao tried to counteract these problems, but he was only moderately successful during his term in office. The economic policy bureaucracy grew and the Development and Reform Commission, which represents conservative views, became more influential. The Chinese government responded to the global economic crisis in 2008 by stabilizing the domestic economy through infrastructure projects (Naughton 2018, pp. 120–121).

In 2012, a new leadership came to power in the form of Xi Jinping, who distinguished himself in a different way. One of the most striking elements was the campaign against corruption. The economy was not initially among the priority reform items. Under Xi, the personalization of politics has increased immensely, attempting to maintain the legitimacy of the CCP (Mühlhahn 2017, pp. 112–118). At the third Plenum of the Communist Party in the fall of 2013, Xi Jinping was elected President of the PRC. Xi expressed the government's will to initiate further reforms. Subsequently, many small leadership groups were formed to solve individual reform problems. However, many reforms have been tackled half-heartedly so far. For example, financial reform was postponed. Reform of the hukou system was initiated, but its abolition was not carried out in the final analysis. Instead, Xi Jinping responded to the growing problems with the "Chinese Dream". Chinese nationalism is increasingly strengthened. In autumn 2017, Xi had himself elected as party secretary for the second time, and in March 2018, age limits for senior cadres were finally lifted (Naughton 2018, pp. 122–123).

Over the past three decades, China has generated almost constant economic growth of 10% of GDP annually, enabling it to rise to become the world's second-largest economic power. Economic growth has brought a certain prosperity to the population, but has negative sides, such as high environmental pollution, the costs of which for the national economy are still not adequately recorded today. Furthermore, economic growth has caused large income disparities within the population, which can become a social problem (Taube 2014, pp. 661–663). The development towards a global society has a flip side: the urban population not only gets rich first, but also gains more from economic growth in the long run, while the rural population only benefits to a small extent. These differences are hardly balanced anymore and lead to a huge social contrast (Vogelsang 2012, p. 610). Finally, it has become clear to the political leadership that a return to a planned economy is no longer possible without jeopardizing economic development. Economic growth, in turn, generates legitimacy for the ruling CCP. If it did not provide economic growth, the question of political participation in society would arise. Overall, state capitalism emerged with a hybrid, mixed economic system based on some state enterprises at the central level and private enterprises at the local level (Mühlhahn 2017, p. 126).

Whether China will change to a Western market economy in the course of economic liberalisation, however, remains questionable, as this includes corporations, cooperatives, legal entities and internationality, but the Chinese understanding of the economy is shaped by the image of the family. Competition and the market are thus defined in a completely different way (Taube 2014, pp. 665–668). Why should the Chinese culture, with its millennia of tradition, its self-assessment, its power of integration and its widespread acceptance among the people, not possess and retain its own understanding of the market and competition (Fikentscher 1993, p. 908)?

References

Davis, Deborah S. 2014. Demographic challenges for a rising China. *Daedalus, the Journal of American Academy of Arts & Sciences* 143 (1): 26–38.

Dikötter, Frank. 2014. Maos großer Hunger. *Massenmord und Menschenexperiment in China.* Stuttgart: Klett-Cotta.

Feng, Wang, Gu Baochang, and Cai Yong. 2016. The end of China's one-child policy. *Studies in Family Planning* 47 (1): 83–86.

Fikentscher, Wolfgang. 1993. Die Rolle von Markt und Wettbewerb in der Sozialistischen Marktwirtschaft der Volksrepublik China: Kulturspezifisches Wirtschaftsrecht. *Gewerblicher Rechtsschutz und Urheberrecht.* Internationaler Teil 12, S. 901–910.

Lin, Justin Yifu. 1992. Rural reforms and agricultural growth in China. *The American Economic Association* 82 (1): 34–51.

Mühlhahn, Klaus. 2017. *Die Volksrepublik China.* Berlin: de Gruyter.

Naughton, Barry. 2018. *The Chinese economy*, 2. Aufl. Cambridge: MIT Press.

Perry, Elizabeth. 2007. Studying Chinese politics: Farwell to revolution? *The China Journal* 57: 1–22.

Taube, Markus. 2014. Wirtschaftliche Entwicklung und ordnungspolitischer Wandel in der Volksrepublik China seit 1949. In *Doris Fischer und Christoph Müller-Hofestede*, ed. Länderbericht China, 645–679. Bonn: Bundeszentrale für politische Bildung.

Vogelsang, Kai. 2012. *Geschichte Chinas*, 2. Aufl. Ditzingen: Reclam.

von Glahn, Richard. 2016. *The economic history of China. From antiquity to the nineteenth century.* Cambridge: Cambridge University Press.

4 State Structure

Barbara Darimont

The structure of the Chinese state is formally laid down in the Constitution of the People's Republic (PR). The institutional organization corresponds to Western countries, but the Chinese Communist Party (CCP) makes all final decisions. While there is an informal policy-making process, it eludes the public. In recent years since Xi Jinping took office, the Party's influence has increased in all areas of life, such as censorship of media – especially social media. The reason for this increased state intervention will be fears of a decline in the CCP's power as the prioritisation of economic growth undermines communist ideology. The unspoken agreement is that the CCP will provide increasing prosperity and in return citizens will stay out of politics. If economic growth declines, this agreement is threatened and the CCP's legitimacy is called into question. While the PRC is a multiparty state, it is de facto led and directed by only one party. A separation of powers into legislative, executive and judicial does not exist (Heilmann 2016, p. 27), but rather an entanglement of powers. This manifests itself particularly in interactions between the central government and the subordinate levels, which is characterized by negotiations and networking (Noesselt 2016, pp. 59–60).

4.1 The Communist Party of China

According to its statute, the CCP is the only ruling party in the People's Republic of China. It was founded in Shanghai in 1921. It unified the country after the civil wars at the beginning of the last century and has ruled ever since. There is no apparent alternative to the

B. Darimont (✉)
East Asia Institute of Ludwigshafen University of Business and Society,
Ludwigshafen am Rhein, Germany
e-mail: darimont@oai.de

B. Darimont (ed.), *Economic Policy of the People's Republic of China*,
https://doi.org/10.1007/978-3-658-38467-8_4

CCP, as it has now guided the country's destiny for 70 years. Therefore, the fate of the PRC is closely linked to the fate of the CCP (Brown 2017, p. 22). Some of the CCP's goals for the future are laid out in great detail in the plan "China 2030: Building a Modern, Harmonious and Creative Society" (Development Research Center of the State Council of PR (China) and World Bank 2013). In this plan for the future, the intention of the Chinese leadership to usher in ecological, technologically innovative development and to become a world power is clearly evident.

At the end of 2014, the CCP had close to 90 million members who came from different walks of life, such as entrepreneurs, businessmen, soldiers and farmers. When three members are together, they have to establish a communist cell. The CCP establishes its own organizations at all levels and in all institutions, including foreign companies (Heilmann 2016, pp. 44–45).

Since 1921, party congresses have been held every 5 years, at which the basic political guidelines are laid down (Noesselt 2016, p. 65). The CCP is led and represented by the General Secretary, who usually also holds the post of President of the PRC. Currently, Xi Jinping is the general secretary of the CCP, state president and supreme commander of the army. The Politburo Standing Committee is the highest decision-making body of the PRC and represents the inner circle of the Party. It currently has seven members. With Xi's inauguration in 2012, the number of members was reduced from nine to seven. These seven individuals simultaneously hold the highest offices in the state: (1) Xi Jinping is President, (2) Li Keqiang is Premier, (3) Li Zhanshu is Chairman of the Standing Committee of the National People's Congress, (4) Wang Yang is Chairman of the Political Consultative Conference, (5) Wang Huning is Secretary of the CCP Central Committee, (6) Zhao Leji is Secretary of the CCP Central Disciplinary Commission, and (7) Han Zheng is First Vice Premier (Cheng 2017).

The Politburo Standing Committee is elected by the Politburo, which has 25 members and is elected for five years by the Central Committee (Noesselt 2016, p. 65). The 205 members of the Central Committee are elected by the nearly 3000 delegates to the Party Congress (Heilmann 2016, p. 46). The Central Committee meets once or twice a year (Noesselt 2016, p. 65). It publishes directives, policies, and guidelines to be implemented by party members at all organizational levels. In addition, personnel or cadre policies are made by the party headquarters, thereby exercising control (Hartmann 2006, p. 71). Many of the CCP's directives take priority over laws and the constitution. Those who are on the Central Committee have many political and economic opportunities, but membership does not guarantee political advancement per se (Brown 2017, p. 24).

The party's administrative bodies include the Disciplinary Control Commission, which has gained influence in recent years as it is largely responsible for fighting corruption. No appeals or reviews are possible against rulings by this commission (Noesselt 2016, pp. 73–76). Bo Xilai, a former rival of Xi Jinping, was accused of corruption by this body and is serving a life sentence. The commission has 130 members and a standing committee of 19 members. The Secretariat of the Central Committee, which consists of six members, has an important function. It has to implement decisions, process documents, ensure consensus within the party, and is responsible for administrative work. It has six departments, namely one each for Organization, Propaganda, Discipline Control Commission, Political Research, General Day Work and the Secretariat. The selection and dismissal of party officials is the responsibility of the Organization Department of the Central Committee (Noesselt 2016, pp. 73–76).

About 200 families share power in China, of which 20 or 30 families are significant. These include the families around Deng Xiaoping, Jiang Zemin and Mao Zedong. They determine who becomes president. Problematically, they use their power to gain economic benefits (McGregor and Utz 2012, p. 201). This leads to resentment among the population. For example, some families are particularly active in certain sectors and defend that turf; for example, the families around Li Peng are active in the energy sector, around Jiang Zemin in telecommunications, and around Zhu Rongji in finance (Brown 2017, pp. 40–41). Xi Jinping has put anti-corruption on the political agenda with the "Strike Hard" campaign, but since politics is above the law and there is no self-government, such a campaign can only be a temporary stop, not a prevention of corruption (Brown 2018, p. 58).

4.2 Xi Jinping

Xi Jinping was elected General Secretary at the 18th Party Congress in 2012 and confirmed in autumn 2017. At the Party Congress in 2017, Xi was able to expand his power and enforce that the age limits for political leadership were lifted, which now makes it possible for him to be President of the PRC until the end of his life. His fate is thus very closely linked to the Party (Brown 2017, pp. 17–19).

During the Cultural Revolution, Xi worked in the countryside. From 1975, he studied chemical engineering at Tsinghua University, after which he studied Marxist philosophy and earned a doctorate in law. Xi first started his professional life in the military in 1979, after which he began his political career in the Beijing area of Hebei Province as deputy party secretary of Zhengding County. This was followed by various posts in Hebei, Fujian and Zhejiang provinces. From 1993, he was a member of the Fujian provincial government. He then went to Ningde and finally to Zhejiang province, where he is credited with the so-called Wenzhou economic model (Fewsmith 2013, pp. 113–117) as an economic achievement. The city of Wenzhou was one of the first cities to allow private enterprises. Xi encouraged private entrepreneurship during his tenure, more specifically companies such as Geely and Alibaba, and created an economic boom throughout Zhejiang province. The last stop was Shanghai (Brown 2017, pp. 68–79). For years, Xi was not in the top spot for president, but Li Keqiang and Li Yuanchao were (Brown 2017, pp. 72–75). What Xi's specific motivations were is anyone's guess, but it will have been power politics rather than charisma. Xi has been called "Jiang Zemin's man" and "Hu Jintao's Li Keqiang." In October 2007, Xi was elected to the Standing Committee of the Central Committee. Bo Xilai, his fiercest rival, was removed from his post in a true political thriller when his wife was implicated in the murder of an Englishman. This happened at the best time for Xi. Either he was lucky or had powerful supporters (Brown 2017, p. 79). Other sources suggest a coup attempt in 2011, as a result of which Xi is installed as a leader by the party elite

(Lu 2019). Xi is not seen as a theorist, but as a practice-oriented politician (Brown 2018, p. 59).

Xi was able to consolidate his power during his first term in office. The period of his second term in office is dominated by the trade conflict with the US. Li Keqiang has so far not been able to strengthen his position significantly and hardly makes an appearance. On the other hand, Wang Qishan, for example, an ally of Xi, is vice president of the PRC, and Liu He, one of the four vice premiers, appears on the international stage in negotiations with Donald Trump. Currently, 11 of the 25 Politburo members are considered friends or allies of Xi. Some of them were only elected to these positions in the last election in autumn 2017 (Cheng 2017).

After the end of the Cultural Revolution in 1976, two camps faced each other, namely the Maoists and the reformers. Xi's father was among the reformers at the time, so his son is not counted among the leftists – the Maoists (Brown 2017, pp. 61–62). Since 1949, radical leftists have been associated with all major disasters. The group of leftists is seen as Xi's greatest enemy (Brown 2017, pp. 128–137). The Maoist leftists represent the most vehement opposition for the ruler. Their criticism is directed against "the bureaucrats who think only of themselves" (Brown 2017, pp. 140–142).

Both Wen Jiabao's family and Bo Xilai's family are associated with riches of over 100 billion RMB, which is not the case with Xi's family. Among the social tasks, Xi must show the people that the CCP does not only want to maintain power in order to enrich its own elites. This intertwining of politics and business, which particularly increased from the mid-1990s, is discussed under the heading of cadre capitalism (Alpermann 2013, pp. 292–294; Heberer 2013, p. 67). Under Hu Jintao criticism had already begun to be voiced, that while the reforms have helped some to become wealthy, the majority of the population is among the losers of the reforms. Although the difference in income is accepted by the population when it becomes clear that someone has done more or has a higher education, many politicians are distrusted because the population assumes that they want to enrich themselves unlawfully (Alpermann 2013, p. 292; Brown 2018, p. 63). Xi needs to build trust and morale in society, otherwise this development may endanger Xi's government (Brown 2017, pp. 189–191, 213).

Xi's political goals are, on the one hand, to reduce the power of state-owned enterprises, in which many selfishly enrich themselves from state property, and, on the other hand, to tax the rich in order to establish social justice. Further urbanization of the population is also on the agenda; however, there are problems with this, as there is a high loss of trust in fellow citizens in the big cities, caused by the high level of anonymity and the decrease in family ties due to high mobility. This loss is attributed to the Party, which is held responsible for it. Furthermore, China is said to become the most digitized nation in the world, which increasingly puts the Party in conflict, more precisely a kind of cyberspace war, with the United States of America (Brown 2017, pp. 153, 163, 175).

The "Chinese Dream" is Xi's political slogan, which implies that the revival of the Chinese nation is the greatest dream of Chinese citizens in modern times. This alludes to the time before the Opium Wars (1839), when China was the Middle Kingdom and took tributes from other states. The dream serves to become a strong and powerful nation (Brown 2017, p. 101). The dream is flanked by Document Number 9, an internal CCP policy document that warns against Western values and their spread in China (Economy 2018, pp. 37–42; Brown 2018, p. 133). While Xi is not the direct author of this document, it does reflect the party line (Buckley 2013).

4.3 National People's Congress

The PRC Constitution was first adopted in 1954, followed by a second constitution in 1976 and the currently valid constitution in 1982, which was revised in 1988, 1993, 1999, 2004 and 2018. Revisions mainly concerned reforms towards a socialist market economy and in 2004, for the first time, the guarantee of private property and human rights (Heilmann 2016, pp. 38–43). The Chinese constitution grants its citizens the basic rights as they are known in Germany: Freedom of Expression, Freedom of Assembly, Freedom of the Person and Personal Dignity. However, there is no constitutional court where these fundamental rights can be invoked. Therefore, Chinese citizens cannot invoke these fundamental rights.

According to the Constitution, the National People's Congress is the legislative body of the PRC. It meets every year in March and is re-elected every 5 years. The first election was held in 1954, as the Political Consultative Conference was initially the legislative body. Since then, the Political Consultative Conference has acted as a platform for multi-party cooperation. It was originally established in 1946 as a joint body by the Nationalists and Communists to stabilize the Second United Front. However, it is no longer mentioned in the 1982 Constitution (Noesselt 2016, p. 62). This Political Consultative Conference still exists today and always meets after the National People's Congress, but it is no longer considered important, especially since it has no state function or competence. The Political Consultative Conference is an advisory body to the state apparatus (Noesselt 2016, p. 61).

There are so-called People's Congresses at the county, district and provincial levels. The members of a People's Congress at the district level are first appointed on the basis of elections. They elect the representatives of the people's congresses at the district and then provincial level. From the provincial-level People's Congress representatives, members are then elected to the National People's Congress. Elections at the lowest level – village and county – are free elections. Deng Xiaoping had planned by 2030 to establish free elections in the PRC. The current leadership has refrained from this goal, and free elections are currently no longer under discussion (Heilmann 2016, p. 87).

The National People's Congress consists of about 3000 members. It decides on all important laws, translates party directives into legislation, elects institutional leaders and receives the government's accountability reports. In addition, according to articles 62–63 of the Constitution, it decides on territorial-administrative divisions and on wars. Decisions are debated at the plenary session of the CCP Central Committee in the fall and adopted at the National People's Congress the following March. Dissenting voices are rarely heard during the votes. Xi Jinping was confirmed as president in March 2018 without a dissenting vote. Exceptions include the vote on the construction of the Three Gorges Dam in 1992, when only 67 percent of deputies voted in favor. Because of these high approval rates, the National People's Congress has been called a "voting machine" (Hartmann 2006, p. 110), which may not be justified (Noesselt 2016, pp. 61–64).

For day-to-day legislative activity, there is a Standing Committee consisting of approximately 180 members, which meets every 2 months. It is in this committee that readings of bills are held. If a bill does not have the necessary approval, it does not even come to a vote. In some cases, laws have been debated for over 10 years. If a law is to be passed quickly but there is no consensus on certain points, the relevant parts are removed from the law, as in the case of the Foreign Investment Law of early 2019, which had 170 articles in the preliminary version and was then reduced to 42 articles in the passed version (Merle 2019). Furthermore, the National People's Congress has the oversight function over the implementation of laws. In the absence of a constitutional court, it is incumbent on it to review the implementation of laws. In practice, this means that 5 years after a law is passed, a group of five to six people will travel to different provinces to inspect the implementation status. This is announced to the relevant authorities a year in advance. If violations are noticed, the inspection group conducts negotiations with the respective officials because the repeal of provincial laws by the National People's Congress would be seen as a loss of face. Although the National People's Congress has the authority to repeal laws, it has never been exercised. There is no provision for sanctions. For Western observers, this leads to the abstruse situation that only about 20 percent of laws are ever implemented (Zhang and Zhang 2000, p. 5).

The National People's Congress adopts the five-year plans, which are understood as a kind of vaguely formulated business plan of the Chinese government. The 13th Five-Year Plan is valid from 2016 to 2020 and envisages, for example, an urbanization rate of the entire population of 60 percent and a gross national income of 6.5% (NDRC 2016). Preparations for the 14th Five-Year Plan are currently (2019) underway, but little content is leaking out to the public (Cheung 2019).

4.4 State Council

The State Council constitutes the central government of the PRC. The Chairman of the State Council is currently (as of late 2019) Premier Li Keqiang. Four politicians are his deputies, Han Zheng, Liu He, Sun Chunlan and Hu Chunhua, who are responsible for different areas. Liu He is in charge of the economic and financial affairs. In addition to the departments of the State Council with their respective State Councils, the ministries, commissions as well as the Court of Accounts are subordinate to the State Council. All leaders are on the CCP Central Committee or Politburo. The prime minister is appointed by the National People's Congress. He then nominates his vice-premiers, state councillors, etc., all of whom are then usually approved by the People's Congress. The State Council has an inner cabinet and an outer cabinet, in which the inner cabinet has ten members: Prime Minister, deputies, state commissioners (Noesselt 2016, p. 65).

Currently, 26 ministries and commissions are under the State Council. The last major restructuring of the state organisation was decided in March 2018. The background was probably that the National Reform and Development Commission (NDRC) had become too powerful and Xi wanted to limit this commission's powers. Since the time of Zhu Rongji, the NDRC has been a think tank that makes political and economic proposals, which have then often been implemented; it also draws up the five-year plans (Tang 2018).

A total of 19 ministries are subordinate to the State Council: Ministry of Foreign Affairs, Ministry of Defense, Ministry of Education, Ministry of Justice, Ministry of Commerce, Ministry of Civil Affairs, Ministry of Finance, Ministry of Human Resources and Social Security, Ministry of Industry and Information, Ministry of State Security (Intelligence), Ministry of Public Security, Ministry of Housing and Urban-Rural Construction, Ministry of Water Resources. Reorganized were: Ministry of Justice, Ministry of Science and Technology, Ministry of Transport and Traffic. Newly established were: Ministry of Agriculture and Rural Affairs, Ministry of Culture and Tourism, Ministry of Natural Resources, Ministry of Ecology and Environment, Ministry of Disaster Management, Ministry of Veterans Affairs. The following commissions exist at the rank of ministry: NDRC, National Commission for Minority Nationalities Affairs and new is the National Commission for Health, as well as the National Audit Office and the Central Bank.

The State Council has a special institution, the State Property Management Commission. It is responsible for the large state-owned enterprises, which often have a monopoly position. In 2003, this special commission was established to oversee state-owned property (Saich 2011, p. 160). It plays a significant role in the restructuring of SOEs by deciding on mergers as well as senior personnel. If SOEs want to make investments, they need approval from this commission. Most of these SOEs are conglomerates, such as the Aviation Industry Corporation of China. A total of 113 state-owned conglomerates are subject to this commission.

New offices introduced in March 2018 under the respective commissions, ministries or directly under the State Council are as follows: National Office of Radio and Television, National Office of Health Care, National Office of International Development Cooperation, National Central Office of Market Supervision Inspection, National Office of Drug Monitoring, National Office of Forestry and Green Areas, National Office of Food and Food Reserves, and National Office of Immigration. Based on the new establishments, certain priorities of the current leadership can be identified. These include health, environmental policy, and food security.

In addition, there are administrative institutions that report directly to the State Council, such as those for overseas Chinese affairs, research and the legal system. They are responsible for coordinating the various departments of the ministries when drafting laws and the like. There are also institutions that report directly to the State Council, such as the Xinhua Publishing House or the Academy of Social Sciences. In the Chinese administrative hierarchy, these institutions are equated to ministries. Finally, the individual ministries and commissions have 16 offices or bureaus that fall under their jurisdiction, such as the Petitions Bureau or the State Tobacco Monopoly Bureau.

4.5 Administrative Divisions and Local Governments

The PRC is divided into six levels: Central, Province, County, District, Township, Village. It has four direct cities, Beijing, Tianjin, Shanghai and Chongqing, two special administrative regions, Hong Kong and Macao, and five autonomous regions, Guangxi, Xizhang (Tibet), Xinjiang, Ningxia and Inner Mongolia. The PRC counts 23 provinces among its territory. Taiwan is also considered a province of the PRC due to the one-China principle. Therefore, in total, the provincial level has 33 or 34 units depending on the point of view, 336 exist at the county level, 2882 at the district level, about 42,000 at the township level, and the number of units at the village level is unknown. This division has been made after the dissolution of the municipalities by the Territorial Organization Law of 1979. The party's direct access ends at the district levels because self-organization prevails below the district level. However, the party intervenes here through other channels, such as the village party secretary (Noesselt 2016, pp. 78–79).

In 1987, the Provisional Organic Law on Village Management Committees established that village management committees are directly elected by villagers. The term of office is 3 years. Re-election is possible and the number of candidates must exceed that of the posts. All persons 18 years of age or older are eligible to vote. The tasks of these village administration committees include, public affairs, welfare services and the support of the municipal government. The problem was and still is that the financial compensation for these tasks is not regulated. After initial experimentation, it was found that over half of the elections were not in compliance with the law. For this reason, the law was revised in 1998 and specific technical measures were prescribed. In addition, the relationship between the village party committee and the village residents' committee was regulated. According to this, the most important post is still the Party cadre, but the CPC has to support self-government (Heberer 2013, pp. 118–123). In 2010, the law was revised again. A set of punitive measures was established by law, which included consequences if there were any irregularities in the elections. Schubert and Heberer (2009) conclude in their study that the elections serve to vote out corrupt cadres. However, for poor areas where there are no opportunities for corruption, the incentive is low. Farmers in wealthy areas are often under the power of a cadre, so their motivation to go to the polls is also rather low. Peasants who have become wealthy are motivated to participate in elections to increase their wealth. The desire for more participation was not evident according to this study (Schubert and Heberer 2009, p. 90). Experiments to vote directly at the municipal level were abandoned again and declared unconstitutional. New forms of e-governance are now supposed to improve administration and expose possible maladministration (Heilmann 2016, p. 87). Elections at other levels are not free, as the people up for vote are predetermined by the CCP in the electoral lists. In 2016, an election scandal rocked the population as a total of 45 businessmen were elected to the National People's Congress through bribery via Liaoning Province (Forsythe 2016).

The structure of local governments is a mirror image of the central government. In concrete terms, this means that people's congresses exist at the subordinate administrative levels, such as non-governmental cities, counties, urban districts, municipalities, minority regions, and in the villages (cf. Art. 95 of the 2018 Constitution). The budget of the respective government must be approved by the relevant People's Congress.

If instructions from the central government and the provincial government contradict each other, those of the central government apply. However, the central government cannot issue binding instructions, only recommendations (Heilmann 2016, p. 66). Central policy is implemented very inconsistently across the provinces (Hartmann 2006, pp. 85–86). Relations among local people's governments are unsettled, as there is no coordinating body in the central administration to organize local people's congresses. For example, if the central units are corrupt, they have no way to stop misconduct. In some cases it is unclear whether the local people's governments have to approve the work of the central administration, so that enclaves have arisen in the central administration that are free of law (Mertha 2005, pp. 792–805).

In principle, the geopolitical landscape can be divided into the coastal provinces, which are economically strong, the northeastern Chinese provinces, which have to contend with high unemployment due to old heavy industry, the poor central provinces and the even poorer western provinces (Hartmann 2006, pp. 118–122). Balancing the different interests is among the challenges faced by the respective political leaders (Heilmann 2016, pp. 72–75). Often these interests are balanced between the different levels through negotiations (Heberer 2013, p. 113).

4.6 Court Organisation

In the case of the PRC, it is generally true that political directives take precedence over laws, so that one speaks of political norms (Eberl-Borges 2018, pp. 81–82). In this system, courts are therefore subordinate to the Party. The court system is divided into four levels: Lower, Middle, Higher and Supreme People's Courts. The latter is the appellate court based in Beijing. The Supreme People's Court is the highest judicial body and supervises the activities of all other courts and interprets the laws (Bu 2009, p. 15). A constitutional court does not currently exist, making it effectively impossible for a citizen to invoke fundamental rights. The absence of a constitutional court also means that disputes over competences among state institutions and government bodies cannot be decided by law or court.

Judges nowadays mostly have a legal education or law degree, as a national judicial exam has been mandatory for lawyers, judges and prosecutors since 2002 (Bu 2009, p. 12). Nevertheless, they are not independent as the salaries of judges are paid by the respective governments. The budget of the courts must be approved by the respective people's congresses. Many judges are in the CCP (Bu 2009, p. 17). In Xi Jinping's era, there is little focus on legal reforms or juridification of governance. For Xi, law is a means of governing, but not a matter in its own right. The PRC is currently still a long way from the rule of law (Economy 2018, pp. 46–48).

4.7 Conclusion

The state structure according to the constitution of the PRC is very similar to that of other states or Western states. However, the CCP has influence on all state decisions and is thus above the law and the state apparatus. The Chairman of the CCP, the President of the PRC and the supreme commander of the military has been Xi Jinping since 2012. All future fortunes of the PRC are tied to him. He is seen more as a reformer, but seems to be timid in pushing through the line of reforms due to internal party infighting.

A major problem in the PRC is that local governments have a great deal of autonomy and use it. This creates a proliferation of legal regulations that can hardly be prevented in the current legal system. A constitutional court would be necessary for this. However, a separation of powers into legislative, executive and judiciary is not envisaged in the PRC, not even in the future.

References

Alpermann, Björn. 2013. Soziale Schichtung und Klassenbewusstsein in Chinas autoritärer Modernisierung. *Zeithistorische Forschungen/Studies in Contemporary History* 10: 283–296.

Brown, Kerry. 2017. *CEO, China. The rise of Xi Jinping*. London: I. B. Tauris.

———. 2018. *China's dream. The culture of Chinese communism and the secret sources of its power*. Cambridge: Polity Press.

Bu, Yuanshi. 2009. *Einführung in das Recht Chinas*. München: C. H. Beck.

Buckley, Chris. 2013. China takes aim at Western ideas. *The New York Times*, 19. August. 2013. https://www.nytimes.com/2013/08/20/world/asia/chinas-new-leadership-takes-hard-line-in-secret-memo.html. Accessed on 08.09.2019.

Cheng, Li. 2017. *China's new Politburo and Politburo Standing Committee.* https://www.brookings.edu/interactives/chinas-new-politburo-standing-committee/. Accessed on 08.09.2019.

Cheung, Gary. 2019. Hong Kong prepares for Beijing's 14th five-year plan earlier than usual, marking proactive approach to role in country's development. *South China morning post.* https://www.scmp.com/news/hong-kong/politics/article/2183802/hong-kong-prepares-beijings-14th-five-year-plan-earlier. Accessed on 08.09.2019.

Development Research Center of the State Council, PR (China) and World Bank. 2013. *China* 2030. Building a modern, harmonious, and creative society. Washington, DC: World Bank.

Eberl-Borges, Christina. 2018. *Einführung in das chinesische Recht*. Baden-Baden: Nomos.

Economy, Elizabeth C. 2018. *The third revolution. Xi Jinping and the new Chinese state*. Oxford: Oxford University Press.

Fewsmith, Joseph. 2013. *The logic and limits of political reform in China*. Cambridge: Cambridge University Press.

Forsythe, Michael. 2016. An unlikely crime in one-party China: Election fraud. *The New York times*, 14. September 2016. https://www.nytimes.com/2016/09/15/world/asia/china-npc-election-fraud-liaoning.html?_r=0. Accessed on 08.09.2019.

Hartmann, Jürgen. 2006. *Politik in China. Eine Einführung*. Wiesbaden: VS Verlag für Sozialwissenschaften.

Heberer, Thomas. 2013. Das Politische System der VR China im Wandel. In Die politischen Systeme Ostasiens. Eine Einführung, Ed. Claudia Derichs und Thomas Heberer, 3. Aufl., 39–231. Wiesbaden: Springer VS.

Heilmann, Sebastian. 2016. Das politische System der Volksrepublik China, 3. Aufl. Wiesbaden: Springer Fachmedien.

Lu, Franka. 2019. Das anpassungsfähigste autoritäre Regime der Welt. *Zeit online*. https://www.zeit.de/kultur/2019-09/70-jahre-volksrepublik-china-mythos-erfolg-autokratie-nationalismus. Accessed on 05.10.2019.

McGregor, Richard, and Ilse Utz. 2012. Der rote apparat. *Chinas Kommunisten*. Berlin: Matthes & Seitz.

Merle, Julia. 2019. *VR China – Neues Gesetz über ausländische Investitionen verabschiedet*. https://www.gtai.de/GTAI/Navigation/DE/Trade/Recht-Zoll/Wirtschafts-und-steuerrecht/recht-aktuell,t=vr-china%2D%2Dneues-gesetz-ueber-auslaendische-investitionen. Accessed on 15.10.2019.

Mertha, Andrew C. 2005. China's centralization: Shifting Tiao/Kuai authority relations. *China Quarterly* 184 (1): 791–810.

National Development and Reform Commission. 2016. *Zhōnghuá rénmín gònghéguó guómín jīngjì hé shèhuì fāzhǎn dì shísān gè wǔ nián guīhuà gāngyào*. http://www.ndrc.gov.cn/gzdt/201603/P020160318576353824805.pdf. Accessed on 08.09.2019.

Noesselt, Nele. 2016. *Chinesische Politik*. Baden-Baden: Nomos.

Saich, Tony. 2011. Governance and politics of China, 3. Aufl. Basingstoke: Palgrave Macmillan.

Schubert, Gunter, and Thomas Heberer. 2009. *Politische Partizipation und Regimelegitimität in der VR China // Politische Partizipation und Regimelegitimität in der VR China. Band II: Der ländliche Raum*, Bd. 2. Wiesbaden: VS Verlag für Sozialwissenschaften.

Tang, Frank. 2018. 'Too big and too powerful': Why xi Jinping is reining in China's economic planning agency. National Development and reform commission will lose a number of key responsibilities under government revamp, leaving it significantly weaker. *South China morning post*, 14. März 2018. https://www.scmp.com/news/china/economy/article/2137043/too-big-and-too-powerful-why-xi-jinping-reining-chinas-economic. Accessed on 24.05.2019.

Zhang, Zhongqiu, and Mingxin Zhang. 2000. Duì wǒguó quánxiàn huàfēn hé lìfǎ quán yùnxíng zhuàngkuàng de guānchá yǔ sīkǎo. *Zhèngfǎ lùntán* 6: 3–6.

Economic Policy Goals and Discourses 5

Johannes Lamade

At the beginning of Xi Jinping's era in 2012, economic policy reforms were announced to provide new impetus. Under Xi, the "New Normal" was proclaimed as a political slogan. This change of direction aims to ensure that the high growth rates of more than 10% of the gross domestic product are no longer practiced. Citizens are to adjust to the fact that prosperity will grow at a slower pace. The New Normal also involves a shift from an export-oriented economy specializing in low-cost products to a domestic-oriented economy with the development of highly technical goods. The economic reforms do not necessarily lead to a liberalization of the economy, but rather underpin the authoritarian state and its ability to intervene. This has given rise to a discussion of the so-called China model, which combines an authoritarian state with high export quotas to generate high economic growth. The trade conflict between the US and China is challenging this China model. Both nationally and internationally, different opinions exist as to whether the model is more beneficial or detrimental to the country's economic development. Moreover, the existence of such a model is sometimes doubted.

5.1 Economic Policy Objectives

In 2013, China's new economic and development policies were presented at the third Plenum of the 18th CPC Central Committee, which traditionally presents the program of the country's incoming leadership. The paper also contains strategies to strengthen the CCP's political power and the country's international orientation (Schucher and Noesselt 2013, pp. 1–2). When the strategy paper was published, the international response was

J. Lamade (✉)
Shaanxi Normal University, Xi'an, China

B. Darimont (ed.), *Economic Policy of the People's Republic of China*,
https://doi.org/10.1007/978-3-658-38467-8_5

initially positive. Thus, according to the CCP, the market was to play a "decisive" role in the future and no longer just a "fundamental" one. This difference was seen as a sign of the party's will to finally tackle the long-awaited economic reforms (Lenz 2013). Doris Fischer (2019), professor at the University of Würzburg, on the other hand, does not recognise any measures that are actually purposeful for further liberalisation of the market when she takes a closer look at the plan.

The 273-page reform document, which was adopted at the 3rd Plenum of the 18th Central Committee, in the drafting of which Liu He, the deputy premier in charge of the economy portfolio, played a major role, can be broken down into three broad main themes: Improving the market economy, changing the role of the state, and carving out innovative enterprise structures. It is divided into eight reform areas: Government, State Monopolies, Land and Land Reforms, Tax Reform, Financial and Fiscal Policy, State Asset Reform, Innovation, and Reforms to Open the Market (The Economics Team 2013). In a report, President Xi highlights that in the five areas of economy, politics, culture, society and ecology, reforms can only be successfully implemented if they support each other. Measures to implement them become much more difficult and their effect is further diminished if effects on other areas are not taken into account (Central Committee of the Communist Party of China 2013). Accordingly, reforms should be seen as a whole package. The reform program is flanked and concretized by the five-year plans.

5.1.1 Five-Year Plans

Every 5 years, the Chinese government adopts new guidelines for its country's economic, social, and political actions: the so-called Five-Year Plan (Huang 2015). The first plan was adopted in 1953 and subsequently regulated the country's course of action until 1957. In the idea and implementation of the first five-year plan, the Chinese government under Mao Zedong was strongly oriented towards the Soviet Union. The leadership under Khrushchev helped intensively in building the Chinese economy during this period, thereby having a strong communist ally. The Soviet government sent thousands of technical advisors to help develop the backward infrastructure (Spence 2001, pp. 642–643).

Today, the five-year plans of the PRC are prepared by the State Development and Reform Commission (NDRC), which is part of the State Council of the PRC. The mission of the NDRC is to shape macro-level control and management of the economy (Lam 2015, p. 101). The NDRC is the successor organization of the State Planning Commission, which elaborated the first five-year plan in 1953 (NDRC n.d.). The current Five-Year Plan sounds less pragmatic than its predecessors; rather, it has an almost visionary character. As almost everywhere in China, modern technology has been brought to the fore. The plan contains 80 chapters. Chapter 6, entitled "Innovation in Technology and Science as a Key Role," is worthy of note because national research is to be strengthened, among other things, by importing technologies from abroad. The expansion of an infrastructure conducive to innovation and better organized networking of individual research centers are being sought. This is to be achieved, among other things, through the establishment of regional innovation centres (NDRC 2016, pp. 22–24). Implementation of a national Big Data strategy is the topic of Chapter 27. An important part in this area is to improve data sharing between government agencies. Furthermore, a national platform for government data is to be launched (NDRC 2016, p. 74). Geopolitical strategies are directly addressed, for example the Belt and Road Initiative (BRI) is to be further advanced. The main focus is on the expansion of so-called economic corridors, which are intended to strengthen international trade. In addition, cultural exchange between the BRI member countries is to be promoted. These exchanges should include education, sport and tourism (NDRC 2016, pp. 147–148).

Last but not least, socio-political topics are addressed in order to react to existing problems in the country. One issue that increasingly concerns the Chinese population, for example, is the rising income inequality. In 2016, the Gini coefficient for China was 0.465 according to the State Statistical Office of the PRC, which means a further increase of 0.003 points compared to the previous year. The United Nations refers to a value of 0.4 or higher as severe inequality of distribution (Leng 2017). Therefore, the political agenda of the People's Republic is to close the national income gap through new redistribution systems (NDRC 2016, pp. 180–182).

In "Perfecting China, Inc." by Scott Kennedy and Christopher Johnson, J Capital's managing director Anne Stevenson-Yang is quoted as saying:

> If you're not a Chinese government official […] then the plan doesn't matter. (Kennedy and Johnson 2016, p. 1)

From these words it can be deduced that the Five-Year Plan plays a much smaller role in the lives of the Chinese people today than it did a good 60 years ago. Commenting on the hope for reform efforts that the plan might spur, Stevenson-Yang said:

> The plan is a manifesto that weakens the faith of those who hoped that under the leadership of strongman Xi Jinping, or through various competing factions, there could be sweeping reforms that would reduce the influence of the state in the economy. (Letts 2017)

5.1.2 New Normal

China's modern history is marked by an economic catch-up process that is unparalleled in the world. Between 2000 and 2010, for example, the country's annual gross domestic product (GDP) growth averaged about 11.4% (World Bank 2019a). Currently, China's economy is barely achieving 7% annual growth. Most recently, growth fell to 6.6% in 2018 (World Bank 2019b). A level that President Xi says is just enough to realize China's 2010 goal of doubling its gross domestic product by 2020. In 2015, Xi cited the need for at least 6.5% economic growth per year (Shao and Yao 2015). However, estimates from 2013 suggest that 7.2% annual growth would need to be achieved by 2020 (Holbig 2018) to begin a smooth downturn.

In November 2014, China's President Xi Jinping publicly announced his country's new policy for the first time, coining the term the so-called "New Normal". At the Asia-Pacific Economic Cooperation (APEC) meeting in Beijing, he spoke of how this adjustment would create new development opportunities for the Middle Kingdom (Xinhua 2014).

In early 2015, Premier Li Keqiang outlined the plan for a new normal in a speech to the National People's Congress. The economy should become more sustainable after more than three decades of rapid growth. The modernization of the country, he said, requires stable and appropriate economic development (Qing and Yao 2015). Moreover, the new strategy aims to prevent China from getting stuck in the so-called middle-income trap, like Thailand or Malaysia, for example. The difficulty here is that, despite an initially rapid upswing, GDP only reaches a medium level (Müller 2015).

However, the idea of lower but more sustainable economic development is not a new idea for Xi Jinping and Li Keqiang. Back in 2012, Wen Jiabao, then China's prime minister and Li Keqiang's predecessor, set a relatively low growth target of 7.5% at the time. This target followed the previously published target of 8% annual growth. He justified the downgrading of the growth targets by saying that he hoped the lower targets would encourage people to make growth more sustainable and efficient (Branigan 2012).

Other sources even assume that the restructuring of the economy has been taking place since around 2010. Accordingly, China's economic growth has been weakening for about 9 years. However, this is not seen exclusively as a weakness of the Chinese economy. Rather, it is argued that it is precisely this creeping process that has played a large part in preventing growth from falling even more sharply in recent years. A much more devastating slump would have been expected if the strong, sometimes double-digit growth figures had been maintained unchecked (Lardy 2019, p. 28).

Political scientist Heike Holbig discusses the possibility that the strategy of the new normal is primarily being used to strengthen the party's legitimacy, rather than any reforms actually promoting healthier and longer-lasting economic growth. Initially, China had portrayed itself as a victim of the international financial crisis, but the crisis would not have caused major problems for the Middle Kingdom because of rapid government intervention. Two years later, by adjusting under the new normal, China would have become the conqueror, rather than the victim, of the global financial crisis. This reinterpretation took place regardless of whether it was actually thanks to CCP policies that the economy stabilized (Holbig 2018, p. 355).

5.1.3 Made in China 2025

Made in China 2025 is described as a plan of action to help China modernize its manufacturing sector. The focus is on ten industries, such as information technology, robotics and the aerospace industry (State Council of the PRC 2015). The expansion of these ten industries has been largely subsidized by state capital. In addition to increased debt (Curran

2018), this has led to international disgruntlement and is said to be the trigger for the trade conflict with the USA.

"Made in China 2025" is based on the German "Industry 4.0" programme and is seen as the Chinese response to the German strategy paper. The closeness of the two programmes in terms of content, especially in terms of industries, is striking. Furthermore, Germany has become a popular target for Chinese corporate acquisitions. In 2015 alone, Chinese companies bought 36 German companies (Deuber 2016). Of course, the increase in Chinese direct investment in Germany and the EU is not exclusively due to the Made in China 2025 project. Policies and other projects, such as the BRI, are also contributing to the increase. Initially, Chinese companies invested mainly in Eastern Europe for cost reasons. However, as China's new economic policies have brought high-tech products further into focus, Germany has become more attractive for direct investment. Company acquisitions in Western Europe require a high investment volume, so it can be assumed that the Chinese state provides financial support for the purchases (Welfens 2017, pp. 18–19).

State intervention in high-tech industry purchases has been a major complaint and reason for US tariff hikes on China (Holland 2019). In order to avoid any further escalation of the already tense situation with the US, Premier Li in March 2019, did not mention the flagship project – despite a nearly two-hour speech to the National People's Congress – with a single word (Giesen 2019). Instead of publicly using the frowned-upon term, media and politicians are now talking about the need to further improve the modernization and quality of production facilities (Holland 2019).

5.1.4 Reform Backlog

Since Xi Jinping took office in 2012, financial and fiscal policy has become more restrictive, despite proclamations of liberalization. In 2013, 57% of loans granted in China went to private companies and only 35% to those under state control. Three years later, the picture is very different. Just 11% of loans extended went to private firms, whereas 83% went to state-owned enterprises (Samuelson 2019). Other sources estimate an increase in loans to state-owned enterprises from 30% to 70% since 2013 (The Economist 2019). This is a clear indication of a renewed strengthening and bettering of state-owned enterprises over their private competitors.

The gradual introduction of free trade zones since 2013 and the 2017 announcement of the establishment of "Free Trade Ports with Chinese Characteristics" were also steps towards liberalization, but many of the measures were not implemented or were later revised. In many cases, the relevant authorities lack the incentive to actually implement the planned economic reforms in the free trade zones (Merics 2018).

Cai Fang (2015, pp. 34–35) identifies a fundamental difference between China and other countries, as reforms pursue different goals in each case. In China, the government does not pursue predefined goals. Rather, he argues, reforms in China are aimed at improving the living standards of the people and ultimately creating a strengthened nation. This is a different view of the implementation of Chinese reforms compared to Western views. The Western press questions whether President Xi, who initially presented himself as a supporter of free global trade, is actually breaking down market barriers, limiting discrimination against foreign companies, or effectively combating technology theft (Deuber 2018).

Since the start of the trade conflict with the US, however, Chinese intellectuals have been hoping that the conflict could lead to an increase in pressure on the Chinese government to the point where the reforms that market liberals are waiting for could be implemented after all (Fischer 2019). Suggestions to this effect were most recently put forward by Liu He, who spoke at the Lujiazui Forum in Shanghai on 13 June 2019 of external pressure – he avoids direct reference to the trade conflict with the US – being helpful in encouraging innovation in China and accelerating planned reforms. Moreover, he calls the pressure now being exerted on China an inevitable test that must be mastered because of China's economic rise (Leng and Wang 2019).

5.2 Discussion of the China Model

The so-called China model is intended to explain China's sustained economic upswing since the beginning of the reform and opening-up policy. Often a positive effect is attributed to the intervention of the Chinese government in the economic processes of the country. There are different opinions on the subject from both international and domestic experts, and there is debate as to whether such a model even exists and whether the 40 years of economic growth can be explained with the help of government intervention. The discussion can be simplistically divided into two camps. One side assumes that the increasing market liberalization of the last 40 years produced the positive result. Others argue that it is mainly the influence of a strong government and the accompanying gradual opening of the market and well thought-out reforms that explain the success of the system. The lack of transferability to other countries is often cited as an argument against the existence of such a model, since China's initial situation before its own rise was unique (Kennedy 2010, p. 461).

Since the beginning of the trade conflict, the question of whether China's actions are compatible with the rules of world trade has become relevant. Interventions by the state, which include the granting of cheap loans to state-owned enterprises, disadvantage both private Chinese and foreign companies. Free market pricing is not possible in the relevant industries and thus the conditions for a free market economy are lacking (The Economist 2019).

5.2.1 Origin of the Term

The US-American Joshua Cooper Ramo (2004, p. 3) was one of the first to take a detailed look at the specifics of Chinese development since Deng Xiaoping and to translate his findings into a model. With his 2004 work, he coined the term "Beijing Consensus". With the term Beijing Consensus, he established a counter-model to the concept established in 1989 by John Williamson (2002) called "Washington Consensus". The Washington Consensus is characterized by a liberal economic policy, whereas the Beijing Consensus is characterized by state intervention. The Washington Consensus owes its name to the fact that in his elaboration Williamson defined ten starting points for advancing the economy in Latin America that everyone in Washington would agree with (Kennedy 2010, pp. 462–463).

According to Ramo, the two main elements that shape the Beijing Consensus are social and economic change. As early as 2004, he stated that China has been changing the international order for years and has shown new possibilities for stable economic development. Ramo describes the Beijing Consensus as a new way of thinking to achieve strong economic growth through just and peaceful means. He describes the system as flexible because no universal solution is sought. Apart from the need to defend borders and self-interest, innovation and the will to experiment play a central role. The system is "pragmatic and ideological at the same time" and is based on Deng Xiaoping's motto of taking small steps rather than great leaps (Ramo 2004, p. 4).

The discussion is currently no longer conducted under the catchword Beijing Consensus, but under the term China Model. This is due on the one hand to the fact that China seems to have weathered the global economic crisis in 2008 better than Western nations, and on the other hand to a book entitled "The China Model" (Zhōngguó Móshì) by Pan Wei, which was published in 2009 on the occasion of the 60th anniversary of the PRC (Pan 2009).

5.2.2 Current Discourse

In 2018, the top Chinese economist Zhang Weiying gave a lecture at Beijing University, which was later translated into German in "Die Zeit" with the title "Die Mär vom China-Modell" (English: "The myth of the China model") (Zhang 2018). The lecture, which appeared in an edited Chinese version on the university's homepage, attracted a lot of attention and was presumably politically undesirable, so that it was deleted from the homepage after only a few days.

In China, there are two prominent players in the current discussion about the China model, Justin Yifu Lin and the aforementioned Zhang Weiying. Both teach at Peking University, but represent entirely different points of view. In 2016, a debate flared up between the two on whether or not industrial policy was positive for the Chinese economy (The Economist 2016). The debate was about the future economic direction of China, a topic that encouraged one million Chinese to watch the three-hour broadcast (Tang 2016).

5.2.2.1 Justin Yifu Lin

Justin Yifu Lin was Vice President of the World Bank from 2008 to 2012, the first Chinese to hold this post. Previously, he studied in Chicago and at Yale University, among other places, and then worked for over 15 years at Peking University, in the China Centre for Economic Research (CCER), which he founded.

Lin is convinced that the China model discussed by foreign and domestic economists exists and that China's actions, at least in part, are transferable to other nations. In a 2011 speech, he answers the question of whether useful conclusions can be drawn from China's development over the past 30 years in the affirmative (Lin 2012, p. 13). According to Lin, industries in which a country has a comparative advantage must be discovered and targeted. This approach, he says, can be applied to any developing country in the world and is the start of rapid and sustained economic growth. If governments adhere to this motto, the competitiveness of their own companies is improved both on the domestic and on the world market. Once the country reaches a new economic level, the comparative advantage changes and the country's infrastructure must be adjusted accordingly. At this point, Lin recommends that the state should intervene. On the one hand, pioneering firms should be compensated for any problems caused by their pioneering work, and on the other hand, the state should help with the sensible coordination of investments and a targeted allocation of resources (Lin 2012, pp. 13–14). He is convinced that markets are inherently flawed and that the state should therefore step in and guide their development. Nevertheless, Lin points out that state-led industrial policy in China has not always been faultless (Tang 2016).

5.2.2.2 Zhang Weiying

Zhang is known for his liberal views, he frequently criticized the – in his opinion – too strong influence of the government on the country's economy. He studied at Oxford University, among other places. His views and positions were strongly influenced by the Austrian economist Friedrich August von Hayek (Zhou 2014). As a result, he is fundamentally critical of government intervention in the economy. He argues that the state cannot act in a completely unbiased manner. Thus, he argues, it is not for the state to use targeted industrial policy to determine economic winners and losers. He cites the failed policy to strengthen the domestic solar energy industry as an example of bad state intervention (Tang 2016).

Zhang distinguishes between two theories that can explain China's rise, namely the China model and the universal model. He himself does not believe in the existence of a China model, but rather that China's rapid economic rise can be explained similarly to that of other countries, such as Germany or Japan. China's rapid economic growth, he said, was due to the fact that China enjoyed the advantages of being a laggard for the past 40 years. Time- and capital-consuming industrial revolutions took place in developed countries. China was able to adopt innovations that emerged elsewhere without having to go through such a phase itself (Zhang 2018).

He illustrates the advantage of being a laggard with a domestic Chinese example. Until 2007, GDP grew much faster in the east of the country than in western China. Since 2007, however, the economy in western China has been growing faster than in the east. To conclude that there is a model in eastern China from which the west of the country can learn is misleading, he says. The explanation for the uneven growth, he argues, is the laggard advantage that the West enjoys over the East (Zhang 2018). According to Zhang Weiying in a speech at Peking University on October 14, 2018, what has ensured economic growth in China over the past 40 years is the following:

> Where the state pulls back and private companies lead the way, the economy grows faster. (Zhang 2018)

In many sectors, the country has already caught up with the West, in some cases even overtaking it. Zhang is alarmed by this realization because, in his view, the potential for growth is declining due to backwardness, which is why the country will have to rely on its own innovations in the future. In this field, he sees a connection between the influence of the state and innovation. At the provincial level, he observes that greater intervention by the state has a negative effect on innovative strength. He sums up that China's economic growth cannot be explained by state capitalism, "but has taken place regardless of it" (Zhang 2018).

Zhang Weiying and Justin Yifu Lin agree in principle on one point, namely that China's growth to date can largely be explained by the country's original backwardness. Lin assumes that clever state intervention has had a positive influence on change and that China's approach over the past 40 years can be applied in this way, or at least in a similar way, to other developing countries. It is the task of the state to intervene in individual stages of its country's development and to strengthen domestic enterprises. Zhang, on the other hand, does not see any particular Chinese characteristics that would explain China's development over the past 40 years. Rather, he believes that a universal model coupled with the opening of markets is responsible for the rapid upswing. In addition, he assumes that the discussion about a China model is harmful for China. On the one hand, because it is often equated with state capitalism in the West, which has led to the current discord with the USA (Gan 2017); on the other hand, because the idea is based on the misinterpretation of facts, leading to the wrong conclusions being drawn for the future direction of the country (Zhang 2018).

5.2.2.3 International Opinions

Internationally, the China model is also controversial. Shaun Breslin (2011, p. 1328) resists the generalization that comes with the term China model. He speaks of many different models of individual provinces in China. Therefore, he refers to the China model as a multitude of individual models that together make up a whole and are thus difficult to copy. In an issue of the US news magazine Time, which appeared in November 2017 and whose title is "China has won" (Zhōngguó yíngle), China's political and economic system is described as better equipped and more sustainable compared to that of the US. The reason given for this statement is the government's easier influence in economic processes. The Chinese government can act almost unhindered, whereas in the U.S. Congress must first be convinced and other hurdles slow down the process. The report concludes by arguing that China is better prepared for the future than the US (Bremmer 2017). Daniel A. Bell

(2015, p. xxii) predicts the demise of the China model if the government does not further liberalize the political system and gradually introduce participation in political decision-making, at least at the local level. In addition, the rule of law must be expanded in order to guarantee a minimum level of human rights.

5.2.2.4 Transferability

One component of a model is its transferability to other countries, which is debatable for the China model. Doris Fischer (2019) names developing countries as possible recipients of a China model, as they are at a similar economic level to China 40 years ago. However, she argues that the starting position of these countries differs significantly from that of China at that time. There is no country that has similarities in politics, education level or population to the conditions of China 40 years ago. Moreover, Hong Kong was an additional building block for China's success, he said, because as a special administrative region it had long acted as a gateway into the country. The city's liberal regulations have allowed the Chinese leadership to partially compensate for the weaknesses of its own system. Last but not least, the size of the country, though not always, had been to some extent advantageous for development. Thus, the government could venture economic experiments in individual provinces and, if successful, transfer them to the entire country; if unsuccessful, the loss was usually limited.

However, there are examples of countries that at least fundamentally resemble the characteristics of China in the late 1970s. The governments of these countries are watching China's development closely and trying to draw lessons for their own country. Of particular interest is the CCP's ability to maintain power despite a variety of crises. To date, the Chinese model has not been clearly defined and is therefore changeable. This very characteristic could represent a central reason for the preservation of power in China. For example, in Cuba, following the 2011 Reform Party Congress, the leading political science journal Temas took up the issue and discussed a transferability to their own country (Noesselt 2012, pp. 249–250).

5.2.2.5 State Capitalism as an Element of the China Model

State capitalism is a mixture of socialism and capitalism; accordingly, it contains characteristics of both systems, such as state shares in economically important companies and, on the other hand, price formation that takes place according to market-economy principles (Bundeszentrale für politische Bildung 2016). It is precisely this interplay between the state and the economy that is cited as the reason for China's stable economic growth in recent decades, and is also sometimes used synonymously with the term China model. The state owns large companies and influences national economic activity through subsidies, cheap lending and the closing off of its own market (Samuelson 2019). State capitalism is not a new or even Chinese invention. Even for earlier emerging economies such as Japan, South Korea, Germany, and even the United States, state capitalism was a means to drive economic growth. Eventually, however, all of these countries came to the conclusion that promoting the economy through state capitalism could only be a temporary solution.

China does not seem to have reached that point yet. Currently, there are still other nations, especially developing countries, such as Brazil or Malaysia, whose economies are heavily influenced by state intervention (The Economist 2012).

But among industrialised nations, a trend towards more state influence in the economy has been evident in recent years. Data from 2015 show that the German state had a stake of at least 25% in 549 companies, an increase of around 150 compared to 2007 (Pennekamp and Schäfers 2015) – a trend that could be directly linked to the 2008 financial crisis. On the one hand, the state provided quite a few banks with rescue packages, and on the other, China served as a role model. The Middle Kingdom survived the economic crisis without any major economic slumps thanks to an economic stimulus programme created by the state, which comprised almost 13% of GDP (Huotari 2018).

5.2.3 Comparable Models

When the China model is discussed, comparisons are often made with other countries in East Asia that have experienced similarly rapid economic growth in the past. As a rule, Japan and the tiger economies are grouped together and their growth model is titled the "East Asian economic model". Hong Kong and Singapore are often excluded in this context, as the difference in size makes a comparison with China inappropriate (Boltho and Weber 2009, p. 267). Nevertheless, similarities with growth in Singapore are discussed, to which a separate model is assigned due to the special features.

5.2.3.1 Singapore Model

Comparing China and Singapore is difficult because the difference in size is a striking contrast. Nevertheless, since the Deng Xiaoping era, China has tried to model itself on the Southeast Asian city-state. Initially, during the early days of the reform and opening-up policy, numerous party cadres went abroad in search of suitable strategies to develop China further. Countries such as the United States, Sweden, and Singapore were the focus of Chinese interest. After the Tiananmen Square incidents in 1989, however, liberal and democratic models were no longer included in the considerations of party officials, which is why Singapore moved to the center of Chinese interest. This interest can be seen in the number of articles dealing with Singapore, as these increased tenfold between 1992 and 2006 (Thompson 2019, pp. 61–63). Following the death of former Singaporean Prime Minister Lee Kuan Yew in 2015, the Singapore debate flared up again. In addition, Xi Jinping's anti-corruption campaign was carried out at this time, and the Chinese government also looked to Singapore for guidance in its execution. The most interest is in the idea that Singapore's economic liberalization could be implemented without a corresponding political liberalization. Moreover, between 2008 and 2015, more than 800 Singapore-related journals were published annually (Ortmann and Thompson 2016, pp. 39–40). In looking at Singapore, China has recognized the importance of cracking down on corrupt officials. However, problematic for Chinese understanding is that in Singapore, the rule of law is an essential component of the fight against corruption. Ultimately, the Chinese leadership came to the realization that a strong and virtuous government was needed to fight bribery (Thompson 2019, p. 66). Others suggest that the CCP leadership may not be interested in learning from Singapore at all. Indeed, the government can use the lessons learned to bolster its own legitimacy and reduce pressure for democratization at home by pointing to the success of another authoritarian-led one-party state. Singapore is not a fully liberal state, yet every citizen in the city-state has basic personal and political rights, which is not the case in China. In addition, Singapore has free elections, although they are not yet comparable to those in other democratically run countries. These elections are nonetheless significant because they require the ruling party to be responsive to the wishes of the people, thereby making them accountable for their policies (Ortmann and Thompson 2016, pp. 42–43).

Most recently, there have been reports of China turning away from the Singapore model. China is now much more self-confident and has confidence in its own growth model. The Chinese political expert Bo Zhiyue suspects that no country now serves as a role model for China, but rather attempts are being made to demonstrate the exemplary nature of its own system to other countries in the world (Gan 2017).

5.2.3.2 East Asian Economic Model

Japan, South Korea and Taiwan are the three economic regions whose rapid upswing has led to the development of the East Asian Economic Model. However, there is no single definition for such a model; five aspects are often mentioned that shaped the development of these three states: a high investment rate, low economic share of the public sector, a highly competitive labor market, expansion of exports, and state intervention in the economy (Kuznets 1988, p. 17). Another common denominator cited is a large labor force, income equality, and authoritarian regimes (Boltho and Weber 2009, p. 267).

Instead of directly making market decisions for companies, governments try to influence companies indirectly with incentives or threats of punishment. This is intended to drive the country's economic growth. It also creates an expectation among the population that the government will intervene to ensure economic growth (Kuznets 1988, pp. 36–37). The government exerts influence with direct government investment, tax incentives and different specifications for depreciation (Boltho and Weber 2009, p. 274), which is how it was practiced in the PRC.

On the one hand, the state promotes exports and at the same time protects its own market from foreign competitors through market entry barriers, so that national companies gain advantages both at home and abroad. As a result, the national economy becomes dependent on exports. East Asian governments aim to strengthen the competitiveness of national firms abroad (Boltho and Weber 2009, p. 268). This approach has similarities with that of the Chinese government.

5.3 Conclusion

The PRC is pursuing a different approach than Russia. The transition from communism to a market economy is not to be abrupt, but gradual (Giesen 2016). In this context, China emphasises that it still needs to protect its own economy in many areas. Therefore, the PRC practices state capitalism, which is criticized by other countries. The Chinese economist Zhang Weiying (2018) argues that the economy grows faster where the influence of the state decreases, which conversely means that state influence in the economy inhibits its growth. He also sees the alienation between the West and China as a direct consequence of the China model, as it is equated with state capitalism, which is not compatible with the international concept of fair trade.

Lin Yifu, on the other hand, sees a model character in what China has achieved in the last 40 years after the start of the reform and opening-up policy. Some international participants consider the China model to be superior to that of Western countries. Still other discussants see the China model as non-existent, primarily because the term "model" suggests a transferability which, in their view, does not exist. Nevertheless, the opponents agree that there are certain Chinese characteristics. However, it is argued that the Chinese rise is based on the same market economy processes as the rise of other industrial nations, such as Japan or Germany after the Second World War.

Under President Xi Jinping, reforms had been announced that pointed to a liberalisation of the economic system, but it turned out that they served to protect China's own economy. The view abroad is that China should open its markets more, as envisaged by the WTO. For the Chinese leadership, it is important to transform its own industry into an innovative one and to become less dependent on labour-intensive production in order to make the leap to an industrial nation.

References

Bell, Daniel A. 2015. *The China model – Political meritocracy and the limits of democracy*. Princeton: Princeton University Press.

Boltho, Andrea, and Maria Weber. 2009. Did China follow the East Asian development model? *The European Journal of Comparative* 6 (2): 267–286.

Branigan, Tania. 2012. *China cuts growth target to 7.5%*. The Guardian. https://www.theguardian.com/world/2012/mar/05/china-cuts-growth-target-7-5. Accessed on 17.09.2019.

Bremmer, Ian. 2017. *How China's economy is poised to win the future*. Time. https://time.com/magazine/south-pacific/5007633/november-13th-2017-vol-190-no-20-asia-europe-middle-east-and-africa-south-pacific/. Accessed on 17.09.2019.

Breslin, Shaun. 2011. The 'China model' and the global crisis: From Friedrich list to a Chinese mode of governance? *International Affairs* 87 (6): 1323–1343.

Bundeszentrale für politische Bildung. 2016. *Staatskapitalismus*. http://www.bpb.de/nachschlagen/lexika/lexikon-der-wirtschaft/20703/staatskapitalismus. Accessed on 17.09.2019.

Cai, Fang. 2015. *Demystifying China's economy development*. Berlin/Heidelberg: Springer.

Central Committee of the Communist Party of China. 2013. *Zhōnghuá rénmín gònghéguó guó yāng rénmín zhèngfǔ. Zhōnggòng zhōngyāng guānyú quánmiàn shēnhuà gǎigé ruògān zhòngdà wèntí de juédìng*. Xinhua News Agency. http://www.gov.cn/jrzg/2013-11/15/content_2528179.htm. Accessed on 17.06.2019.

Curran, Enda. 2018. *China's debt bomb*. Bloomberg. https://www.bloomberg.com/quicktake/chinas-debt-bomb. Accessed on17.09.2019.

Deuber, Lea. 2016. Was hinter Chinas Kaufrausch in Europa steckt. *WirtschaftsWoche*. https://www.wiwo.de/unternehmen/industrie/syngenta-kraussmaffei-und-co-was-hinter-chinas-kaufrausch-in-europa-steckt/12964438.html. Accessed on 17.09.2019.

———. 2018. Rambopolitik statt Reformen. *Süddeutsche Zeitung*. https://www.sueddeutsche.de/politik/china-rambopolitik-statt-reformen-1.4258114. Accessed on 17.06.2019.

Fischer, Doris. 2019. *Chinas Wirtschaftspolitik: Einblicke und Diskussion*. Podcast Nummer 18. Weithmann consulting. https://weithmann.com/chinas-wirtschaftspolitik/. Accessed on 17.06.2019.

Gan, Nectar. 2017. Has China outgrown its need for Singapore as a role model? *South China Morning Post*. https://www.scmp.com/news/china/diplomacy-defence/article/2095310/has-china-outgrown-its-need-singapore-role-model. Accessed on 17.09.2019.

Giesen, Christoph. 2016. China im Zweikampf der Systeme. *Süddeutsche Zeitung*. https://www.sueddeutsche.de/wirtschaft/samstagsessay-zweikampf-der-systeme-1.3234649. Accessed on 17.09.2019.

———. 2019. Warnung und Eigenlob. Schulden runter, Militärausgaben rauf: Premierminister Li Keqiang gibt beim Volkskongress ambitionierte Ziele vor. *Süddeutsche Zeitung*. https://www.sueddeutsche.de/politik/china-warnungen-und-eigenlob-1.4355438. Accessed on 17.09.2019.

Holbig, Heike. 2018. Whose new normal? Framing the economic slowdown under Xi Jinping. *Journal of Chinese Political Science* 23: 341–363.

Holland, Tom. 2019. Beijing's 'made in China 2025' plan isn't dead, it's out of control. *South China morning post*. https://www.scmp.com/week-asia/opinion/article/3004900/beijings-made-china-2025-plan-isnt-dead-its-out-control. Accessed on 17.06.2019.

Huang, Cary. 2015. How China's five-year plan, an overhang from the Soviet era, has evolved. *South China Morning Post*. https://www.scmp.com/news/china/policies-politics/article/1866736/how-chinas-five-year-plan-overhang-soviet-era-has. Accessed on 17.09.2019.

Huotari, Mikko. 2018. *China in der Weltwirtschaft*. Bundeszentrale für politische Bildung. http://www.bpb.de/izpb/275583/china-in-der-weltwirtschaft?p=all. Accessed on 17.09.2019.

Kennedy, Scott. 2010. The myth of the Beijing consensus. *Journal of Contemporary China* 19 (65): 461–477.

Kennedy, Scott, and Christopher K. Johnson. 2016. *Perfecting China, Inc*. Washington, DC: Center for Strategic International Studies.

Kuznets, Paul W. 1988. An east Asian model of economic development: Japan, Taiwan, and South Korea. *Economic Development and Cultural Change* 36 (3): 11–43.

Lam, Willy Wo-Lap. 2015. *Chinese politics in the era of Xi Jinping*. New York/London: Routledge.

Lardy, Nicholas. 2019. *The state strikes back*. Washington, DC: Peterson Institute for International Economics.

Leng, Sidney. 2017. China's dirty little secret: Its growing wealth gap. *South China Morning Post*. https://www.scmp.com/news/china/economy/article/2101775/chinas-rich-grabbing-bigger-slice-pie-ever. Accessed on 17.09.2019.

Leng, Sidney, and Orange Wang. 2019. Chinese Vice-Premier Liu He says 'external pressure' can actually help China's economy. *South China Morning Post*. https://www.scmp.com/economy/china-economy/article/3014301/chinese-vice-premier-liu-he-says-external-pressure-can. Accessed on 17.09.2019.

Lenz, Moira. 2013. *Pressespiegel zum Kommuniqué des 3. Plenums: Macht und Markt.* Beijing Rundschau. http://german.beijingreview.com.cn/german2010/mt/2013-11/15/content_578353.htm. Accessed on 17.09.2019.

Letts, Stephen. 2017. *China's next five-year plan on the way and it is a big deal for Australia.* ABC News. https://www.abc.net.au/news/2017-10-15/china-communist-party-congress-should-we-worry/9050032. Accessed on 17.09.2019.

Lin, Justin Yifu. 2012. *Demystifying the Chinese economy.* New York: Cambridge University Press.

Merics. 2018. *China update 17/2018.* Mercator Institute for China Studies. https://www.merics.org/de/newsletter/china-update-172018. Accessed on 17.06.2019.

Müller, Matthias. 2015. China am Scheideweg. *Neuste Zürcher Zeitung.* https://www.nzz.ch/wirtschaft/oekonomische-literatur/china-am-scheideweg-1.18554217. Accessed on 17.09.2019.

NDRC. 2016. *The 13th five-year plan – For economic and social development of the People's Republic of China.* http://en.ndrc.gov.cn/newsrelease/201612/P020161207645765233498.pdf. Accessed on 18.09.2019.

———. n.d. *Main functions of the NDRC.* National Development and Reform Commission. http://en.ndrc.gov.cn/mfndrc/. Accessed on 17.09.2019.

Noesselt, Nele. 2012. *Governance Formen in China – Theorie und Praxis des chinesischen Modells.* Wiesbaden: Springer Fachmedien.

Ortmann, S., and M.R. Thompson. 2016. China and the "Singapore model". *Journal of Democracy* 27 (1): 39–48.

Pan, Wei. 2009. *Zhōngguó móshì: Jiědú rénmín gònghéguó de 60 nián.* Beijing: Zhōngyāng biānyì chūbǎn shè.

Pennekamp, Johannes, and Manfred Schäfers. 2015. Deutschland lässt den Staatskapitalismus blühen. *Frankfurter Allgemeine Zeitung.* https://www.faz.net/aktuell/wirtschaft/wirtschaftspolitik/ruecklaeufiger-trend-deutschland-laesst-den-staatskapitalismus-bluehen-13734501.html. Accessed on 17.09.2019.

Qing, Koh Gui, and Kevin Yao. 2015. *China signals 'new normal' with higher spending, lower growth target.* Reuters. https://www.reuters.com/article/us-china-parliament-idUSKBN0M103W20150305. Accessed on 17.06.2019.

Ramo, Joshua Cooper. 2004. *The Beijing consensus: Notes on the new physics of Chinese power.* London: The Foreign Policy Centre.

Samuelson, Robert. 2019. Why China clings to state capitalism. *The Washington Post.* https://www.washingtonpost.com/opinions/why-china-clings-to-state-capitalism/2019/01/09/5137c6d4-141e-11e9-b6ad-9cfd62dbb0a8_story.html?noredirect=on&utm_term=.9869acbad572. Accessed on 17.06.2019.

Schucher, Günter, and Nele Noesselt. 2013. *Weichenstellung für Systemerhalt: Reformbeschluss der Kommunistischen Partei Chinas.* German Institute of Global and Area Studies 10. https://www.giga-hamburg.de/de/publication/weichenstellung-f%C3%BCr-systemerhalt-reformbeschluss-der-kommunistischen-partei-chinas. Accessed on 17.09.2019.

Shao, Xiaoyi, and Kevin Yao. 2015. *China economy shows signs of steadying, more policy support needed.* Reuters. https://www.reuters.com/article/us-china-economy-activity-idUSKBN0TV05U20151212. Accessed on 17.09.2019.

Spence, Jonathan D. 2001. *Chinas weg in die moderne.* München: Deutscher Taschenbuch.

State Council of the PRC. 2015. *'Made in China 2025' plan issued.* http://english.gov.cn/policies/latest_releases/2015/05/19/content_281475110703534.htm. Accessed on 17.09.2019.

Tang, Frank. 2016. To embrace market or state: Chinese economists debate the country's future. *South China Morning Post.* https://www.scmp.com/news/china/economy/article/2045438/embrace-market-or-state-chinese-economists-debate-countrys-future. Accessed on 17.09.2019.

The Economics Team. 2013. *China's third plenum may be more than just talk.* U.S.-China Economic and Security Review Commission. https://www.uscc.gov/sites/default/files/Research/Staff%20 Backgrounder_China%E2%80%99s%20Third%20Plenum%20May%20Be%20More%20 Than%20Just%20Talk%20%281%29.pdf. Accessed on 17.09.2019.

The Economist. 2012. The rise of state capitalism. *The Economist.* https://www.economist.com/ leaders/2012/01/21/the-rise-of-state-capitalism. Accessed on 17.06.2019.

———. 2016. Plan v market – China's industrial policy. *The Economist.* https://www.economist. com/finance-and-economics/2016/11/05/plan-v-market. Accessed on 17.06.2019.

———. 2019. Pandas can fly – The struggle to reform China's economy. *The Economist.* https:// www.economist.com/leaders/2019/02/21/the-struggle-to-reform-chinas-economy. Accessed on 17.09.2019.

Thompson, Mark Richard. 2019. Authoritarian modernism in East Asia. In *Security, development and human rights in East Asia.* New York: Palgrave Macmillan.

Welfens, Paul JJ. 2017. *Chinas Direktinvestitionen in Deutschland und Europa.* Hans-Böckler-Stiftung. https://www.boeckler.de/pdf/p_mbf_report_2017_36_ci_welfens.pdf. Accessed on 13.06.2019.

Williamson, John. 2002. *What Washington means by policy reform.* Peterson Institute for International Economics. https://piie.com/commentary/speeches-papers/what-washington-means-policy-reform. Accessed on 17.06.2019.

World Bank. 2019a. *GDP growth (annual %).* https://data.worldbank.org/indicator/NY.GDP.MKTP. KD.ZG?end=2017&locations=CN&start=2001. Accessed on 17.06.2019.

———. 2019b. *GDP growth (annual %).* https://data.worldbank.org/indicator/NY.GDP.MKTP. KD.ZG?locations=CN. Accessed on 03.06.2019.

Xinhua. 2014. *Xi JInping shouci xitong 'xin changtai'.* Xinhua News Agency. http://www.xinhua-net.com//world/2014-11/09/c_1113175964.htm. Accessed on 17.06.2019.

Zhang, Weiying. 2018. Die Mär vom China-Modell. *Zeit Online.* https://www.zeit.de/2018/47/ china-staatskapitalismus-privatunternehmen-wirtschaft-wachstum. Accessed on 17.09.2019.

Zhou, Zhenghua. 2014. *Zhāng Wéiyíng: Wǒ shì yǒu zìyóu yìzhì de rén shuōhuà yào duì dé zhù liángzhī.* Sina Finance. http://finance.sina.com.cn/leadership/crz/20140820/120620066996. shtml. Accessed on 17.09.2019.

State-Owned Enterprises

6

Mirko Zumholz

With the opening-up and reform policies, the state sector in China has become less important in recent decades, as private enterprises are increasingly profitable and create jobs. Moreover, a large proportion of Chinese state-owned enterprises (SOEs) have been transferred to the private sector or merged. The profitability and competitiveness of many SOEs have declined, and some would cease to exist without substantial government support. The Chinese government artificially keeps SOEs alive so that they can compete against private enterprises. In recent years, there has been renewed support for SOEs because they perform other economic and political functions in addition to their commercial function. They act as a tool for the government to maintain control over employment and the direction of the economy.

This brings up the question of how the state sector will develop in the future. Will the closure of SOEs be further delayed, so SOEs have to surrender to the mechanisms of the market economy in the long run, or will the government revive the traditional sector?

6.1 Nationalisation of Enterprises

In the years following the proclamation of the PRC in 1949, the process of economic development was by no means structured. The period was marked by regulatory upheavals in which the relationship between the Chinese state and the country's companies changed radically on several occasions, but the economic sector was always dominated by SOEs. The first steps toward a centrally administered economic order became apparent as early as 1950. Handicraft, industrial and commercial enterprises were nationalized, and as a

M. Zumholz (✉)
Ludwigshafen University of Business and Society, Ludwigshafen am Rhein, Germany

B. Darimont (ed.), *Economic Policy of the People's Republic of China*,
https://doi.org/10.1007/978-3-658-38467-8_6

result, privately managed industrial enterprises no longer accounted for even 20 percent of gross industrial production. With the introduction of the planned economy and the first five-year plan (valid from 1953 to 1957), which mainly provided for the strengthening of heavy industry, nationalization was pushed ahead. The CCP dissolved the remaining private sector structures in the secondary sector and took direct control of industrial enterprises. In addition, agriculture was collectivized. By merging sovereign farmers into production cooperatives, the rural population was easier to control. By the end of 1956, over 96 percent of farmers already belonged to such a production cooperative (Taube 2014, pp. 648–650).

The Chinese economy suffered a major setback during the "Great Leap Forward" from 1958 to 1961. Mao Zedong's ambition was to accelerate industrialization – especially in Chinese heavy industry (Giese and Zeng 1993, p. 176). A few years earlier, the hitherto mainstay agricultural sector had already been downsized in favor of industrialization. In 1958, government fixed investment, most of which went to heavy industry, was increased by 85 percent. As a result, the number of people employed in the state industrial sector tripled from 7.5 million to over 23 million (Taube 2014, pp. 649, 651). From 1957 to 1978, there were only two forms of enterprise ownership in the People's Republic: SOEs and collective enterprises (Qing 1996). It was not until the death of Mao Zedong in 1976 that change came to economic policy. Policymakers turned away from the values that had prevailed until then.

The decisive factors for China's rapid development from 1978 onwards were the transition from a centrally administered economy to a market economy and the establishment of market-oriented institutions. Special economic zones were initially established, in which market-based regulations were tested. The development of these areas was by and large left to foreign companies (Taube 2014, pp. 659–665).

6.2 Characteristics of State-Owned Enterprises

SOEs are empirical concepts for which no precise definitions are available as yet. The Chinese state does not specify what characteristics an enterprise must have to be considered state-owned. However, enterprises that are owned by the state or the people, as well as those that are managed or controlled by the state, can be considered SOEs. Since the state is considered the representative of the people, there is practically no difference between the forms of ownership just mentioned. The extent of state intervention in decision-making processes determines the degree to which an enterprise is state-owned (Wei and Li 2019, p. 45).

According to Sebastian Heilmann (2016, p. 199), Chinese companies can essentially be divided into four different basic company types. Type 1 and type 2 are companies whose ownership and control rights are predominantly state-owned. In Type 3 and Type 4, on the other hand, these rights are in the hands of private individuals. The two SOE types differ in terms of their positioning concerning the state and the market. Type 1 is state-based and is thus strongly supported by the state. In contrast, Type 2 is market-based, pursues competitive strategies, and is less reliant on government support. The same distinction applies to type 3 and type 4, type 3 is close to the state, and type 4 is market-based.

The first type of company is predominantly found in industries that are directly controlled by the central government due to their high strategic relevance. These include sectors such as aerospace, oil, and gas, but also essential infrastructures such as electricity and telecommunications networks. Companies of the second type are mainly medium-sized firms. They are usually owned by local governments and are therefore only indirectly controlled by the central government. Type 2 companies are found in industrial sectors that are not of particularly high strategic importance. There, however, because they are market-based enterprises, they have to face competition from Chinese and foreign private enterprises (Heilmann 2016, pp. 199–200; China Institute at the University of Alberta 2018).

In addition to state-owned and private enterprises, there is another group of companies in the Chinese economy: collectively owned firms. Collective enterprises are firms that are officially owned by the workers or management employed there (Pigott 2000, p. 17). These emerged shortly after Mao Zedong came to power and initiated the collectivization of agriculture (Heilmann 2016, p. 253). They are hardly relevant in modern times (Pigott 2000, pp. 17, 72).

6.2.1 Internal Structure of State-Owned Enterprises

In Chinese SOEs, a strong hierarchical-authoritarian structure prevails. Guidelines and instructions are delegated without involving employees in the decision-making process. Questioning or disobeying orders is not an option. The top of a state-run enterprise is always made up of two executives, a president, and a party secretary. The president of a SOE acts as the chief executive officer and is responsible for the operations of the enterprise. This person ensures that targets set by the government as well as the party are met or achieved (Slegers and Atzler 2017, pp. 89–90). The party secretary is not directly involved in the day-to-day business, he assumes a supervisory role. He is responsible for monitoring whether party policies are implemented. Similar to the role of the president, this is a balancing act, as he must ensure compliance with party directives while not burdening operations. The next highest authority after the president and party secretary are the supervisory board and executive board, in which the company's top management is represented (Slegers and Atzler 2017, p. 94).

The starting point for a career in the upper management of a state enterprise is first of all a promotion within the party. Only then is the assumption of a senior position in the economy. It is characteristic of top management officials that they only stay in their positions for a relatively short time, usually between 1 year and 5 years (Slegers and Atzler 2017, p. 89).

6.2.2 Influence Mechanisms of the State

Despite reform and opening-up policies, the political influence on Chinese companies remains immense. In general, the government can draw on various mechanisms of influence and intervention. According to Heilmann, these can be divided into six mechanisms (Heilmann 2016, p. 201).

Probably the greatest influence is exerted through ownership. As central or local governments have ownership rights over companies, they are responsible for hiring managers and defining strategic goals. In addition, profit transfer and reinvestment decisions are reserved for the government (Heilmann 2016, p. 201). Large corporations subject to the central government are supervised by the State Assets Control and Administration Commission (SASAC). The purpose of this institution is to ensure the competitiveness and profitability of firms in international comparison, as well as to prevent the outflow of state funds. In addition, since 2010, the Commission has been more involved in tasks such as restructuring, regulation, and supervision, as well as the strategic organization of the state sector. Since then, the Ministry of Finance has been responsible for managing the dividends of SOEs that are subject to the central government. However, local governments at the provincial or city level are responsible for the majority of Chinese SOEs. These also have administrative apparatuses similar to SASAC (Heilmann 2016, pp. 61–62).

Influence happens through officials in leadership positions who implement party and state interests (Slegers and Atzler 2017, p. 89). The CCP, but especially SASAC, is heavily involved in personnel selection within SOEs. It is responsible not only for selecting managers but also for filling positions at lower staff levels. There are regular transfers of personnel between state-affiliated enterprises and the party. Executives from business are transferred to government posts and later back to the management level of SOEs (Heilmann 2016, p. 201). This creates a close personnel network between the state and companies.

In large SOEs, the chairmen of the board of directors often simultaneously hold office as deputy ministers and, in rare cases, even as ministers. In SOEs, there are hardly any employees in the middle and upper management levels who do not belong to the party. Since the Chinese state has authority over both the equity and debt markets, companies are heavily dependent on the government for investment. In theory, necessary approvals could be denied, thus limiting companies' room for maneuver. In addition, it is up to the government to determine subsidy measures and issue licenses. Influence can also be exerted over this. However, since the government sets goals, targets, and room for maneuver of SOEs, it is unlikely to deny licenses to such enterprises (Slegers and Atzler 2017, pp. 89–94).

In the People's Republic of China, there have been several market regulation authorities over the years, such as the NDRC and the Ministry of Foreign Commerce (MOFCOM). Originally, these regulatory authorities were supposed to reduce direct intervention in the economy, to ensure fair market access and fair competition on a neutral basis. De facto, however, they are used to enforce political interests. Not least through its procurement policy, the Chinese state influences companies. The state is considered the largest purchaser in the People's Republic and can thus specifically promote or deliberately neglect certain industries (Heilmann 2016, p. 202).

6.3 Number and Productivity of State-Owned Enterprises

Compared to the years before the turn of the millennium, SOEs have become less relevant in the Chinese economy, although their importance has been increasing again for the past 2 or 3 years. In 1998, SOEs still provided 60 percent of total employment in China; by 2010, this share had fallen to 30 percent. The share of SOEs in the total number of industrial enterprises fell from 40 percent to 10 percent over the same period (International Trade Administration U.S. 2017; Heilmann 2016, p. 200).

In total, there are about 150,000 SOEs. The central government is the owner of about a third of them. The rest are owned by provincial and municipal governments. More than 100 of these SOEs are managed directly by the central government, or more precisely by SASAC. Of these, in turn, nearly two-thirds are listed domestically or abroad (International Trade Administration U.S. 2017; Heilmann 2016, p. 61). SOEs can be found in any sector of the economy. The share of SOEs in the electricity supply and coal extraction sector is 15 percent. This is followed by the aerospace industry and the military sector with 10 percent. Close behind are machinery, information technology, real estate, and building materials, each with 9 percent. The agricultural sector comprises just 4 percent and the iron and steel industry only 3 percent (China Institute at the University of Alberta 2018).

6.4 Economic Efficiency of State-Owned Enterprises

SOEs are controversial in their function; on the one hand, they form the backbone of the economy in many areas, on the other hand, they are subject to public criticism because of state intervention. The Party's efforts to reform the state sector and make it more competitive are contrasted with the social functions that SOEs still fulfill (Wei and Li 2019, pp. 37–38).

6.4.1 Competitiveness

SOEs compete with private firms in China (Heilmann 2016, p. 199). It has already been shown in the past that most SOEs cannot compete with private firms in terms of productivity. The average return on assets is 4–6 percent higher for private firms than for state-owned firms (Trivedi 2018). Even during the years of high economic growth until the financial crisis of 2008, both productivity growth and profitability of SOEs were significantly lower than those of the private sector. Furthermore, various studies indicate that return on assets as well as return on capital are significantly lower for SOEs than for private enterprises (Cheng et al. 2018, pp. 2, 22).

This is consistent with the findings of other studies, which indicate that private enterprises, in particular, have contributed to China's economic recovery. The economic output of private enterprises has risen from zero percent before the reform and opening up to the current level of 52 percent of total GDP. However, it is striking that this share has stagnated in recent years, even though the return on assets of private enterprises is at a high level and these enterprises are extremely productive, implying high growth potential (Walk 2019, p. 2).

6.4.2 State Intervention

In the Chinese market, competition is distorted enormously. Heavily subsidized SOEs and competitive private enterprises are pitted against each other (Heilmann 2016, p. 199). Both domestically and internationally, the close relationship between the government and SOEs is criticized (China Institute at the University of Alberta 2018, p. 16; Wei and Li 2019, p. 37).

The 2008 financial crisis prompted the Chinese government to provide SOEs with cheap loans to contain the damage caused by the crisis. However, they used the capital provided to them to make investments in the real estate market. This has resulted in a sharp increase in the debt of Chinese SOEs. Although the financial crisis has been overcome, SOEs continue to be given preferential treatment when it comes to lending. Not only do they receive loans from state-controlled banks more easily than private companies, but they also receive better terms (Fan and Hope 2013, p. 11). In 2016, 83 percent of all loans went to SOEs (Walk 2019, p. 1). Private enterprises are disadvantaged; they do not have the same financing options as SOEs (Cheng et al. 2018, p. 25). Furthermore, there are barriers to entry into a wide variety of markets for which SOEs obtain licenses much more easily than private enterprises (Fan and Hope 2013, p. 14). In addition, SOEs have strategic resources such as land and natural resources, which gives them a monopoly position (Cheng et al. 2018, p. 18).

In 2006, SASAC defined seven strategic industries in which Chinese companies should retain absolute control. These sectors are national security, energy supply, petroleum and petrochemicals, telecommunications, coal mining, aerospace, and shipping. In addition, SASAC has identified five other sectors that should remain under state control, namely machinery, information technology, the automotive industry, the construction sector, and the extraction and production of steel, base metals, and chemicals (Fan and Hope 2013, p. 13). What is problematic about the influence is that overcapacities are created, such as in the case of solar panels or steel (China Institute at the University of Alberta 2018, p. 16).

6.4.3 Social Responsibility

SOEs not only pursue economic interests, but also take on social tasks (Wei and Li 2019, p. 37). These are usually difficult for private companies to implement due to their competitive orientation. These include, for example, strategies to make the country competitive in key industries, maintaining economics and social stability, and providing public goods (Wei and Li 2019, pp. 102–103). In this regard, three types of enterprises can be distinguished. First, there are the normal commercial enterprises whose main focus is on profit maximization. In addition, there are SOEs that operate in strategically important sectors and thirdly, there are public service enterprises that are supposed to focus on the quality of their services rather than their profitability (Asia Society and Rhodium Group 2018).

Social benefits, education, health care, pensions, and job security were the responsibilities of SOEs before the 1978 reforms (Fan and Hope 2013, p. 2). Currently, over 17 million retired workers still receive their pensions through SOEs. In addition, job retention is a high priority. For example, some SOEs hired workers even though there was no need for them (China Institute at the University of Alberta 2018, p. 11). The average number of employees in SOEs exceeds the average number in private enterprises by a factor of three (Cheng et al. 2018, p. 13).

The development of infrastructure and the promotion of urbanization fall within the scope of activities of SOEs operating in the public sector. For example, the strong state presence in the telecommunications sector has ensured that the country's network expansion has been realized in the less developed regions. Similarly, state-owned corporations guarantee energy supply for the population and businesses (China Institute at the University of Alberta 2018, p. 11). SOEs must operate profitably while ensuring the stability of the national economy, providing public goods, and keeping employment rates high (Wei and Li 2019, p. 4).

6.4.4 Zombie Companies

The State Council defines zombie enterprises as firms that have made losses for three consecutive years, cannot meet environmental and technology standards, are not in compliance with national industrial guidelines, and rely on government support or banks (Lam et al. 2017, p. 6). According to official data from SASAC, there were a total of 2041 zombie companies in 2016, many of which were in the coal and steel sectors. Due to overproduction, the corporations rely on loans to continue their day-to-day operations (Hao 2019).

Zombie companies are a major problem for the Chinese economy. For years, these companies received loans, preventing more meaningful investments elsewhere (Woo and Zhang 2019). China's zombie companies contribute significantly to corporate debt as well as the low productivity of SOEs. While the government hopes to revive the firms through capital investment, it often achieves the opposite. However, zombie companies become dependent on their backers. Thirty percent of enterprises are not viable after 5 years (Lam et al. 2017, pp. 4, 16).

The government plans to eliminate this problem by 2020 (Woo and Zhang 2019). Local authorities have been tasked with drawing up lists of zombie companies. They will either be closed down or efforts will be made to rebuild the companies. Measures beyond capital provision, such as stopping excess production, are to be established (Ouyang and Wang 2018). However, SASAC has not disclosed exact details so far (Hao 2019).

Probably the greatest concern for those responsible is the loss of jobs. After all, SOEs, especially in the hard-hit coal and steel sector, employs countless people. A medium-sized state coal mine employs between 2000 and 8000 workers. Many of them have been employed in the sector for generations so that in the event of a closure, entire families would be left without work. Unrest, which could become a direct threat to the CCP's leadership position, would be inevitable (Hao 2019). Furthermore, in some cases workers would have to be relocated to find new work, as there are usually no other employment opportunities in the regions where the large corporations are located (Woo and Zhang 2019).

Closures of zombie companies represent a loss for financial institutions. Banks that have provided loans to these companies for a long time fear for their existence if the loans are not repaid. The reasons for local government officials' concern are many. Some simply have a good relationship with the companies, while others fear unrest that could be an obstacle to their careers. In addition, in the event of closures, local governments would have to forgo tax revenues from these companies in subsequent years (Hao 2019).

6.5 State Enterprise Reforms

Since the financial crisis in 2008, China's SOEs have been in crisis and their underperformance in the economy is negatively affecting the growth of the entire country. This is prompting the government to initiate reforms to improve the efficiency of weakening SOEs (Lardy 2019, p. 81). Reform of Chinese SOEs began as early as the late 1980s (Taube 2014, pp. 659–660). According to Lardy, in recent years the Chinese government has adopted six basic approaches to revitalizing the state sector: Socialization, merger, shared ownership, debt-bond transfer, governance reforms, and financial reforms (Lardy 2019, p. 81).

6.5.1 Socialisation

Through socialization, China's traditional SOEs are transformed into joint-stock companies and limited liability companies. Company formations are necessary to be able to list SOEs on the national and international stock exchanges. The first steps socialization were taken in the early 1990s, and a draft law was passed in 1993. A year later, an enterprise law was passed that started the process of transforming SOEs (Pigott 2000, p. 53). By 2011, approximately 106,000 SOEs, or about 40 percent of all SOEs, had been transformed into jointcompanies or limited liability companies. With the new corporate form, firms have shareholders for whom shareholder meetings must be held. Furthermore, a list of shareholders must be maintained and share certificates issued. This increases the transparency of the activities of SOEs (Lardy 2019, pp. 81–82).

The long-term impact of the reform can best be seen in state-owned industrial enterprises. In 1995, there were approximately 120,000 state-owned industrial enterprises. After the Enterprise Law came into force in 1994, a significant decline in SOEs to 42,000 was first seen in 2000. However, the decline cannot be explained by socialization alone; a large number of state-owned industrial enterprises will have been privatized or closed down because of the state enterprise reforms by then Premier Zhu Rongji. In 2000, just 11,000 industrial SOEs had been socialized (Lardy 2019, pp. 82–85).

6.5.2 Mergers

Mergers are intended to improve the performance of the largest SOEs (Lardy 2019, p. 86). The primary goal is to reduce central SOEs (Bloomberg 2016). Some have been merged to increase their influence and reduce competition in the relevant sectors (Wildau 2017). These "megamergers" also aim to increase their market share at the global level (Wildau 2017). For example, in 2015, the government merged two train manufacturers into one megamerger group to increase the number of orders from abroad (Bloomberg 2016). In doing so, the state is preparing domestic companies for the global "One Belt One Road" infrastructure project (Wildau 2017). Mergers have a further benefit through their representative effect, as in this way the reform of the state sector appears to be progressing (Wildau 2017). Mergers are implemented by SASAC. From 2013 to 2017, the original nearly 200 SOEs under the central government were reduced to 101 (Wildau 2017). Among the merged companies were large oil companies, telecommunication companies, most of the state-owned power generators and electricity providers, steel companies, and the largest airlines (Lardy 2019, p. 86). When two state-owned companies merge, they remain 100 percent state-owned.

Despite the mergers by SASAC, success did not materialize. The transformed firms became larger as a result, but no more efficient than before (Trivedi 2018). The returns of the firms involved decreased (Bloomberg 2016). Indirectly, SASAC admitted the negative trend in asset returns because it reported a profit increase of 30 percent in the years from 2012 to 2016 to the previous period. At the same time, since total assets increased by 80 percent from 2011 to 2016, this indicates that profits decreased relative to assets (Lardy 2019, pp. 88–89).

Mergers between two industrial giants in the same sector create monopolies. This reduces competition in the relevant industrial sectors, which in turn leads to low innovation potential (Lardy 2019, pp. 96–97). In this way, large SOEs become even larger and absorb their competitors, so that competition ceases (Wildau 2017).

6.5.3 Mixed Ownership

By mixing ownership, SOEs are to be transformed. With the implementation of this state program in 2013, the possibility was created to invest in SOEs with private and collective capital. As early as 1997, the foundation for this was laid by Jiang Zemin, then general secretary of the CCP. He had the idea that publicly owned forms of the enterprise should not be limited to state-owned and collective enterprises, but should be allowed to mix with private ownership (Lardy 2019, p. 91). In 2013, there was a crucial innovation, since then the principle of mixed ownership works both ways. Private capital is tolerated in SOEs and state capital can be invested in private enterprises. Before 2013, only private capital was allowed to invest in SOEs (Lin 2018). Under the reforms, the government is increasing the number of enterprises with shared capital and supporting the private sector (Nan 2018). Already in 2017, two-thirds of enterprises under the central government and 50 percent of subsidiaries of these enterprises were at least partially financed by private capital (Lardy 2019, pp. 91–92). Between 2013 and 2017, private firms invested a total of about RMB 1.6 trillion in SOEs. At the same time, local SOEs invested more than RMB 600 billion in private sector firms (Nan 2018).

It is speculated that this reform step only serves to relieve the state. This is because by bringing in private capital, the government has to invest less to prop up SOEs (Lardy 2019, pp. 91–92). Furthermore, this allows the government to increase its influence in the private sector, while private entrepreneurs lose their voice. This is because although private firms are equal to their state counterparts in theory, in reality, inequality prevails. The majority of the companies concerned continue to be controlled by the state after the reforms, and positive changes are manageable. In some cases, the situation of the companies is worsening. Due to price differences in the two sectors, profits can be made from the transfer of assets between state and private companies. Mixed ownership has made such machinations much easier, and entrepreneurs have been virtually encouraged to take advantage of this (Lin 2018).

6.5.4 Debt-Bond Transfer

Another way to reform state-owned firms is to transfer debt and bonds. Corporate debt should be transferred directly to investors as bonds (García-Herrero and Ng 2018). In other words, the bank forgives the outstanding debt of the indebted firm and is awarded company shares in exchange, which are then to be resold to investors. In practice, the whole process is quite complicated. The procedure described above is hampered by a law that prohibits banks from owning shares in companies that do not belong to the financial sector. To get around this regulation, the bank invests capital in a legally completely separate subsidiary that acts as a link. This then buys the loans from the SOEs from the bank and exchanges them for shares in the enterprise with the original borrower. This does not violate the Banking Act, since the intermediary company belongs to the financial sector and the bank is thus allowed to own shares in it. Nevertheless, the risk remains with the banks because in the event that the intermediary firm has to sell the bonds for less than the value of the original debt, the value of the intermediary firm will also fall. This would then be reflected on the bank's consolidated balance sheet (Lardy 2019, p. 93). In recent years, the Chinese government has been trying to promote the transfer program and make it more attractive to investors. Some changes have been made, and foreign investors are now also to be encouraged to participate (Zhang 2019).

Implementation is difficult due to the enormous sums of money involved and the fact that the value of the companies concerned is determined on a market basis. Very few of these companies are listed on the stock exchange, which makes price negotiations difficult. Furthermore, it is planned that investors from the non-financial sector will be involved. These are to buy up appropriate shares first and then enter into the management of the company. The problem is that outside investors would like to have control over the companies. However, this is not in line with the government's interests. Not least because of this, few of these transfers have been implemented in reality to date (Lardy 2019, pp. 93–94). However, the program only focuses on large SOEs and therefore only allows the country's largest banks to participate (García-Herrero and Ng 2018).

6.5.5 Corporate Governance Reforms

With the introduction of enterprise management reforms in 2013, the Chinese government started to divide the SOEs under its control into two different categories. This concerns all state-owned companies that do not operate in the financial sector. They were divided into commercial and public services. Since it can be assumed that most companies will be classified as commercial, these reforms are a further step towards increasing the profits of SOEs. Companies classified as commercial are to be judged on their financial performance. As a result, companies that generate losses will no longer be granted loans and other sources of funding will also be eliminated if they perform negatively. In the long run, this will either lead to bankruptcy of these companies or takeover by other companies. However, there are special rules for commercial SOEs belonging to the national security sector, for example. In addition, public sector firms that operate in sectors where the government sets prices cannot be valued based on their financial performance. Examples of such firms can be found in the public transport sector, where transport fares are fixed. Therefore, these companies are instead judged on variables such as the quality of products or services (Lardy 2019, p. 94).

SASAC was commissioned to categorize the companies and completed this back in 2014. However, for unknown reasons, this categorization was rejected by the government. It took until 2017 to create a new classification that included all SOEs. However, information on the classification was never made public, suggesting that some influential large corporations were able to reject the classification as a commercial enterprise to avoid having to give up privileges in the future (Lardy 2019, p. 95).

6.6 Return to a Planned Economy

The PRC uses large SOEs to maintain economic output (Trivedi 2018). Although private enterprises have become more important after China's opening and the independence of state-run enterprises has increased (Ankenbrand 2017), since about 2017, officials favoring the old system of state control over the economy have gained influence (Hornby 2018). The CCP's support of SOEs has increased, and the government's authority has also been strengthened (Ankenbrand 2017). Especially in times of stagnant economic growth, the CCP resorts to the time-honored system of large SOEs to ensure stability (Trivedi 2018).

These changes in the planned economy can be seen in the percentage of the country's total profits accounted for by the profits of state-owned companies. This fell from over 30 percent in 2012 to just 15 percent in 2016 but has since increased again so that the percentage share was over 30 percent again in 2018 (Trivedi 2018). Moreover, the focus on the state sector is evident in the People's Republic's economic programs. In the awarding of projects under One Belt One Road, it is clear that the promotion of SOEs has come to the forefront for China's policymakers. The ten strongest Chinese companies operating abroad are all state-owned (Lardy 2019, p. 122).

The coastal region is economically progressive and dominated by private companies. In the interior, on the other hand, state-owned heavy industry is often present. Unbalanced development of the different areas could lead to social instability (Ankenbrand 2017). Since 2011, the Chinese government has been trying to find a solution to this problem that is as business-friendly as possible (Fan and Hope 2013), but until then it has relied on SOEs as a factor of stability in China's western and northeastern regions.

6.7 Conclusion

Many reforms aim to increase the autonomy of enterprises by mixing ownership or changing the form of the enterprise. However, the Chinese government is unwilling to reduce or even relinquish control over SOEs. This reduces the benefits of many of the changes brought about by the reform measures of recent years.

Promoting the profitability of SOEs is vital for the CCP's survival. This is because the CCP's authoritarian leadership is legitimized by the steady rise in the population's income. Without economic growth, this cannot be guaranteed in the long run, which could lead to a threat to the CCP's position of power. The government needs to launch reforms that make the state sector more profitable while ensuring a high level of employment. It is conceivable that in the context of the trade conflict with the US, SOEs will act as economic stabilizers. Whether this phenomenon is short-lived will be seen in the coming years.

References

Ankenbrand, Hendrik. 2017. China will wieder mehr Staat. *Frankfurter Allgemeine Zeitung*. https://www.faz.net/aktuell/wirtschaft/chinas-staatschef-setzt-wieder-verstaerkt-auf-planwirtschaft-15252876.html. Accessed 18 Sep 2019.

Asia Society and Rhodium Group. 2018. State-owned enterprise. *The China dashboard*. https://aspi.gistapp.com/fall-2018/page/state-ownedenterprise. Accessed 03 Feb 2020.

Bloomberg. 2016. *Why China's $1 trillion merger makeover could fail*. https://www.bloomberg.com/news/articles/2016-09-07/china-s-1-trillion-makeover-of-bloated-soes-attracts-skeptics. Accessed 18 Sep 2019.

Cheng, Hong, Hongbin Li, and Tang Li. 2018. *The performance of state-owned enterprise: New evidence from the China employer-employee survey*. Standford Center on Global Poverty and development. Working Paper No. 1037. https://kingcenter.stanford.edu/publications/performance-state-owned-enterprise-new-evidence-china-employer-employee-survey. Accessed 19 Sep 2019.

China Institute at the University of Alberta. 2018. *State-owned enterprises in the Chinese economy today: Role, reform, and evolution*. https://cloudfront.ualberta.ca/-/media/china/media-gallery/research/policy-papers/soepaper1-2018.pdf. Accessed 18 Sep 2019.

Fan, Gang, and Nicholas C. Hope. 2013. *The role of state-owned enterprises in the Chinese economy*. US-China 2022: Economic Relations in the Next 10 Years. China-United States Exchange Foundation. https://www.chinausfocus.com/2022/wp-content/uploads/Part+02-Chapter+16.pdf. Accessed 18 Sep 2019.

García-Herrero, Alicia, and Garry Ng. 2018. *China's "matryoshka" approach for debt-to-equity swaps could be good for banks, but bad for investors*. http://bruegel.org/2018/03/chinas-matryoshka-approach-for-debt-to-equity-swaps-could-be-good-for-banks-but-bad-for-investors/. Accessed 18 Sep 2019.

Giese, Ernst, and Gang Zeng. 1993. Regionale Aspekte der Öffnungspolitik der VR China. *Geographische Zeitschrift* 81 (3): 176–195.

Hao, Nicole. 2019. Chinese authorities begin the complicated task of shutting 'Zombie' businesses. *The Epoch Times*. https://www.theepochtimes.com/chinese-authorities-begin-the-difficult-task-of-shutting-down-zombie-firms_2789145.html. Accessed 18 Sep 2019.

Heilmann, Sebastian. 2016. *Das politische System der Volksrepublik China*. Wiesbaden: Springer Fachmedien.

Hornby, Lucy. 2018. China renews focus on state-owned sector as US trade war builds. *Financial Times*. https://www.ft.com/content/f1ca06b2-c837-11e8-ba8f-ee390057b8c9. Accessed 18 Sep 2019.

International Trade Administration U.S. 2017. *China – state owned enterprises*. https://www.export.gov/article?id=China-State-Owned-Enterprises. Accessed 19 June 2019.

Lam, Raphael W., Alfred Schipke, Yuyan Tan, and Zhibo Tan. 2017. *Resolving China's zombies: Tackling debt and raising productivity*. IMF Working Paper – Asia and Pacific Department. https://www.imf.org/en/Publications/WP/Issues/2017/11/27/Resolving-China-Zombies-Tackling-Debt-and-Raising-Productivity-45432. Accessed 18 Sep 2019.

Lardy, Nicholas R. 2019. *The state strikes back*. Washington, DC: PIIE Peterson Institute for International Economics.

Lin, Zhang. 2018. The three dangers of China's mixed-ownership reform. *South China Morning Post*. https://www.scmp.com/news/china/economy/article/2158036/three-dangers-chinas-mixed-ownership-reform. Accessed 18 Sep 2019.

Nan, Zhong. 2018. Mixed ownership reform to expand. *China Daily*. http://www.chinadaily.com.cn/a/201811/21/WS5bf499aca310eff303289f70.html. Accessed 18 Sep 2019.

Ouyang, Shijia, and Yanfei Wang. 2018. Major measures to cut zombie firms. *China Daily*. http://www.chinadaily.com.cn/a/201812/05/WS5c0721e0a310eff30328f16d.html. Accessed 18 Sep 2019.

Pigott, Charles A. 2000. *Reforming China's enterprises*. Paris: Organisation for Economic Co-Operation and Development (OECD).

Qing, Zhao Youg. 1996. The company law of China. *Indiana International & Comparative Law Review* 6 (2): 461–492.

Slegers, Arnd, and Peter Atzler. 2017. *Chinesische Staatsunternehmen verstehen*. Wiesbaden: Springer Gabler.

Taube, Markus. 2014. Wirtschaftliche Entwicklung und ordnungspolitischer Wandel in der Volksrepublik China seit 1949. In *Doris Fischer und Christoph Müller-Hofestede*, ed. Länderbericht China, 645–679. Bonn: Bundeszentrale für politische Bildung.

Trivedi, Anjani. 2018. *China's imperial growth delusion just won't die*. https://www.bloomberg.com/opinion/articles/2018-09-11/china-turns-to-state-companies-to-bolster-economic-growth-again. Accessed 18 Sep 2019.

Walk, Edgar. 2019. *Opfert China sein langfristiges Wachstumspotenzial?* https://www.fundresearch.de/fundresearch-wAssets/partnercenter/metzler-asset-management/docs/Marktaktuell_KW06_2019.pdf. Accessed 18 Sep 2019.

Wei, Qingong, and Hanlin Li. 2019. *Entities and structures in the embedding process*. Singapore: Springer.

Wildau, Gabriel. 2017. China prepares fresh round of state-orchestrated megamergers. *Financial Times*. https://www.ft.com/content/e3972f54-62e2-11e7-91a7-502f7ee26895. Accessed 18 Sep 2019.

Woo, Ryan, and Lusha Zhang. 2019. *'Zombie' enterprises hampering China's economic transformation: Chinalco*. https://www.reuters.com/article/us-china-economy-zombiefirms/zombie-enterprises-hampering-chinas-economic-transformation-chinalco-idUSKBN1QR05V. Accessed 18 Sep 2019.

Zhang, Yue. 2019. More efficient debt-to-equity swaps urged. *China Daily*. http://www.chinadaily.com.cn/a/201905/23/WS5ce5facaa3104842260bd4a3.html. Accessed 18 Sep 2019.

Private Company

7

Enrico Cordes, Konstantin Krone, and Viktoria Paul

Private enterprises are companies in which private individuals have predominant ownership and control rights. The state does not provide the managerial staff and profits generated remain in the companies. This raises the question of whether private companies have any right to exist in the PRC – a communist state – since under communism all property is common property. The People's Republic walks a tightrope in allowing private enterprises. In thc PRC, a distinction can be made between those companies that are close to the state and receive state support, and those companies that act predominantly market-based and competition-oriented (Heilmann 2016, pp. 199–201).

In addition, there are a large number of mixed forms and gradations, each with a different legal framework, such as collective enterprises, joint stock companies with state and private investors, or joint ventures with foreign partners. Since it is difficult to draw a clear distinction between enterprises with varying degrees of state influence, most economists divide enterprises into state-owned as well as non-state-owned enterprises and do not pursue further differentiation (McGregor 2010, p. 200). Non-state-owned enterprises include collective enterprises and foreign enterprises (Lardy 2019, p. 58).

China had a total of more than 18 million companies in 2017. Of these, more than 300,000, or just under 2 percent, were state-owned. Just under 250,000, or just under 1.5 percent, were collective enterprises, over 16 million, or almost 90 percent, were privately

E. Cordes
University of International Business and Economics, Beijing, PR China

K. Krone (✉)
University Mannheim, Mannheim, Germany

V. Paul
KPMG, Düsseldorf, Germany
e-mail: viktoriapaul@kpmg.com

B. Darimont (ed.), *Economic Policy of the People's Republic of China*,
https://doi.org/10.1007/978-3-658-38467-8_7

owned, and over 100,000, or more than 0.5 percent, were foreign-owned. Over one million, or more than 5 percent, were 'other', although it remains unclear what is meant by this (National Bureau of Statistics of China 2018). Individual businesses were not included here.

Among the various forms of enterprises in China, private enterprises generate nearly 70 percent of China's gross domestic product (GDP) and account for about 90 percent of new jobs created (Lardy 2014, p. 122). Networks provide advantages to private enterprises: Entrepreneurs' networks often include officials, and direct relationships between entrepreneurs and state officials play an important role, in creating a symbiotic relationship between government and business.

In the following, we will first clarify what characterizes private enterprises in the PRC. This is followed by an explanation of the interdependence between private enterprises and the state. In recent years, the Chinese government has encouraged the establishment of companies, which is why the extremely virulent start-up scene will be examined afterward. When many companies are founded, the question arises as to how failures are dealt with. Therefore, the chapter concludes with an inventory of bankruptcies in the PRC.

7.1 Emergence of Entrepreneurship

In the 1980s, private enterprises emerged in the form of individual businesses (chin.: gètǐ gōngshāng hù). Typically, these were family businesses that were not allowed to hire more than eight employees outside the family. However, many companies exceeded this limit (Dickson 2008, p. 34). Although private ownership was first mentioned in a state document regulating employment in private enterprises as early as 1984 (Dickson 2008, p. 35), it was not until 1988 that the National People's Congress passed an amendment to the constitution that allowed private enterprises in the first place (Tse 2015, p. 43).

At the end of the 1980s, various problems arose as part of the general reform and opening-up policy. The growth spurt resulted in severe inflation, and there was also resentment towards corruption by officials this eventually led to protests among the population (Tse 2015, p. 44). After the 1989 demonstrations, which were not supported by the vast majority of private entrepreneurs (Dickson 2008, p. 17), reform stalled for the time being. In 1989, private entrepreneurs were banned from joining the CCP, which had previously been at least tolerated (Dickson 2008, p. 3). Therefore, most entrepreneurs registered as collective enterprises, which were publicly owned and supervised by the local government (Dickson 2008, p. 37). However, these enterprises were run like private businesses, they were called "red hat collectives". Concrete statistics suggest that the number of private enterprises at the time was relatively small: In 1990, only 72 million people (8 percent of all workers) worked in private enterprises (Lardy 2014, p. 19).

In 1992, in the course of Deng Xiaoping's trip to South China, an economic boom began. Deng justified the market economy he intended with the following words: "A market capitalism is not capitalism, because there are also markets under socialism" (Dickson 2008, p. 38). In this context, the CCP urged its members to "jump into the sea" (chin.: xiàhǎi) of private enterprise. In this way, Party members were to show others how they could take the lead in getting rich (Dickson 2008, p. 19). After 1992, the number of private enterprises increased by 35 percent annually (Dickson 2008, p. 38). In 1993, many state-owned enterprises were privatized. Limited liability companies were introduced in 1994 and just three years later 48 percent of all registered private companies were limited liability companies, rising to 65 percent in 2004. Previously, private enterprises could only be wholly owned with unlimited liability of private property (Nee and Opper 2012, p. 116).

Jiang Zemin, as President of the PRC in 1997, promoted mixed ownership: "Public ownership must not only be state-owned or collective, but it also includes mixed ownership." However, mixed ownership is often criticized for serving as an end to access to private capital (Lardy 2019, p. 91). In 2000, the party government finally deviated from its hitherto restrictive attitude towards private enterprises. Instead of "guide, monitor and control", "support, encourage and lead" became the guiding principle (Dickson 2008, p. 39). On 01 July 2001, Jiang Zemin promulgated the "Three Representatives" principle (Hong 2015, p. 51). Besides peasants and workers, China was now represented by "advanced productive forces". The status of private entrepreneurs improved significantly (Dickson 2008, p. 28). Eventually, some private entrepreneurs were crowned "model workers" (Dickson 2008, p. 44). In 2002, the CCP's constitution was amended to allow private entrepreneurs to join the party before 2002, private entrepreneurs had joined the party illegally (Dickson 2008, p. 3).

The party had realized that it needed representatives of new technologies (Hong 2015, p. 51). From 2002 onwards, party structures were established within private companies, thus continuously working on the involvement of private companies in the political system (Heilmann 2017).

7.2 Characteristics of Private Companies

The number of private enterprises has increased by 146 percent during the last five years (Xinhua 2018). Accordingly, the importance of the private sector increased. Private companies already generate more than half of the tax revenue (Sommer 2019, p. 67).

7.2.1 Importance for the Economy and the Labour Market

The exact size of the private sector is unknown, as it is difficult to define the original ownership structure. Estimates of the private sector's contribution to China's GDP vary. The most significant difference in estimates was in 2005 when the Hong Kong-based investment group CLSA estimated that GDP generated by private enterprises was 70 percent, while the Swiss bank UBS assumed that the private sector contributed no more than 30 percent to GDP, regardless of the indicators (McGregor 2010, p. 199). Reasons for the discrepancies in estimates of private companies and their economic power in China vary. For example, Huawei, the largest private company in China, has never published a full breakdown of its ownership structure. In addition, when asked whether their companies are "sīyíng", i.e. "privately run", private entrepreneurs prefer to answer "mínyíng", literally: "run by the people" (McGregor 2010, p. 200).

Private companies are important for the labor market in China; estimates suggest that 60–85 percent of the total labor force is employed by private companies. Moreover, private enterprises generate almost all of the growth in the labor market. They contribute 90 percent of all new jobs and are responsible for all growth in the labor market since 1978 (Lardy 2014, p. 122).

7.2.2 Sectoral Focus of Private Companies

Since 2010, the state has specifically promoted some sectors by supporting private companies in these areas (Fischer 2017, p. 188). As a result, private companies are overrepresented in some sectors, such as retail, while the state continues to dominate in other sectors. Here, private companies are either not represented at all or underrepresented, such as in the tobacco industry. According to the chairman of SASAC (State Assets Control and Administration Commission), Li Rongrong, China has set itself the goal of maintaining absolute control in seven strategic sectors: Military industry, electricity generation, and distribution, oil, petrochemicals, telecommunications, coal, civil aviation and shipping (Lardy 2014, p. 54). In addition, the state controls other "products of a resource nature" such as water and gas (Lardy 2014, p. 12). Furthermore, except shadow banks, the capital market is one of the most restrictive markets. In 2012, only 15 percent of all bank investments could be considered private (Lardy 2014, p. 20). Due to the monopoly position of the state in some sectors where private companies can't enter dictated and above-average returns can be achieved (Lardy 2014, p. 24).

Among 14 different services, China is only as open as the OECD average in three sectors, namely architecture, engineering, and IT. In other sectors such as banking, insurance, telecommunication,s and logistics, China is two to four times more restrictive (Lardy 2019, p. 101). Nevertheless, most private companies or self-employed individuals are active in the service sector, with wholesale and retail trade dominating. In 2009, out of over seven million private enterprises, 68 percent or over five million were in the service sector, of which again over two million were in wholesale and retail. Thus, 80 percent of retail trade was privately owned (Lardy 2014, pp. 72, 80). While private firms could operate retail and restaurant businesses early in the reforms, modern service sectors often remained closed to them. The share of investment by private firms in scientific research and technical services increased from 35 percent in 2012 to 53 percent in 2015. Across the service sector, investment by state-owned enterprises was only 45 percent of all investments in 2015 (Lardy 2019, pp. 39–40).

In industry, the share of investment by state-owned enterprises declined significantly. In 2011, the share of industrial investment by private companies was almost three times that of state investment (Lardy 2014, pp. 140–141). The decline in investment is reflected in the share of industrial production. While state-owned enterprises were responsible for 78 percent of output in 1978, the share fell in comparison to 26 percent in 2011 (Lardy 2014, p. 72). The difference in the total returns to the capital of private and state-owned enterprises is particularly striking. This difference fell across the industry in 2016. SOEs' 2.5 percent return on assets was significantly lower than private firms' 10.5 percent (Lardy 2019, p. 57). In the service industry, private firms have been able to show much higher returns on assets. Thus, they are much more efficient (Lardy 2019, p. 39).

7.2.3 Success of Private Companies

In China, a total of 408 IPOs were recorded in 2017. Of these, 377 were private companies and 31 were SOEs (Lardy 2019, p. 111). The latter are much more indebted than the former. In 2016, the debt-to-equity ratio of private companies was 105 percent, whereas the debt-to-equity ratio of state-owned enterprises was 160 percent (Lardy 2019, p. 61). Private companies are significantly more productive and successful than many state-owned enterprises, which can be attributed to several reasons. For example, private companies are much more market-oriented, which brings significant advantages to free markets (Lardy 2014, p. 49). What are the success factors of private companies in China?

7.2.3.1 Networks

The success of private entrepreneurs in China is heavily dependent on functioning networks and relationships. As in other countries, there are representations of the interests of economic actors in politics. Private entrepreneurs represent their interests through informal channels, such as lobbying associations, business clubs, etc., or formal organizations, such as the People's Congresses, the Political Consultative Conference of the CCP, and business associations (Heberer and Schubert 2017, p. 100).

On the one hand, there are entrepreneur clubs such as the Chinese Entrepreneur Club (CEC), in which almost all of China's important private entrepreneurs are members, or the China Entrepreneur Forum (CEF), which has several hundred members. There are also numerous local and regional associations, such as the General Association of Zhejiang Entrepreneurs. In addition to these smaller privately founded associations, there are official associations. The largest is the All-China Federation for Industry and Commerce (ACFIC), a state-owned and party-affiliated organization that does not play a major role among private companies (Hirn 2018, pp. 31–32). Nonetheless, the ACFIC is described as "the bridge that connects the party and government with people in the non-state business sector" (Hsu and Hasmath 2013, pp. 68–69).

In many business associations, government officials play an intermediary role. Official statistics from 2004 show that in Zhejiang, the province with the largest private sector, 35 percent of all business associations appointed government officials-already retired or nearing retirement-as board chairmen, while 46 percent of private companies asked government officials to become secretaries. Many other associations have local officials either as honorary directors or as members of the advisory committee (Hsu and Hasmath 2013, pp. 76–77).

7.2.3.2 Interdependence Between Government and Private Enterprise

The Chinese economic model has been described by Lardy (2014, p. 121) as "relationship" (guānxi) capitalism. There is a strong "we-feeling" between private entrepreneurs and state employees based on guanxi – good relations – and mutual material benefits (Yan and Huang 2017, p. 41). For a long time, Chinese private entrepreneurs sought contact with the Party. For example, by the end of 2014, a full 53 percent of all private enterprises (1,579,000 in numbers) had established contacts with the CCP. At the same time, this figure was even over 95 percent for officially defined large enterprises (Yan and Huang 2017, p. 38). Good relations with the state are important because China is not yet a constitutional state. Political guidelines stand above the laws and the CCP or the state can change them in favor of individual entrepreneurs (Hirn 2018, p. 38).

However, since the PRC's accession to the WTO, a shift towards a formal legal order can be discerned (Ahl 2005, p. 30). This is evident from the rhetoric of President Xi Jinping, who said the following about building the rule of law at the 19th Party Congress of the CCP in 2017:

> Scientific legislation, strict law enforcement, just exercise of justice, and law observance by the whole people were comprehensively advanced. The building of the rule of law, the establishment of the government based on the rule of law and the establishment of a society based on the rule of law complemented each other. The rule-of-law system of Chinese-style socialism was increasingly perfected and the legal consciousness of the whole society was greatly strengthened. (Xi 2017, p. 4)

Contacts with the government and a good guanxi relationship with government officials brought many advantages. But it should not be neglected that not only good relations are important.

> Entrepreneurs can use the government to butter their bread, but they must somehow earn their bread from the start. (Ma et al. 2012, p. 103)

Party Membership of Private Entrepreneurs

According to Dickson (2008, p. 94), the first step towards a profitable relationship with the Chinese state is joining the CCP. Party membership offers entrepreneurs advantages such as easier access to credit and discretionary powers when issuing licenses and permits. In addition, there is goodwill in the implementation of workplace safety and environmental regulations as well as protection against competition and unfair political measures. Furthermore, membership in the CCP allows easier access to material resources as well as financial and tax advantages. A company that maintains a good relationship with the authorities has better chances in public tenders. In public contracts, the payment of the purchase price is guaranteed, unlike in private contracts (Ma et al. 2012, p. 18). Moreover, especially in the past, good relationships with government agencies helped overcome legal and institutional failures. A 2002 study that surveyed 2324 private firms nationwide found that a private firm's party membership status significantly contributed to the firm's profitability (Li et al. 2008, pp. 283–299).

According to Dickson (2008, p. 2), private entrepreneurs in China who are active in politics or are party members are called red capitalists. Among the red capitalists, a distinction can be made between the so-called xiahai (xiàhǎi) entrepreneurs, who were already Party members when they went into business, and co-opted entrepreneurs, who joined the Party only after establishing their business. The former constitute the majority of Red Capitalists, using their pre-existing contacts to become economically successful. Reasons for joining the Party can be of various origins. Lu Guoqiang, an entrepreneur from Zhejiang, thought about himself that he could not fight the Party outside the system and decided to change the system from within (Dickson 2008, p. 95). The desire for political participation is clear, even if it is mainly motivated by entrepreneurship in the case of co-opted entrepreneurs.

Private entrepreneurs are not satisfied with party membership. For example, a 2009 study showed that over 50 percent of private entrepreneurs are on party committees or the Chinese People's Political Consultative Conference (PPCC). At the 12th National People's Congress (NPC), private entrepreneurs were the second largest group of delegates at 23 percent (Heberer and Schubert 2017, p. 110). For example, Lei Jun, the founder of Xiaomi, was a member of the NPC, and Robin Li, the founder of Baidu, was a member of the PPCC (Tse 2015, p. 81). Wang Jianlin, the founder of the Wanda Group of Companies, aptly described the importance of the government, "If anyone says they can ignore the government in this business, I say that is impossible." Or to put it in Jack Ma's words, "Fall in love with the government, but don't marry it – respect it" (Hirn 2018, p. 38).

Influence of Companies on Politics

A company's performance is determined by forces and factors external to the market. The government can influence company performance by publishing or changing policies and regulations (Ma et al. 2012, p. 5). Conversely, companies need to develop their strategies to deal with these external factors. Businesses and the government come into contact through different channels in China. A company can approach the government on two levels, either as a whole company or through the entrepreneur as an individual. Likewise, influence can be exerted on the government as an institution (individual government agencies) and individual officials. Ma et al. (2012) identify various forms of influence, which are described below.

When individual officials are to be influenced, this is often done through a direct handover of money or benefits in kind. On a long-term basis, companies may regularly gift officials in various ways. While this practice has become more difficult under Xi Jinping's corruption campaign, it has not been abolished. An indirect way of exerting influence is for a business owner to actively support government work, for example by becoming a spokesperson or active participant in government-initiated movements. A more direct option is for companies to sell shares to government investors (asset management). In this way, the state becomes a shareholder and the private company signals power and status to the government by wearing the so-called "red hat". Then, all the preferential treatments that otherwise only state-owned enterprises have are available to the company (Ma et al. 2012, p. 71). Furthermore, entrepreneurs can join the party and occupy government posts. In these posts, they can influence draft legislation. These strategies of private companies to influence the government are mostly dependent on individuals. As a result, Chinese private companies do not seek mergers with other companies when exerting political influence but act individually (Ma et al. 2012, p. 101).

Influence of the Party on Private Companies

Hans Joachim Fuchs (2007, p. 33) described the situation as follows:

> Chinese companies are not only managed, subsidized, and controlled by business economics, but always by state economics as well. Their success or failure in China and abroad is decided not only by the markets and the management but also by the central government in Beijing and the powerful provincial governments.

Today, the state no longer controls the economy through direct ownership of enterprises, but much more through recruiting private entrepreneurs into the Party, thereby exerting indirect control (Lardy 2014, p. 59). This was also reflected in Xi Jinping's 2018 address to private entrepreneurs, "They should take the Party seriously and take its concept to their hearts" (McGregor 2018). That the Chinese government wants private entrepreneurs in the Party and congresses such as the Political Consultative Conference is reinforced by the statement of Mr. Yin, chairman of Lifan Group: "When I was elected vice chairman of the PPCC of Chongqing Municipality, I wanted to sell all my private assets and become a professional politician. But I was told that if I sold Lifan Group, I would lose all representation of the POE sector and thus would no longer be qualified for the position on the PPCC" (Ma et al. 2012, p. 31). Most private companies deny that political connections play a role (Jourdan and Ruwitch 2018). Nonetheless, telecommunications companies, such as Huawei, are directly obligated by laws such as Article 28 of the Cybersecurity Law to provide technical assistance and succor "in the context of national security" (Feng 2019).

In addition, officials, most of whom are members of the CCP, have an interest in gaining influence over private enterprises. Officials who controlled resources, land use rights, or bank loans were able to convert their power into capital by taking advantage as well as supporting entrepreneurial pursuits for additional returns, so-called "rent-seeking" (Hong 2015, p. 28). However, progress in the development of China's legal system, such as the impact of anti-corruption campaigns under President Xi, should not be ignored. Zhang (2018, pp. 392–393) examined the impact of the campaign on corporate fraud and concludes that from 2004 to 2014, due to increased enforcement activities, companies cheated by nearly 4 percent less. However, anti-corruption campaign activities have recently waned due to the trade conflict with the US.

Party Cells

Setting up Party cells or CCP grassroots organizations in private enterprises is another way of exerting political influence on businesses. Xi Jinping said in March 2012 on the establishment of party cells:

> The immediate goal was to increase the CCP's organizational presence in the nation's POEs to strengthen its influence over this increasingly important sector. (Yan and Huang 2017, p. 38)

Although the state already attached importance to party representation in private companies in the years before 2012, it focused primarily on large private companies and ignored smaller ones. While in 2000 there were party cells in just 17 percent of all private companies, by 2006 there was 35 percent. In 2014, 53 percent of all private enterprises had CPC grassroots organizations (Heilmann 2016, p. 44).

Yan and Huang (2017, pp. 46–55) illustrate the growth and importance of party cells using the example of China's Anhui province. There were 350,000 private enterprises there in 2013. These accounted for 60 percent of GDP, 70 percent of taxes, and 80 percent of urban employment in the province. At the beginning of the period under study in 2011, party cells existed in 42 percent of private enterprises. A year later, after setting the CCP's goal of having more organizational presence in all private enterprises, the percentage of private enterprises with Party cells in Anhui increased to 91.6 percent. To achieve this rapid increase, the CCP's Anhui Committee dispatched Party trainers to expand Party cells. Each instructor had to support at least three enterprises, of which at least one had to have employed 50 people. This work was supervised by the Non-State Economic Organizations and Social Organizations Working Committee (NEOSOWC). As a result, all counties and municipalities in Anhui established their NEOSOWC branches. However, many employees, including those with Party membership, were unaware of the existence of Party branches in their workplace. For example, one Party education expert said that these Party branches existed only "on paper." In many companies in Anhui, the position of Party secretary was held by the business owner, and other senior Party posts were held by the owner's immediate family members. Therefore, party cells follow the will of the owner. Against this background, party cells are of little importance.

Reasons for the CCP's Influence on Private Enterprises

The legitimacy of the CCP is based on the implementation of goals in various areas, such as the steady economic development of the PRC, social balance, and national unity (Zhu 2011, p. 124). According to Bloomberg, the most important contribution to China's economic development comes from private enterprises. Thus, the Chinese government must avoid losing its legitimacy base to private entrepreneurs to prevent an otherwise imminent loss of power. By exerting influence, the CCP creates a dependency on private enterprises. Conversely, the CCP's legitimacy depends on a thriving economy. Moreover, influence is intended to prevent private entrepreneurs from forming an opposition to the CCP, which could threaten the CCP's hold on power (McGregor 2018).

Direct dependence can mean cooperation. Technology companies, such as Tencent and its subsidiary WeChat, are subject to restrictive state regulations. WeChat must follow state censorship requirements, according to which it deletes videos, censors texts, and blocks links. In May 2018, a new Chinese Federation of Internet Companies (CFIS) was formed. Pony Ma, Jack Ma, and Robin Li, among others, were appointed as vice presidents. CFIS is under the direction of the Cyberspace Administration of China (CAC). One of the first commitments of CFIS is to "diligently study and implement the spirit of Xi Jinping's thoughts on building a cyber superpower" (Cook 2018).

7.3 Start-Up Scene

Until now, Silicon Valley in the US has been considered the go-to place for high-tech start-ups. But Taiwanese start-up investor Kai-Fu Lee, a former high-ranking official at Apple, Microsoft, and Google describes Silicon Valley as follows:

> I can tell you that Silicon Valley looks downright sluggish compared to its competitors across the Pacific. (Kai-Fu Lee 2018, p. 15)

With this statement, he refers to the start-up scene in China. Alongside the start-up scene in the USA, this is considered to be the largest and most promising in the world. It is seen as a key component of China as a location for innovation and thus as a major factor influencing China's development into the world's innovation leader. Western media frequently report on China's superiority in many future technologies, and artificial intelligence, in particular, is often mentioned. It is questionable whether the Chinese start-up landscape corresponds to this image.

At this point, it must be mentioned that, due to the topicality of this subject, there is little academic literature on the subject so far, so numerous newspaper articles and publications from on Internet had to be consulted. The book "Entrepreneurship in China: The Emergence of the private sector" by Andrew Atherton and Alex Newman from 2018 forms the basis of this section.

7.3.1 Definition of Start-Up Companies

The term start-up stands for newly founded companies that are still in the early stages of their development, i.e. in the phase in which they are not yet established on the market. However, it is not only the age of a company that determines whether it is a start-up according to the definition; after all, not every newly founded craft business is considered a start-up. Other characteristics of a start-up are an above-average degree of innovation, an exceptionally high potential to grow, and a scalable business model (Gründerszene 2018b). For the most part, start-ups have only limited financial resources at their disposal at the beginning. As a consequence, founders are often dependent on investors and external financing, such as venture capital and seed capital, at an early stage (Achleitner 2018). So-called bootstrapping, in which outside financial support is dispensed with entirely, is the exception (Gründerszene 2018a). In general, the development of a start-up can be divided into four phases. There is the early financing phase (seed phase), the start-up phase, the growth phase (emerging growth), and the expansion phase. In the early financing phase, the acquisition of financial resources plays a key role. The start-up phase begins with the founding of the company, and the product or service is optimized and launched on the market. In addition, production, sales, and marketing strategies are developed. In the growth phase, the market is to be penetrated across the board. The expansion phase is characterized by the enlargement of the company (Gründerszene 2018b).

There is no clear guideline after which period a start-up ceases to be a start-up, but the status as a start-up expires at the latest when the company is established on the market.

7.3.2 Presentation of the Chinese Start-Up Scene

The establishment of modern private companies began in China more than 30 years ago. In the meantime, a lively start-up scene has emerged.

7.3.2.1 Development of the Start-Up Scene

In the mid-1980s, in the course of the reform and opening-up policy, numerous Chinese set up their businesses. This led to the highest wave of start-ups. The focus was particularly on the manufacturing sector, but the founding of today's technology giants Huawei and Haier also occurred during this period. During the second wave of start-ups in the early to mid-1990s, the start-up atmosphere changed. The government actively encouraged private start-ups, the market was partially deregulated, and start-ups became easier. The third wave of start-ups emerged after the beginning of the Internet boom in the late 1990s and early twenty-first century. It was often educated Chinese who had lived and studied abroad who brought foreign corporate and business models to China and adapted them to the Chinese market. The focus on the startup scene shifted from the manufacturing industry to the service industry during this period. The internet giants Baidu and Tencent, which today together with Alibaba dominate the Chinese internet market, were founded at that time (Atherton and Newman 2018, pp. 10–11).

At the beginning of the twenty-first century, the Chinese government at the time realized that focusing exclusively on low-wage, labor-intensive work would not ensure sustained and steady economic growth. To keep growth and employment high in the long term, the state began to promote modern technologies and innovation (Atherton and Newman 2018, p. 42). It was in this context that funding methods such as venture capital first emerged to facilitate the financing of start-ups of all stripes. Before this, almost all public funding was for small businesses specializing in future technologies (Atherton and Newman 2018, p. 86). This funding largely ran through the Torch Program, established in

1988. This program for the promotion of innovation and technology laid the foundation for the start-up scene in its current form as early as 1988 (Molloy 2016).

Particularly since 2014, the Chinese start-up scene has developed so rapidly that it is now one of the leading global scenes (Abele 2019). This boom was fuelled by various factors. On the one hand, the Chinese government established an entrepreneurship and innovation program, under which start-ups received a whole bundle of support measures. Second, after decades of growing the Chinese economy, there were a large number of consumers willing to pay for new solutions and technologies. In addition, a critical mass of know-how and resources had been reached, so the Chinese government began to consider mass entrepreneurship as the next step in the development of global innovation leadership. As a result, several policies and measures were proclaimed to encourage entrepreneurship as well as startups (Covestro 2018, p. 3). For example, in May 2014, the general start-up requirements were significantly simplified. This measure alone had a major impact; in the same year, according to the State Administration for Industry and Commerce (SAIC), the total number of start-ups increased by almost 50 percent compared to the previous year. This can be seen as a clear signal that start-ups were becoming an increasingly viable option for the Chinese (Atherton and Newman 2018, p. 122). Furthermore, new financing methods for small businesses and the lowering of capital requirements in many key industries helped to simplify market entry for start-ups and further develop the scene (Covestro 2018, p. 3).

7.3.2.2 Scope of the Start-Up Scene

The exact size of the scene is not known. Although the SAIC collects data on new companies, it remains unclear how many of these are start-ups. Moreover, dissolutions and liquidations of startups are not statistically recorded in China, so only estimates are possible. According to Chinese information platforms, there were more than 100,000 startups across the country in 2018, of which nearly 65 percent were located in the metropolitan areas of Beijing, Shanghai, and Guangdong province (Abele 2019). More than 100 Chinese start-ups were listed on stock exchanges worldwide in 2017 (Covestro 2018, p. 3).

The number of so-called unicorns can be used as an indicator of the strength of a start-up scene. Unicorns are start-ups whose market valuation exceeds USD 1 billion (Bendel 2019). Although the number of these companies varies greatly from source to source, it can nevertheless be assumed that there are more than 160 such companies in China with a total value of almost USD 630 billion. This already exceeds the number of unicorns in the US start-up scene, although the scene there is still considered a benchmark worldwide. However, Chinese companies tend to be overvalued by international standards, so the figures may be slightly skewed. The People's Republic has also caught up with the US in venture capital. In 2018, USD 107 billion of venture capital was invested in China, compared to USD 113 billion in the US, which was only slightly more (Abele 2019). Total investment in the Chinese start-up scene is already higher than in the USA (Sieren 2018, p. 10).

7.3.2.3 Priority Sectors

Leading technology companies form an important backbone of the Chinese start-up scene. Entire start-up systems have emerged due to the active support of companies like Tencent, Alibaba, Baidu, Lenovo, Fosun, etc. This is due to their active support of startups. They provide ideas, infrastructure as well as financial resources and thus contribute significantly to shaping the start-up scene; after all, the majority of unicorns are backed by such a company. It is also thanks to these corporations that the focus of the Chinese start-up scene is primarily on the high-tech industry (Covestro 2018, p. 4).

In general, Chinese entrepreneurs see artificial intelligence (36 percent), life sciences and healthcare (16 percent), Big Data (11 percent), fintech (9 percent), and robotics (5 percent) as the most promising industries (Silicon Valley Bank 2018, p. 9). China is already a global leader in artificial intelligence, due to massive government and private investment in the field (Lee 2018, p. 3). In 2017, 48 percent of venture capital invested globally in artificial intelligence was invested in the People's Republic. In the USA, it was only 38 percent at the same time (Abele 2019).

Nevertheless, there are by far the most start-ups in the area of e-commerce. This is due on the one hand to the huge target group and on the other hand to the fact that the technical and financial hurdles for developing a good product and for entering the market are lowest in the area of e-commerce (Mattheis 2014). On the other hand, most investments flow into internet services, smart hardware, and B2B services. At the same time, the Chinese start-up scene is experiencing the highest growth in the automotive, artificial intelligence, and, logistics sectors (Covestro 2018, p. 11).

7.3.2.4 International Influences

On the one hand, foreign influences in China's startup scene are growing. An increasing number of founders and co-founders are from abroad or are Chinese nationals who have spent parts of their studies or careers abroad. Sometimes foreign founders are at the forefront of China's most successful startups. Foreign founders often operate in multinational management teams. This is to compensate for a lack of knowledge about China and local conditions, as a deep understanding of the Chinese market is required due to its fast pace and constant development (Abele 2019).

In addition, international corporations are increasingly entering the Chinese start-up scene. For example, companies such as Mercedes Benz, Bayer, and BASF are establishing their local accelerators and incubators with the aim of attracting promising business models and talent as early as possible (Abele 2019).

On the other hand, more and more Chinese players are moving abroad in search of international talent, products, and business models for the Chinese market. In recent years, several state- and privately-funded Chinese accelerators and incubators have set up shop in Silicon Valley in the US. Furthermore, Chinese startups themselves are increasingly making the move abroad. A prominent example is the scooter manufacturer Ninebot (Abele 2019).

7.3.3 Policy Guidelines

In the course of the last few years, the Chinese government has announced several policy guidelines, strategic orientations, and programs of measures that influence the start-up scene to a large extent. Both the Made in China 2025 strategic direction and the 13th Five-Year Plan of the People's Republic of China emphasize the promotion of the startup scene. This is most evident in the 13th Five-Year Plan, valid from 2016 to 2020, which specifies support measures for Chinese startups as part of an innovation-driven development of the economy. It calls for the start-up industry to flourish. The population is to be encouraged to set up businesses through specific incentives. In addition, start-up centers are to be established, certain regions are to be further developed by attracting start-ups, and jobs are to be created, especially for highly qualified personnel such as university graduates, by pushing the scene forward (National Development and Reform Commission 2016, pp. 8, 15, 71, 120).

At the same time, the Made in China 2025 program is modernizing the Chinese industry and specializing it in certain key areas. These include high-tech industries such as automotive, robotics, IT, etc. (Wübbeke et al. 2016, p. 6). Start-ups are expected to deliver innovations specifically for these industries and are supported accordingly.

7.3.3.1 Torch Program

The Torch Program was initiated by the Ministry of Science and Technology in 1988 and represents the basis of today's start-up scene (Molloy 2016). From the beginning, the program served to promote innovation and technology. Within this framework, the establishment of high-tech development zones was envisaged, where science, technology, and industry were to come together and technology was to be transformed into market-ready products (Embassy of the People's Republic of China in Ireland n.d.). Incubators and research parks were implemented in these high-tech zones, which focused on future technologies such as information technology from the very beginning of the program. Since the inception of the Torch program, 156 high-tech zones have been established and are home to over 50,000 companies (Molloy 2016). In 2017, nearly 45 percent of all business investment in research and development nationwide was attributable to companies from these zones (Abele 2019). The total volume of investment that has gone into the Torch program is over $25 billion (Molloy 2016).

7.3.3.2 Mass Entrepreneurship Programme

The mass entrepreneurship program, established in 2009, aims to stimulate entrepreneurship and thereby innovation. Initially, it was subsidized with nearly $6.5 billion annually, and by 2016, a total of more than $56 billion had been poured into the program (Reshetnikova 2018, p. 508). In 2014, this endeavor received a boost when Chinese Premier Li Keqiang emphasized the importance of entrepreneurship to Chinese economic policy at the World Economic Forum in Davos (Xinhua 2014). Meanwhile, mass entrepreneurship is considered a stated goal of the government to tap the innovation potential of the population. A document issued by the State Council in 2017 explicitly requires several ministries to take measures to promote entrepreneurship and innovation (State Council of the PRC 2017). Central and provincial governments devote huge amounts of resources primarily to promoting innovative start-ups (He et al. 2019, p. 564).

The mass entrepreneurship program is still ongoing. In September 2018, the State Council published a revised version that addressed the latest developments in the Chinese economy (China Innovation Funding 2018). The revision includes 34 specific tasks to be completed by relevant authorities in the short to medium term. Management and administration are to be better adapted to the program, and more decision-making powers are to be distributed to lower levels. In addition, support measures for innovative companies and other players in the start-up scene are to be extended, inter alia by involving start-ups more in public procurement, especially when it comes to high-tech. Other priorities include improving services for entrepreneurs and start-ups, creating jobs, promoting technology transfer between research, development, and industry, integrating international resources, and centralizing mass innovation and entrepreneurship in high-tech zones (State Council of the People's Republic of China 2018).

In addition, the NDRC, the Ministry of Science and Technology, and the Ministry of Industry and Information Technology, together with other relevant authorities, are to promote international cooperation in the field of innovation and entrepreneurship. This is to be done especially along the New Silk Road as well as in cooperation with other organizations, such as ASEAN (State Council of the PRC 2018). In early June 2019, Li Keqiang reiterated how important mass entrepreneurship is for the People's Republic (Xinhua 2019).

Looking at all these programs, it is obvious that Chinese policy has a strong focus on promoting the domestic startup scene. The measures taken illustrate the direction that the Chinese government has taken and intends to continue in the future.

7.3.4 Promoting the Start-Up Scene

At the beginning of the twenty-first century, start-ups often found it difficult to attract external funding (Ahlstrom and Ding 2014, p. 613). In the meantime, however, the Chinese start-up scene benefits from a wide-ranging system of support measures and financing methods. Privately or publicly funded incubators and accelerators, start-up cafés, business angels and other venture capitalists all mingle in this system. Roughly speaking, a distinction can be made between public and private support and financing, although mixed forms also exist.

7.3.4.1 State Support Instruments

The start-up scene receives massive support from the state, provincial and local governments. Investment in the creation and growth of innovation and high-tech companies is widespread. Indirect measures mainly aim to advance science and technology, modernize the industry, and build a bridge between academic research and practical application

(Covestro 2018, p. 5). Special attention is paid to building and developing an academic base and public investment in research and development. The number of academic works from China cited worldwide is growing rapidly, and China will be one of the most cited countries in the world in terms of scientific research. This is a crucial step in establishing a scientific base that will be transferred into the innovation and development of technology. In some respects, the investment seems to be paying off. Funding in recent years has created high-tech giants such as Xiaomi, whose success in turn encourages other entrepreneurs to start up (Atherton and Newman 2018, p. 125). Direct action will improve institutional infrastructure to facilitate entrepreneurship. Incentives and attractive financing options are offered to founders and talent is also promoted and encouraged to start a business (Covestro 2018, p. 5).

To achieve these goals, the various government agencies resort to various means. These include, for example, direct funding, tax incentives, legislative changes, development of entrepreneurial infrastructure, etc. These funds are mainly used in a sector- and location-specific manner. Especially since 2015, high-tech areas have been strategically promoted. These include industries such as electric mobility, information and communication technology, Big Data, artificial intelligence, Internet Plus, etc. (Liu 2016, p. 54). Especially under the Made in China 2025 project, many of these industries are provided with individual and specific support measures (Abele 2019). This leads to a large disparity between funding for high-tech start-ups and start-ups without a technological focus (Fuller 2010, p. 452).

There are numerous so-called start-up hubs. These are start-up hubs that are home to various players in the start-up scene. Incubators, accelerators, maker spaces, start-ups themselves, and other institutions of the scene come together here to exploit synergy effects. Most of these start-up hubs are located in high-tech zones, i.e. in large industrial parks that are already home to a large number of established and innovative companies. In addition, research parks are often attached. These contain research and development centers of private companies as well as university facilities. The aim is to transform scientific results into market-ready products.

Start-ups receive special support from provincial and local governments. For example, the Shanghai government has initiated measures to develop the local start-up scene to compete with Beijing. Rental costs for start-ups have been lowered. They are now only half as high as in Beijing, and hiring foreign employees has also been made easier. The scene in Shanghai is considered the most international in the People's Republic. The combination of elite universities, research, and development centers of global corporations, and international accelerators guarantees a developed infrastructure, both for Chinese start-ups and start-ups from abroad. Despite these support measures and Shanghai's status as an international financial center, the entire start-up support system there lags behind other national and international cities. Lacking venture capital and business angels, in particular, Shanghai has few options for providing sufficient capital to young early-stage start-ups (Atherton and Newman 2018, p. 44). Nevertheless, fintech companies in particular have a strong presence in Shanghai (Abele 2019). In addition, Zhangjiang Hi-Technology

Park is home to many international pharmaceutical and biotechnology companies. The Shanghai Technology Innovation Center, a large technology incubator, is where many tech-oriented start-ups come together and take advantage of the commercialization and development strategies there (Atherton and Newman 2018, p. 44).

In Zhejiang Province, especially in its capital Hangzhou, a variety of support measures have been taken aimed at subsidizing innovative entrepreneurship and high technology, thus particularly affecting start-ups. For example, several high-tech zones were established in 2015, one of which is "Internet Town" in Hangzhou, which alone provides space for 2000 start-ups. With Hangzhou, the provincial government is aiming for a technology center that can operate on an equal footing with the centers of Shanghai and Beijing. The success of the local Alibaba Group and its subsidiaries Taobao, Alipay, et, serve as a basis (Atherton and Newman 2018, pp. 44–45). To achieve this goal, the provincial government spares no expense or effort. Numerous incentives have been put in place for entrepreneurs. On the one hand, these include common subsidy measures such as free office space, tax breaks, free use of IT infrastructure, etc. On the other hand, unconventional measures and services are offered. A government office has been established in a technology park, which guarantees the registration of start-ups and the filing of patents within one week. In addition, the provincial government, in cooperation with the Bank of Hangzhou, has launched a start-up fund to help university graduates start up. This will provide seed funding with loans of up to RMB 50,000. Interest rates on the loans can be reduced by as much as 50 percent in some circumstances (Ernst and Young 2013, p. 5). Hangzhou benefits from the local internet giant Alibaba, which revolutionized the Chinese e-commerce market a few years ago with its online retailer Taobao. Accordingly, the focus of the start-up hub in Hangzhou is on e-commerce (Abele 2019). Similar support measures have been set up in other cities in the province, but not to the same extent (Atherton and Newman 2018, p. 45).

The scene in Shenzhen is strongly characterized by incubators and accelerators that primarily support start-ups specializing in hardware. It is not for nothing that the scene there is sometimes referred to as the "Silicon Valley of Hardware" (Abele 2019). Because of its manufacturing and innovation infrastructure, Shenzhen, in combination with Hong Kong, is now listed as the second most important innovation cluster in the world behind the Tokyo-Yokohama cluster (Bergquist 2018), making it decidedly attractive for companies from the information, communication technology, and robotics sectors. Technology giants such as Huawei and Tencent ensure a local imprint on the start-up scene there (Abele 2019). However, the state intervenes to a large extent. The local government has established a variety of policies to encourage innovation-driven entrepreneurship. For example, in 2015, an RMB 200 million fund was established to provide financial support to 50,000 hardware startups over five years. Business registrations have also been simplified, and a system has been developed to guarantee financing and loans for startups (Atherton and Newman 2018, p. 46).

Alongside Beijing, Shanghai, Hangzhou, and Shenzhen are by far the largest start-up centers in the People's Republic, which is mainly due to the enormous support provided at the provincial and local levels. In other areas of the country, similar support measures are widespread on a smaller scale. Major cities such as Nanjing or Chengdu, for example, are also building technology centers to promote start-ups (Atherton and Newman 2018, p. 46). The internal competition within the country stimulates cities and provinces to invest more and more in the start-up scene.

In general, companies that establish themselves in such centers have good chances of benefiting from direct and indirect measures by government authorities. With this locational advantage, they often qualify for tax breaks, for the receipt of research funds as well as for money for science projects from the Ministry of Science and Technology (Abele 2019).

7.3.4.2 Government and Private Support Instruments

Some support measures cannot be categorized into state support and private support, as both levels are intertwined.

Venture Capital

In venture capital, venture capitalists contribute equity capital to a company without fixed interest and repayment obligations; only in exceptional cases does this involve majority stakes (Röhr 2018, p. 7). This usually runs for a limited period so that the venture capitalist can sell his shares profitably after the company has grown (Schefczyk 2006, pp. 10–11).

Venture capital was first actively promoted in China at the beginning of the twenty-first century. Meanwhile, some of the world's largest private equity firms and venture capitalists, such as Sequoia Capital and Tiger Global, have set up shops in China's high-tech hubs and are investing heavily in start-ups. At the same time, there are numerous Chinese private equity firms and venture capitalists also making financial commitments (Atherton and Newman 2018, p. 46). Leading technology companies are increasingly acting as venture capitalists. The government itself is also involved in venture capital (Wang et al. 2013, p. 16). In 2007, the Ministry of Finance, together with the Ministry of Science and Technology, introduced regulations that determine how government agencies at the provincial and regional levels may cooperate with venture capital firms. Four investment options were highlighted through which government agencies can invest financial resources in high-tech start-ups in cooperation with private investors. Government agencies can raise seed capital for venture capital funds, invest directly in high-tech companies in cooperation with venture capital firms, subsidize private venture capital firms, and provide risk protection for private venture capital firms (Atherton and Newman 2018, p. 87).

This has resulted in a rapid increase in the venture capital industry. From 2008 to 2014 alone, the number of venture capitalists almost doubled, from 694 to 1334 (PWC 2015, p. 10). It is now easier to access venture capital in China than in the G20 countries (Ernst and Young 2013, p. 6). In terms of the total amount of venture capital raised and invested, China is now the second largest market after the USA (Covestro 2018, p. 3).

International Talent Promotion

More and more Chinese talents who have studied abroad or preferably worked in Silicon Valley are returning to China because Chinese institutions provide incentives for them to return. Foreign talent is also being recruited (Abele 2019). In 2008, the Thousand Talent Scheme was launched to encourage foreign-educated talent and scholars to work in the high-tech industry and start their own companies. By 2019, the scheme had attracted over 6000 of these individuals. Academics in particular are paid large sums of money for private use and in the form of start-up packages (Yang and Liu 2019). Experts from the high-tech industry are in high demand. These benefit from temporary tax exemptions, rent exemptions, etc. if they return. The Chinese government has thus found an effective means of bringing back talent from abroad and increasing China's competitiveness. However, application processes for these programs are sometimes opaque, so some places are awarded through good relationships. As a result, not only the elite are recruited. Independent of the Thousand Talent Scheme, there are other incentives in this direction. University incubators and partly private or state-owned high-tech and science parks, such as Zhongguancun in Beijing, are often responsible for such incentives (Abele 2019). In 2011, the People's Republic began building 150 incubators, specifically as an incentive for Chinese study abroad students to start a startup in China. In addition, those students are offered advice on startup-related topics such as financing, human resource management, etc. In total, these incubators work with around 8000 companies and 20,000 students (Ernst and Young 2013, p. 4).

Private Funding Instruments

Young start-ups cannot often obtain sufficient loans from banks for early-stage financing, as they usually do not have a credible credit history or other collateral. As a result, start-up entrepreneurs often look to private funders such as venture capitalists and business angels. This is to drive faster development and expansion, in part to protect themselves from copies of their business model. Due to high competition, rapid growth and market establishment are essential for many young start-ups (Atherton and Newman 2018, p. 46).

Business Angels

Compared to venture capital, the volume of angel investment seems very small, but it should not be ignored. Just like the allocation of venture capital, the number of angel investments has increased significantly in recent years. Business angels are usually more risk-averse and less picky than venture capitalists. They usually invest in the early stages of a start-up, when the greatest growth is still to come. In 2008, there were only 72 cases of angel investment in the whole of China, but in 2013, there were as many as 262 and the trend is rising. Most angel investors are found in the major high-tech zones on the East Coast and build their incubators locally to support the start-ups they have invested in (Atherton and Newman 2018, p. 89).

Companies as Sponsors

Leading technology companies, such as Alibaba or Tencent, are aware that they cannot rely solely on the state to subsidize the burgeoning start-up scene and develop start-ups into internationally competitive high-tech companies. Increasingly, they are acting as venture capitalists or starting up themselves. In this, they see an opportunity to get closer to customers by delivering value to them through start-ups (Atherton and Newman 2018, pp. 46–47). Alibaba, for example, on the one hand, indirectly invests in local high-tech and small businesses and national technology hubs through subsidiaries (Atherton and Newman 2018, p. 45). On the other hand, Alibaba invests directly in emerging internet start-ups. Similarly, Tencent invests in young service start-ups to offer their services in its apps WeChat and QQ. Partnerships like these seem to be a trend in China, as they bring great benefits to both. For the big tech corporations, investing in startups also offers them the chance to establish improved business models, new working methods, and new technologies within their own companies. Moreover, they thereby assume a degree of social responsibility and improve their reputation in society and politics (Atherton and Newman 2018, p. 47). In addition, the leading corporations hold start-up competitions in which the winners receive promotional measures.

International companies also support the Chinese start-up scene. Microsoft set up its accelerator in Beijing back in 2012 and supported over 200 start-ups free of charge with a relatively high success rate. This resulted in a total of 15 company takeovers and four IPOs. German corporations operate similar support programs. For example, Daimler, BASF, Porsche, and, other companies have jointly launched the "Startup Autobahn" initiative to promote start-ups dealing with the topic of mobility. Several companies were provided with premises and specialist support, and contacts were also established with potential investors (Hsu 2018b).

Other Institutions and Organizations

Many institutions and organizations are focused on supporting start-ups in their early stages before they have found institutional investors, i.e. investors apart from family and friends. Not every start-up has the same requirements and needs, this often depends on the industry, the founders, and the stage the start-up is in. Nevertheless, almost every founder goes through the so-called pre-idea phase, the pre-incorporation phase, and the pre-funding phase (Hsu 2018a).

In the pre-idea phase, there is often no concrete idea that can be implemented, only the will and the decision to found a start-up. In this case, there are various ways to find out about the latest trends. Numerous specialized media and magazines, such as TechNode Media, 36 Kr, and Cyzone, report on new technology trends, opportunities for acquiring venture capital, etc. In addition, there are numerous events, especially for founders and people who want to become one. Startup Salad, AngelHack, Startup Grind, etc. organize talks, startup competitions, and other meetups. Such events are intended to bring together founders, investors, and other actors in the scene (Hsu 2018a).

In the pre-incorporation phase, there is already a concrete idea that is to be implemented. In this phase, the start-up must be registered, personnel recruited and the business model refined. To support this phase, special competitions are held to bring together potential founding partners, obtain feedback from independent experts, and possibly even gain funding. Several institutions offer various training courses on how to run a start-up business, some free of charge, some at high prices. Typically, such training is designed to improve creativity and soft skills (Harris 2019, p. 140). For example, Lenovo and the Chinese Academy of Sciences support such course programs (Hsu 2018a). In addition, there are numerous incubators, accelerators, FabLabs, as well as other platforms that offer various services to promote start-ups, some of which collaborate with large companies and local government or are directly run by renowned universities. In addition, there are so-called coffee shop incubators that offer entrepreneurs jobs, fast internet, and social events (Hsu 2018a).

In the pre-funding phase, the registered start-up must prove itself with its new employees and a mature business model, and investors must be convinced. Various players and programs support the search for investors. Some of them are fee-based, and some require shares in the start-up. The sponsors pursue different intentions with the support. These sponsors include the state, global start-up companies, and international companies that are active in this field for strategic reasons. Through this, both government and private organizations and companies seek to form strategic partnerships from which they can benefit in the future. Other programs, such as Plug and Play China, a platform that specializes in bringing together startups and investors, themselves inject equity into the startups (Hsu 2018b).

The actors described above serve start-ups and their founders in different ways, thus enabling the success of the Chinese start-up scene. After these three phases, further major challenges lie ahead for Chinese start-ups. In this area, other actors are again there, such as venture capital firms, they also do not disinterestedly stand by the start-ups (Hsu 2018b).

Microfinancing, Peer-to-Peer Lending, and Crowdfunding
Microfinance plays an important role in China's rural regions in particular. This mainly involves smaller loans being granted to entrepreneurs with low incomes and no collateral. Although there are now over 10,000 microfinance providers in China, this type of financing plays only a secondary role in the development of the start-up scene. While microfinance does lead to people from rural households spending more time on entrepreneurial activities, these are rarely innovative start-ups (Atherton and Newman 2018, pp. 90–91).

Peer-to-peer lending and crowdfunding, on the other hand, have developed strongly in recent years, mainly due to the huge growth of the e-commerce sector (Atherton and Newman 2018, p. 91). Crowdfunding for start-ups is even part of the 13th Five-Year Plan (National Development and Reform Commission 2016, p. 15). The Chinese peer-to-peer industry is now the largest in the world and continues to grow (Aveni 2015, p. 9). Large companies such as Alibaba have entered the peer-to-peer market, offering accounts, loans, etc. online, sometimes even to the tune of several million RMB to entrepreneurs who cannot show any collateral. Another internet giant, JD.com or Jingdong, offers a crowd-funding platform to raise money for microfinance (Atherton and Newman 2018, p. 92).

State Interventions

The state intervenes massively in the promotion of the start-up scene, especially in the allocation of venture capital, and has thus coined the English term "venture communism". Through investments in venture capital firms, joint investments with private venture capital firms, state venture capital guidance funds, and risk hedging of venture capital firms, it is ensured that sufficient venture capital flows into technology start-ups (Abele 2019). These measures have already pumped USD 1.8 billion into the market, distorting competition (Feng 2018). In addition, the state supports venture capital firms by, for example, lowering taxation bases. At the end of 2017, there were approximately 2300 venture capital institutions nationwide, with numbers dropping slightly the following year due to the economic downturn. Nevertheless, a record $100 billion was invested in 2018 (Abele 2019). This compares to just under six billion USD in 2010 (Fannin 2019).

7.3.5 Practical Example: Zhonguangcun (Beijing)

The Zhongguancun Science Park in Beijing is China's largest and oldest high-tech park. It was founded in 1988 and consists of a total of 16 smaller parks covering a total area of 488 km^2 in several districts of the city. In total, more than 20,000 high-tech companies are located there (Kooperation International n.d., p. 1) and reportedly employ more than 1.5 million people (Atherton and Newman 2018, p. 43). Zhongguancun has its development plan. It aims to create a system that promotes independent innovation capability, research and development, tech talent, and entrepreneurship (China Daily 2011). The close proximity to the elite universities Qinghua University and Peking University as well as to the Chinese Academy of Sciences has always been a great location advantage to link research and application (Kooperation International n.d., p. 1). In total, there are over 50 universities and colleges and more than 100 research institutions nearby (Cai et al. 2009, p. 6). Close partnerships are maintained and companies distribute research contracts to universities or enter into concrete collaborations. Conversely, universities and research centers invest in companies (Kooperation International n.d., p. 12).

On the one hand, Zhongguancun attracts major international corporations such as Google, Microsoft, and IBM; on the other hand, local start-ups such as Lenovo have grown into major international companies (Eesley 2009, p. 14). The focus of companies in Zhongguancun is often on information and communication technology, biotechnology, and the pharmaceutical industry, as well as mechanical engineering and renewable energy (Kooperation International n.d., p. 10). In recent years, the high-tech zone has become more of a Chinese hub for internet start-ups, comparable to Silicon Valley (Atherton and Newman 2018, p. 43). Especially in the fields of artificial intelligence, blockchain, and tech play a major role. Beijing is a national leader in these areas, not least thanks to targeted investments in AI start-ups (Abele 2019).

The administration of the high-tech park is under the control of the local government. It is trying to make it as easy as possible to set up start-ups in Zhongguancun, and there are a large number of incentives for companies there. For example, setting up a company is extremely quick, with the process taking just over four days on average. Generally, only a reduced corporate income tax of 15 percent has to be paid, and newly founded start-ups even pay no corporate income tax at all for the first three years. Furthermore, the hukou system has been relaxed to attract talent from all over the country. Employees of the high-tech park are allowed to register their residence in Beijing (Eesley 2009, p. 14). In 2014, the so-called InnoWay was established, an entire street exclusively for start-ups. There are co-working spaces, start-up cafés, accelerators, and venture capitalists, most of which are state-supported. This is a clear indicator that the local government in Beijing is assisting the local start-up scene in its development (Atherton and Newman 2018, p. 43).

Companies wishing to locate in Zhongguancun are subject to certain requirements. For example, half of the company's revenues must come from high-tech projects, expenditure on research and development must account for at least 3 percent of sales, and at least 20 percent of employees must have a university degree (Kooperation International n.d., p. 7).

Subsidies, investment, and proximity to academic research create a hugely start-up-friendly environment (Atherton and Newman 2018, p. 43). This is reflected in the high number of patent applications and overseas returnees. Over 10,000 Chinese who were abroad for educational purposes found work in Zhongguancun and founded a total of 420 start-ups (Kooperation International n.d., p. 5).

7.3.6 Special Features of the Chinese Start-Up Scene

The People's Republic has a huge talent pool. In recent years, the number of academics in China has increased enormously. While in 2004 there were still around four million Chinese graduating with a bachelor's or master's degree, in 2017 there were almost 13 million (Unesco 2019). The labor market has not been able to keep up with this rapid development. The unemployment rate for graduates was 7.5 percent in 2015, well above the official Chinese average of 4.1 percent (DAAD 2017, p. 23). Due to the sometimes difficult labor market situation, founding a start-up company appears to be an attractive alternative. In addition, the high social reputation of successful founders, such as Alibaba's Jack Ma, plays a role for young founders (Covestro 2018, p. 3).

Furthermore, there is a lack of investment opportunities for private assets in China. The Chinese housing market is overheated and the stock market is subject to both enormous fluctuations and state control. The transfer of capital abroad has become much more difficult in recent years, making investments in venture capital funds or directly in young start-ups all the more interesting. In addition, Chinese culture plays an important role. Three-quarters of Chinese entrepreneurs are convinced that their culture encourages them to start up (Ernst and Young 2013, p. 7). To this day, entrepreneurship is strongly influenced by Confucianism. The idea that success can be achieved through hard work and the acquisition of knowledge (Atherton and Newman 2018, pp. 98–99) is a major factor behind the booming start-up scene. Moreover, the Chinese are often very risk-averse and rarely shy away from entrepreneurial risk. Buddhism is also in some ways conducive to the success of start-ups, as Buddhist values promote long-term orientation, social responsibility, and social and political capital (Liu et al. 2019, pp. 723–724). Not to be neglected is the great enthusiasm and affinity for technology that runs through all strata of the population from young to old (Abele 2019).

Despite the great success of Chinese start-ups in recent years, the scene has various weaknesses. Some start-up founders complain of being isolated in global start-up networks, which makes it much more difficult to recruit international talent (Covestro 2018, p. 3). In addition, there is a large investment gap outside the high-tech zones. Only within these zones have funding opportunities for tech startups greatly improved. Start-ups outside these centers often struggle to obtain adequate funding (Atherton and Newman 2018, pp. 48–49). In general, it has become more difficult to access private venture capital in recent years as overall investment in this area has declined (Covestro 2018, p. 3). As a result, university graduates in particular are finding it difficult to turn their start-up ideas into reality due to a lack of financing options (Liu et al. 2015, p. 109).

Despite this decline in venture capital investment, the market for innovation-driven start-ups is highly competitive. The battle between start-ups for the same consumers and market share leads to price reductions and, as a consequence, the bankruptcy of many start-ups. In addition, this intense competition leads to business mergers both within the same industry and across different industries to take advantage of economies of scale. The best examples are the apps Meituan, a food delivery company, and Dianping, a group-buying app (Atherton and Newman 2018, p. 48).

There are general doubts about the effectiveness of programs designed to promote national innovation, especially the Torch Programme (Heilmann et al. 2013, p. 896). There is also criticism of the innovative strength of Chinese start-ups. Compared to US start-ups, they lag be some areas. Founders in the high-tech sector are generally more risk-averse than their international colleagues. Moreover, it is not uncommon for successful business models to be imitated by start-ups from abroad instead of developing their own (Abrami et al. 2014). However, this does not apply to all high-tech sectors. Furthermore, there is often a lack of translation of scientific knowledge into practical applications (Atherton and Newman 2018, p. 125). Furthermore, there are discriminatory phenomena in the allocation of subsidies. Ethnic minorities, especially Uygurs, often find it difficult to receive the same funding support as Han Chinese. They are often subject to certain restrictions that do not affect Han Chinese. However, this problem is not specific to start-up funding, but can be found in other areas of the economy as well (Howell 2019, p. 710).

7.4 Insolvencies of Companies

When the People's Republic of China was founded, there was initially no insolvency law, as there were only state-owned enterprises that were financed by the government in the event of insolvency (Fehl 2008, p. 325). However, as China's reform and opening-up policies progressed and the country's economy grew rapidly, it became clear that ailing companies would have to be wound up. In 1988, an insolvency law was passed, which initially only applied to state-owned enterprises (Weng 2011, p. 107). However, the increasing number of private companies led to the need to pass a new insolvency law, which finally happened in 2006 (Falke 2006, p. 399). This law was largely received from Western insolvency law texts (Hetzel and Blumer 2014, p. 7). The wording of the law is ambiguous in many places. Thus, not only for foreign but also for local companies, the outcome of a filed insolvency petition is not exactly predictable. This leads to uncertainty, which ultimately results in a reluctance to use the new insolvency law on the part of companies. Many companies are withdrawing from the market under company law rather than with the help of the court. Foreign investors have to ask themselves whether their assets are protected enough in case of insolvency or whether investments are too risky. Although this Chinese law is now strongly aligned with the corresponding laws of Western countries, this does not apply to the eventual application of the law.

In Chinese law, and especially in insolvency law, cultural influences play a role that should not be neglected. The Chinese population's dismissive attitude towards debt or insolvency, which can be traced back to imperial times, differs greatly from Western culture. This is due to Chinese cultural traits such as loss of face and Confucian tradition. Insolvency is seen as a defeat and therefore leads to loss of face, which must be avoided. Therefore, debts have often been passed on. This loss of face, however, does not only apply to individuals. The government would also lose face with the closure of a state-owned enterprise and the resulting loss of many jobs, which is why insolvency has been seen as a last resort in the past as well as today (Martin 2005, pp. 71–72).

The teachings of Confucius, whose values still guide many Chinese to this day, play a special role in the application of insolvency law. Confucius always emphasized the natural harmony within society. Based on this body of thought, Chinese society ultimately developed towards collectivism and, for cultural reasons, rejects individual rights.

> If one governs by edicts and orders by punishments, the people evade and have no conscience. If one leads by force of nature and orders by custom, the people have a conscience and achieve the good. (Confucius, Chinese philosopher, 551 B.C. to 479 B.C.)

Confucianism encourages citizens to harmony and to resolve conflicts among themselves or through respected arbitrators. Thus, according to Confucianism, the law only serves to secure the hierarchical social ord to sanction any disruption of the ruling apparatus through criminal law and organize the state administration (Zweigert and Kötz 1996, p. 284).

Moreover, it is contrary to Confucian doctrine to push someone into insolvency, which is why the filing of insolvency by creditors is perceived as immoral (Martin 2005, p. 73).

7.4.1 Bankruptcy Law

The 2006 Bankruptcy Act consists of 126 articles, which in turn are divided into twelve chapters. One of the most important goals of the law is the equal treatment of state-owned and private enterprises (Münzel 2007, p. 47). Furthermore, both the USA and the EU did not recognize the PRC as a fully-fledged market economy for a long time. One of the reasons given by the PRC's two largest trading partners was the inefficient bankruptcy laws (Parry et al. 2010, p. 14). The desire to gain the confidence of foreign investors, and thus attract them to the country, eventually led to the introduction of new bankruptcy law (Martin 2005, p. 70). Security for international investors was created on two levels. First, liquidation committees, which often consisted of only marginally experienced corporate management functionaries (Godwin 2007, p. 760), were replaced by qualified insolvency administrators. Second, the preferential treatment of secured creditors was included in the Chinese Bankruptcy Law for the first time in 2006 (Jiang 2014, p. 587).

Since there is no uniform bankruptcy procedure, each petitioner must choose between bankruptcy liquidation, reorganization, or composition. The law contains a general section covering all three types of proceedings and three special chapters containing the specific provisions relating to each type of proceedings.

Furthermore, all other creditors were granted more rights by the new law. For example, the 2006 law gave creditors themselves the ability to apply for reorganization. The pressure was particularly high to end state investment in inefficient SOEs and to restructure or remove these companies from the market (Bufford 2017, p. 1). The numerous, heavily indebted SOEs are employers of many millions of Chinese citizens. Therefore, the People's Republic of China had to find a way to renew state-owned enterprises without spending billions of dollars from the treasury, but equally avoiding job losses. The reorganization procedure introduced a solution to this acute problem (Xu 2013, p. 106). Furthermore, the new law was intended to protect both the order of the market and workers' rights (Hetzel and Blumer 2014, p. 13).

7.4.1.1 Scope of Application

The Bankruptcy Act applies to all companies that constitute a legal entity. This is irrespective of state or private company management and irrespective of the domestic or foreign nationality of the shareholders. Of interest to foreign investors is that representative offices, branches, partnership firms, and sole proprietorships do not fall within the scope. On the other hand, other companies with foreign participation, such as equity joint ventures, companies with exclusively foreign participation, and foreign limited liability investment companies, are fully covered by the law (Fehl 2008, p. 326). One of the criticisms has been that institutions such as foundations, non-profit institutions, sole proprietorships, etc. remain in a grey area. It is unclear whether the Bankruptcy Act covers them (Münzel 2007, p. 48). For financial institutions, such as insurance companies, banks, etc., the Council of State can establish special provisions so that they are excluded from the scope. The possibility of private insolvency was also not included in the law. Private individuals fall exclusively within the scope of the General Principles of Civil Law (Hetzel and Blumer 2014, p. 18).

7.4.1.2 Insolvency Application

In principle, there is no obligation to file for insolvency in the PRC. An exception exists if a liquidator in extrajudicial liquidation determines that the assets are insufficient to cover the claims. In this case, he is obliged to file a petition. The debtor may file a petition for reorganization, bankruptcy liquidation, or the composition, provided that the grounds for opening the bankruptcy proceedings exist. The petition must be filed with the court in whose district the debtor company has its registered office. If an investor owns more than 10 percent of the debtor's registered capital, the investor may also file for reorganization during the liquidation proceedings requested by the creditor. Furthermore, any insolvency creditor may file a petition for insolvency if the debtor is unable to pay debts that are due. For this purpose, the creditor only has to prove the insolvency of the debtor company, as the filing of a petition by a creditor would otherwise hardly be possible due to the lack of access to the debtor's balance sheets. Once the petition has been filed, the court must decide within 15 days whether to accept the bankruptcy petition for processing. If the court rejects the bankruptcy petition, the petitioner has the opportunity to appeal to the next higher people's court within ten days. In addition, the petition may be withdrawn at any time before the opening of proceedings (Art. 7–9 Bankruptcy Act).

7.4.1.3 Reasons for Insolvency

Upon receipt of the petition, the competent court must check whether there is a ground for bankruptcy. On the one hand, there is insolvency, which exists "if the claim has arisen under the relevant law, the payment deadline has expired and the debtor has not fully satisfied the claim" (Fehl 2012, p. 213), and on the other hand, there is overindebtedness, which exists "if the debtor's balance sheets or other accounting documents show that its assets are not sufficient to repay all its liabilities" (Fehl 2012, p. 213). If the creditor files for insolvency, it is sufficient if he can prove insolvency. The presentation of an enforcement order and proof of unsuccessful enforcement is sufficient for this purpose. If the debtor files for insolvency, he must prove not only insolvency but also overindebtedness. To facilitate the entry into insolvency nevertheless, the so-called "obvious over-indebtedness" was introduced. In the case of "obvious over-indebtedness", it is sufficient to prove that it is obviously impossible for the debtor to pay the liabilities with the available assets (Art. 7–8 Bankruptcy Act).

7.4.1.4 Insolvency Administrator

The insolvency administrator has to represent the interests of both the creditors and the debtor and has to find a compromise between the two. The insolvency administrator can be a non-natural person, such as a law firm, an accounting firm, a bankruptcy liquidation company, or other qualified institution. In simple bankruptcy proceedings, natural persons are usually appointed as insolvency administrators. In complex cases with foreign participation, the courts usually appoint an institution. Bankruptcy trustees are vetted by the court based on some regulations and guidelines set by the Bankruptcy Law and then placed on a list (Yang 2008, pp. 533–534). If the petition is accepted, the court is required to select a receiver from the bankruptcy trustee list. This selection is random and cannot be influenced by either party. However, the random selection of the insolvency administrator is highly controversial. Some Chinese insolvency law specialists claim that the administrator should be chosen by the creditors so that they can already express their interests when the administrator is selected (Tang and Au 2007, p. 48). Others see the problem with a random selection that the suitability of the particular receiver for the particular case cannot be guaranteed. It is demanded that the creditors' meeting should determine the bankruptcy trustee, as they are most closely familiar with the debtor company and thus with the case. The creditors' assembly cannot independently remove the insolvency administrator but must apply to the court for the replacement. However, if the court refuses the replacement, there is no legal remedy for the creditors to enforce their application anyway (Li 2008, p. 127).

7.4.1.5 Creditors

Creditors in bankruptcy proceedings are divided into different groups. The first group is the ordinary bankruptcy creditors. Their claims were justified at the time of acceptance of the bankruptcy petition. The second group is the mass creditors. Their claims arise as a result of the bankruptcy proceedings. These include directly the court costs, the costs of administration, realization, and distribution of the bankruptcy estate, the remuneration of the bankruptcy trustee, etc. There are also segregation and segregation-related claims. There are also creditors entitled to separate satisfaction and creditors entitled to separate satisfaction. Creditors entitled to segregation can claim certain items that do not belong to the debtor's assets but their own and receive them back directly from the bankruptcy trustee. Creditors entitled to separate satisfaction receive preferential satisfaction based on security interests. Furthermore, the new Bankruptcy Act has no provisions for subordinated bankruptcy creditors (Art. 44–49 Bankruptcy Act).

7.4.1.6 Meeting of Creditors

Only creditors who have filed their claims in due time may attend the creditors' meeting and are entitled to vote. Employees cannot register their claims arising from the employment relationship and are therefore not entitled to attend the creditors' meeting. They can only bring in their interests through representatives or trade unionists who are entitled to participate. Probably the most important task of the creditors' meeting is to decide on the continuation of the business. A voting member is appointed by the court as chairman of the creditors' meeting, who also chairs the meetings. Creditors entitled to separate satisfaction are entitled to vote at the creditors' meeting on a plan of arrangement or distribution only if they have waived their rights to preferential satisfaction. A double majority is generally required for a resolution at the creditors' meeting. This means that a majority of the creditors present and entitled to vote and a majority of all claims, not just those present, must be in favor. Once a resolution has been passed at the creditors' meeting, it is binding on all creditors and can only be set aside by the court if the interests of individual creditors have been harmed. Also, if the creditors' meeting fails to reach an agreement regarding the administration, realization, or realization plan, the court may make a decision (sections 64–69 of the Bankruptcy Act). In addition, a creditors' committee may be appointed by the creditors' meeting both in the liquidation case and in the reorganization case to supervise the administration and disposal of the estate's assets, to supervise the distribution of the insolvency assets, to convene the creditors' meeting, etc. (Art. 59–69 Bankruptcy Act).

7.4.1.7 Insolvency Court

As a general rule, the People's Court at the registered office of the debtor company has jurisdiction over the relevant bankruptcy case. The registered office of a legal entity is defined as the place of the center of business activity or the head office. In particularly complex cases or cases involving foreign participation, the Intermediate People's Court shall have jurisdiction. Upon approval of the insolvency proceedings, the insolvency court must immediately appoint an insolvency administrator. The courts have wide-ranging powers in Chinese insolvency proceedings. At the outset of the insolvency proceedings, the consideration of the grounds for insolvency is predominantly at the discretion of the courts. All these powers of the court affect the insolvency proceedings. It is criticized that due to the strong role of the courts, the influence of the government on the proceedings is serious and this can have a negative impact on procedural practice (Jiang 2014, p. 569).

7.4.1.8 Insolvency Proceedings

After the formal and material requirements have been examined and confirmed by the court, the proceedings are opened by the court in the requested type of proceedings. Upon acceptance of the petition, the debtor initially loses the right to manage the assets, which immediately pass to the insolvency administrator appointed by the court. The creditors must file their claims within a time limit set by the court. The list of claims must be submitted to the first meeting of the creditors' assembly. If there are objections by the creditors or the debtor to one or more claims, the court must decide on them. Furthermore, upon acceptance of the petition, all claims not yet due become due (Art. 107–110 Bankruptcy Act).

Like all other proceedings, bankruptcy liquidation begins with the court's declaration of bankruptcy. All rights of disposal go to the insolvency administrator and cannot be transferred back to the joint and several debtors. The liquidator is responsible for drawing up a realization and distribution plan, which he must first submit to the creditors' meeting for a vote and then to the court for approval. The assets to be realized include all assets of the debtor company, both domestic and foreign (Bufford 2017, p. 21). Creditors with a right to separate satisfaction can access the corresponding claims immediately after the opening of proceedings. Creditors with security interests may also seek preferential satisfaction of their rights from the bankruptcy trustee outside the bankruptcy proceedings after the declaration of bankruptcy. In China, the liquidation of insolvency assets is generally to be carried out by public auction. If the insolvency administrator wishes to use a different liquidation procedure, he must obtain the consent of the creditors' meeting. After the realization of the debtor's assets, a realization plan drawn up by the insolvency administrator must be submitted to the creditors' meeting for confirmation. In principle, the distribution takes place according to a legally defined scheme. First of all, the costs of the proceedings and the costs of the insolvency estate are satisfied. The costs of maintaining contracts, wage costs arising from the continued operation, etc. belong to the costs of the insolvency estate. Immediately after satisfaction of the procedural and mass costs, the wages, pensions, compensation, etc. owed to the employees are paid. Thereafter, social security costs will be satisfied first and general bankruptcy claims last. If the bankruptcy estate is insufficient to satisfy all claims, satisfaction is made on a pro-rata basis. Unless otherwise decided by the creditors' meeting, the bankruptcy estate is distributed in the form of cash (Articles 107–119 of the Bankruptcy Law). Until 2008, the Chinese government implemented a program that promoted the liquidation of loss-making state-owned enterprises. In doing so, the Chinese state was particularly concerned with compensating workers who had relied on lifetime employment in the state-owned enterprise. This included the issuance of loans by the state to ensure adequate compensation for the workers. After the completion of this program in 2008, the number of state enterprise liquidation cases dropped significantly. A gather compensation of employees in state-owned enterprises remains one of the biggest problems in liquidation processes.

The right to apply for the composition procedure lies only with the debtor himself. The debtor may apply for a composition agreement after the bankruptcy petition has been accepted and after the bankruptcy proceedings have been opened. Together with the application for a composition, the debtor must submit a draft composition agreement. If the court approves the composition proceedings, it submits the draft composition agreement to the creditors' meeting for discussion. Security creditors may enforce their rights against the debtor immediately after the court approves the composition proceedings. The composition procedure is a common type of procedure for small, at most medium-sized companies with a manageable number of creditors and a rather low level of debt (Hetzel and Blumer 2014, p. 38).

In China, not only the debtor itself but also any creditor or investor of the debtor whose shares account for at least 10 percent of the total assets is entitled to file an application. Secured creditors may only assert their rights during the reorganization proceedings if there is a risk that the security interests will be jeopardized by impairment or damage. Upon the opening of proceedings, all powers of disposal and administration shall pass to the insolvency administrator. With the approval of the people's court and under the supervision of the insolvency administrator, the debtor himself may manage the assets and the business during the reorganization proceedings. Liquidation proceedings may be converted into reorganization proceedings with the approval of the people's court to save the enterprise from liquidation. Both the debtor and any investor with at least 10 percent of the total assets of the debtor enterprise have the option to apply to the court to change to reorganization proceedings (Bao 2006, pp. 599–602).

7.4.2 Practical Application

The number of new company registrations has increased from ten million in 2008 to almost 20 million in 2014 (Finder 2016). An even stronger increase in newly registered companies since 2013 is predominantly related to the simplified conditions for opening a company from the revised Companies Act of 2013. Limited liability companies, one-man companies, and joint stock companies have no minimum capital requirements, and cash deposits are not required. Further, there are no deposit requirements, no requirements regarding capital contributions, and no registration of paid-in contributions in the trade certificate. The government intended this new regulation to encourage new business start-ups and thus innovation in general. The assumption is obvious that due to these low entry barriers to setting up a company, many dubious companies also enter the market, which ends the business activities again after only a short time and possibly even goes bankrupt.

The latest figures from the Supreme People's Court say that 9642 new bankruptcy cases were filed in Chinese courts in 2017. This is a 68 percent increase compared to the previous year. As recently as 2015, there were only 3569 cases, according to the Supreme People's Court. Insolvency figures are also forecast to continue rising in the coming years. This increase would be explained, among other things, by the fact that application rules of the new Insolvency Law are gradually being published by the Supreme People's Court, making the outcome of insolvency proceedings more predictable (Finder 2016). Although it is almost impossible to determine the exact insolvency figures for the People's Republic of China, a trend can be identified. For example, the already low insolvency figures initially fell after the new insolvency law was passed. But as clarity grows about the application of the new law, the number of companies filing for insolvency has been rising for the past two to three years. This is illustrated in more detail by the following examples.

7.4.2.1 Taizinai

Taizinai, a dairy products producer funded by Goldman Sachs and Morgan Stanley, among others, has operating units in Hunan, Beijing, Hubei, Sichuan, and Jiangsu, as well as a holding company in the Cayman Islands. Most of the company's international capital was provided through the Cayman Islands holding company. In 2008, the company ran into financial difficulties for the first time. In the same month, the government reached an agreement with another company, allowing Taizinai to borrow RMB 20 million. After further attempts to rescue the company, the local court announced in July 2010 that a reorganization process had been opened over Taizinai. For this purpose, Beijing Deheng Law

Firm was appointed as the administrator. It is worth mentioning here that a large law firm was appointed as administrator in this cross-border case, which is internationally active and experienced through operating units in Hong Kong, Seoul, Berlin, Sydney, etc. At the end of the case, the resolutions made in the Cayman Islands were recognized by the Chinese courts and Taizinai Group was sold in large parts to two domestic investors. This case was considered one of the first representative cases of Bankruptcy Law regarding litigation with foreign participation. The appointment of an internationally experienced law firm as insolvency trustee and the recognition of foreign court orders was seen as a major confidence boost for foreign investors (Godwin 2012, pp. 182–185).

7.4.2.2 East Star Airlines

In the 2009 case of East Star Airlines, based in Wuhan, it was not the debtor itself but a group of six creditors that filed a petition for liquidation with a court in Wuhan. The debtor, on the other hand, wanted to enforce the reorganization of the company. In this case, the creditors received the support of the local government and were ultimately able to obtain the liquidation of the company. The court indicated that the plan of reorganization filed by East Star Group and China Equity, which promised to invest $200 million to $300 million in the reorganization, was not feasible and thus the conditions for accepting the reorganization proceeding did not exist. The government tried to push through a takeover by state-owned Air China, which East Star Airlines rejected. In this case, the local government sided with the creditors, some of whom were foreign, despite the loss of jobs (Yan 2009).

7.4.2.3 Circumvention Possibilities

The number of insolvency cases suggests that it is common practice in China to withdraw from the market in a disorderly manner. One reason for this is the imprecise law, on which further interpretative guidance is awaited from the Supreme People's Court. As a result, most companies tend to choose to dissolve their business under corporate law in the event of insolvency, rather than filing for bankruptcy with a court. The Bankruptcy Law and Restructuring Research Center of the Chinese University of Politics and Law have collected figures showing that 780,000 companies exited the Chinese market in 2008. According to the Supreme People's Court, this compares to only 3139 bankruptcy cases. Of these 780,000 cases, 380,000 had exited the Chinese market through simple deregistration and 400,000 through license cancellation, it said. Chinese enterprises have two ways to exit the market under the Chinese Enterprise Law, Insolvency Law, and other, relevant regulations. One is to file for insolvency with the court and the other is to dissolve under enterprise law without court involvement. Dissolution under corporate law can be either voluntary under the company's articles of association or involuntary through the cancellation of the business license by the China Bureau of Industry and Commerce. Even in the case of corporate dissolution without court involvement, there is a prescribed dissolution process. This states, among other things, that the company must be deregistered before losing its legal independence. However, before this deregistration, it must go through the complete dissolution process. Problematically, many companies deliberately bypass this dissolution process to be (Li and Wang 2010, pp. 1–2).

7.5 Conclusion

So far, private entrepreneurs and official bureaucrats have benefited from mutual interaction and have had little incentive to change existing economic and political structures (Yan and Huang 2017, p. 40). Ten Brink (2014), p. 696) calls this relationship a "private-public growth coalition." According to Hirn (2018, p. 38), entrepreneurs and politicians are interdependent, the latter depending on the success of the former.

With the government's anti-corruption measures, entrepreneurs are more likely to be forced to comply with laws and re-evaluate their activities (Hong 2015, p. 50). It is becoming increasingly difficult for private entrepreneurs and corrupt officials to conduct profitable unofficial business. Nevertheless, fewer hurdles should stand in the way of private enterprise development and start-up creation in the future. China's government recognizes the importance of entrepreneurship and innovation and encourages it. Another look at the rhetoric of leading Chinese politicians confirms this: Premier Li Keqiang mentioned "innovation" 59 times and "entrepreneurship" 22 times during the 2016 National People's Congress (Tse 2016).

China has managed to create a unique start-up environment in recent years, and the local scene has really blossomed. Various policy measures and strategic orientations, such as the Torch Program or the Mass Entrepreneurship Program, have contributed significantly to this. The start-up scene has been heavily promoted at all levels, both by the private sector and the government. Numerous incubators, accelerators, and investors converge in innovation hubs and play a major role in shaping the scene. Thanks to huge investment packages and subsidy measures, the Chinese start-up scene has become the global market leader in some areas such as artificial intelligence.

However, the Chinese start-up scene has significant weaknesses. It remains questionable whether the goal of the enormous support measures for start-ups, namely to increase China's innovative power, is being achieved at all. It seems that in some industries and sectors there is a lack of efficient allocation of resources as well as a lack of transformation of innovation into market-ready products. In addition, the bulk of startups benefits only marginally from the overwhelming subsidies, as they are mainly intended to support high-tech and innovation. However, most Chinese startups are in e-commerce and other traditional industries. In addition, startups outside the metropolitan regions of Beijing, Shanghai, Hangzhou, and Shenzhen are often neglected, creating strong regional disparities.

Undeniably, great progress has already been made in high-tech areas such as internet services, artificial intelligence, and consumer electronics. Nevertheless, among all the positive results and reports on the successes of the Chinese start-up scene, there are negative aspects mentioned, such as the lack of efficiency in the use of the enormous financial resources and the transformation condition of research results into market-ready products. This hardly plays a role in their perception. It will be exciting to continue to follow the Chinese start-up scene on its way to the innovation leadership it is striving for.

In the last two to three years, there have been some developments in Chinese insolvency law that simplify the transparency of proceedings and their application. These include, among others, the already partially published application regulations of the Supreme People's Court. These judicial interpretations concern, for example, the entry into the proceedings including the grounds for bankruptcy and the court's acceptance of the petition, and the, election of the bankruptcy trustee (Fehl 2012, p. 212). Another important step was the inauguration of an insolvency information platform by the Supreme People's Court in August 2016. The website (http://english.court.gov.cn/) publishes the latest annual reports, but also details of insolvency cases. Insolvency administrators are required to post their decisions and actions on the website. This also allows parties to insolvency proceedings to always have the latest information on their insolvency case. The only problem is that the website is currently only available in Chinese. With the publication of all insolvency cases, the pressure on Chinese courts to adhere closely to the text of the law and the interpretations of the Supreme People's Court is increasing.

References

Abele, Corinne. 2019. *Chinas Hightech-Start-ups mischen an der Spitze mit.* https://www.gtai.de/GTAI/Navigation/DE/Invest/business-location-germany.html. Accessed 18 June 2019.

Abrami, Regina M., William C. Kirby, and F. Warren McFarlan. 2014. Why China can't innovate. *Harvard Business Review.* https://hbr.org/2014/03/why-china-cant-innovate. Accessed 18 June 2019.

Achleitner, Ann-Kristin. 2018. *Start-up-Unternehmen.* https://wirtschaftslexikon.gabler.de/definition/start-unternehmen-42136/version-265490. Accessed 18 June 2019.

Ahl, Björn. 2005. China auf dem Weg zum Rechtsstaat. *Die Politische Meinung* 423: 25–30.

Ahlstrom, David, and Zhujun Ding. 2014. Entrepreneurship in China: An overview. *International Small Business Journal* 32 (6): 610–618.

Atherton, Andrew, and Alexander Newman. 2018. *Entrepreneurship in China. The emergence of the private sector.* New York: Routledge Taylor & Francis Group.

Aveni, Tyler. 2015. New insights into an evolving P2P lending industry: How shifts in roles and risk are shaping the industry. *Positive Planet.*

Bao. 2006. Comparative studies of China's enterprise bankruptcy law and the US bankruptcy law. *Norton Journal of Bankruptcy Law and Practice* 2: 599–602.

Bendel, Oliver. 2019. *Einhorn.* https://wirtschaftslexikon.gabler.de/definition/einhorn-119178. Accessed 18 June 2019.

Bergquist, Kyle. 2018. *GII 2018: Localized innovation. World intellectual property organization.* https://www.wipo.int/pressroom/en/stories/gii_2018_localized_innovation.html. Accessed 18 June 2019.

Bufford, Samuel L. 2017. China's bankruptcy law interpretations: Translations and commentary. *American Bankruptcy Law Journal* 91 (1): 1–54.

Cai, Hongbin, Yasuyuki Todo, and Li-An Zhou. 2009. *Do multinationals' R&D activities stimulate indigenous entrepreneurship? Evidence from China's "Silicon Valley"*. NBER working paper series 2009. http://www.nber.org/papers/w13618. Accessed 18 June 2019.

China Daily. 2011. Outline of the development plan for Zhongguancun Science Park (2011–2020) selected edition II. *China Daily*. http://www.chinadaily.com.cn/m/beijing/zhongguancun/2011-11/17/content_14113993.htm. Accessed 18 June 2019.

China Innovation Funding. 2018. *Opinions on promoting high-quality development and the establishment of an upgraded version of "mass innovation and entrepreneurship"*. China Innovation Funding. http://chinainnovationfunding.eu/dt_testimonials/china-opinions-on-upgraded-mass-innovation-and-entrepreneurship/. Accessed 18 June 2019.

Cook, Sarah. 2018. Tech firms are boosting China's cyber power. Cooperation with the CCP, often mandatory, carries risks as well as benefits. *The Diplomat*. https://thediplomat.com/2018/09/tech-firms-are-boosting-chinas-cyber-power/. Accessed 05 February 2019.

Covestro; Kairos Future hg. 2018. *China's start-up landscape (and how to engage with it)*. https://press.covestro.com/news.nsf/id/2018-177-EN/$file/KAIROS_ENG.pdf. Accessed 23 June 2019.

DAAD. 2017. *China: Daten & Analysen zum Hochschul- und Wissenschaftsstandort | 2017*. https://www.daad.de/medien/der-daad/analysen-studien/bildungssystemanalyse/china_daad_bsa.pdf. Accessed 18 June 2019.

Dickson, Bruce J. 2008. *Wealth into power. The communist party's embrace of China's private sector*. Cambridge: Cambridge University Press.

Eesley, Charles. 2009. *Entrepreneurship and China: History of policy reforms and institutional development, 2009*. https://pdfs.semanticscholar.org/6bdd/8e0d1f51004150c80f04b00ca2f442b783dd.pdf?_ga=2.59014826.607143882.1560779987-1470356820.156077998. Accessed 18 June 2019.

Embassy of the People's Republic of China in Ireland. n.d. *Torch Programme*. Embassy of the People's Republic of China in Ireland. http://ie.china-embassy.org/eng/ScienceTech/ScienceandTechnologyDevelopmentProgrammes/t112843.htm. Accessed 18 June 2019.

Ernst and Young. 2013. *The power of three. The EY G20 Entrepreneurship Barometer 2013*. G20 Young Entrepreneurs' Alliance and Ernst & Young.

Falke, Mike. 2006. Chinas neues Gesetz für den Unternehmenskonkurs: Ende gut, alles gut? *Zeitschrift für Chinesisches Recht* 4: 399–403.

Fannin, Rebecca. 2019. *China rises to 38% of global venture spending in 2018, Nearing US Levels*. https://www.forbes.com/sites/rebeccafannin/2019/01/14/china-rises-to-38-of-global-venture-spending-in-2018-nears-us-levels/#2ddb6f05a5c9. Accessed 18 June 2019.

Fehl, Elske. 2008. Das neue Insolvenzrecht der VR China – Mehr Schutz für ausländische Investitionen? *Zeitschrift für Chinesisches Recht* 4: 325–331.

———. 2012. Auf dem Weg zu einem vorhersehbaren und geordneten Konkursverfahren in China: Die neuen Interpretationen des Obersten Volksgerichts zum Konkursgesetz der VR China. *Zeitschrift für Chinesisches Recht* 3: 212–216.

Feng, Emily. 2018. China's state-owned venture capital funds battle to make an impact. *Financial Times*, 23 December 2018. https://www.ft.com/content/4fa2caaa-f9f0-11e8-af46-2022a0b02a6c. Accessed 18 June 2019.

Feng, Ashley. 2019. *We can't tell if Chinese firms work for the party. Huawei claims to be an independent firm, but China's laws mandate a different reality*. https://foreignpolicy.com/2019/02/07/we-cant-tell-if-chinese-firms-work-for-the-party/. Accessed 05 January 2019.

Finder, Susan. 2016. *Ramping up China's bankruptcy courts, the latest data*. https://supremepeoplescourtmonitor.com/2016/05/18/ramping-up-chinas-bankruptcy-courts-the-latest-data/. Accessed 25 May 2019.

Fischer, Doris. 2017. Neuartige Innovationsmuster in der chinesischen Industrie – Entrepreneurship in China. In *China Innovationsstrategie in der globalen Wissensökonomie*, ed. Joachim Freimuth and Monika Schädler, 179–203. Wiesbaden: Springer.

Fuchs, Hans Joachim. 2007. *Die China AG: Zielmärkte und Strategien chinesischer Markenunternehmen in Deutschland und Europa*. München: FinanzBuch.

Fuller, Douglas B. 2010. How law, politics and transnational networks affect technology entrepreneurship: Explaining divergent venture capital investing strategies in China. *Asia Pacific Journal of Management* 27 (3): 445–459.

Godwin, Andrew. 2007. A lengthy stay? The impact of the PRC enterprise Bankruptcy Law in the rights of secured creditors. *UNSW Law Journal* 30 (3): 755–773.

———. 2012. Corporate rescue in Asia – Trends and challenges. *Sydney Law Review* 158 (34–1): 163–187.

Gründerszene. 2018a. *Bootstrapping. Gründerszene.* https://www.gruenderszene.de/lexikon/begriffe/bootstrapping?interstitial. Accessed 18 June 2019.

———. 2018b. *Startup. Gründerszene.* https://www.gruenderszene.de/lexikon/begriffe/startup?interstitial. Accessed 18 June 2019.

Harris, Tom. 2019. *Start-up*. Cham: Springer International Publishing.

He, Canfei, Jiangyong Lu, and Haifeng Qian. 2019. Entrepreneurship in China. *Small Business Economics* 52 (3): 563–572.

Heberer, Thomas, and Thomas Schubert. 2017. Private entrepreneurs as a "strategic group" in the Chinese policy. *China Review* 17 ((2), Special Issue: Evolving State-Society Relations in China): 95–122.

Heilmann, Sebastian. 2016. *Das politische System der Volksrepublik China*, 3. Aufl. Wiesbaden: Springer VS.

———. 2017. *How the CCP embraces and co-opts China's private sector*. https://www.merics.org/de/blog/how-ccp-embraces-and-co-opts-chinas-private-sector. Accessed 05 February 2019.

Heilmann, Sebastian, Lea Shih, and Andreas Hofem. 2013. National planning and local technology zones: Experimental governance in China's torch programme. *The China Quarterly* 216: 896–919.

Hetzel, Ludwig, and Maja Blumer. 2014. *Chinesisches Insolvenzrecht. Deutsche Übersetzung des chinesischen Gesetzes über die Unternehmensinsolvenz mit einer Einleitung*. Stäfa: Fajus.

Hirn, Wolfgang. 2018. Chinas Bosse. *Unsere Unbekannten Konkurrenten*. Frankfurt/New York: Campus.

Hong, Zhaohui. 2015. *The price of China's economic development. Power, capital and the poverty of rights*. Lexington: The University Press of Kentucky.

Howell, Anthony. 2019. Ethnic entrepreneurship, initial financing, and business performance in China. *Small Business Economics* 52 (3): 697–712.

Hsu, Daniel. 2018a. *A curated guide to Beijing's startup scene, part 1*. https://technode.com/2018/06/02/a-curated-guide-to-beijings-startup-scene-part-1/. Accessed 18 June 2019.

———. 2018b. A *curated guide to Beijing's startup scene, part 2*. https://technode.com/2018/06/16/a-curated-guide-to-beijings-startup-scene-part-2/. Accessed 18 June 2019.

Hsu, Jennifer Y.J., and Reza Hasmath. 2013. *The Chinese corporatist state. Adaption, survival and resistance*. New York: Routledge.

Jiang, Yujia. 2014. The curious case of inactive Bankruptcy Practicein China: A comparative study of U.S. and Chinese Bankruptcy Law. *The Northwest Journal of International Law & Business* 34 (3): 559–582.

Jourdan, Adam, and John Ruwitch. 2018. *Alibaba's Jack Ma is a Communist Party member, China state paper reveals*. Reuters. Shanghai. https://www.reuters.com/article/us-alibaba-jack-ma/

alibabas-jack-ma-is-a-communist-party-member-china-state-paper-reveals-idUSKCN1NW073. Accessed 05 February 2019.

Kooperation International. n.d. *Cluster Zhongguancun*. Kooperation International. https://www.kooperation-international.de/laender/hightech-regionen/zhongguancun/#c23502. Accessed 18 June 2019.

Lardy, Nicholas R. 2014. *Markets over Mao. The rise of private business in China*. Washington, DC: Peterson Institute for International Economics.

———. 2019. *The state strikes back. The end of economic reform in China?* Washington, DC: Peterson Institute for International Economics.

Lee, Kai-Fu. 2018. *AI superpowers. China, Silicon Valley, and the new world order*, International. Aufl. Boston/New York: Houghton Mifflin Harcourt.

Li, Yongjun. 2008. Die Stellung des Gläubigers im Konkursverfahren. In *Chinesisches Zivil- und Wirtschaftsrecht aus deutscher Sicht. Neuere Entwicklungen im Sachen-, Konkurs-, Arbeits-, Wettbewerbs-, Gesellschafts- und Kapitalmarktrecht*, ed. Bu Yuanshi, 115–134. Tübingen: Mohr Siebeck.

Li, Shuguang, and Zhofa Wang. 2010. Review of the PRC Bankruptcy Law in 2009. *INSOL International, Technical Series Issue* 11 (3): 1–2.

Li, Hongbin, Li-An Zhou, Lingsheng Meng, and Qian Wang. 2008. Political connections, financing and firm performance: Evidence from Chinese private firms. *Journal of Development Economics* 87 (2): 283–299.

Liu, Sylvia Xihui. 2016. Innovation design: Made in China 2025. *Design and Management Institute: Review* 27 (1): 52–58.

Liu, Xinzhi, Yushan Liu, and Zhuxin Wan. 2015. Difficulties of college students' business startups under economic new normal and their countermeasures. *Cross-Cultural Communication* 2015: 107–112.

Liu, Zhiyang, Zuhui Xu, Zhao Zhou, and Yong Li. 2019. Buddhist entrepreneurs and new venture performance: The mediating role of entrepreneurial risk-taking. *Small Business Economics* 52 (3): 713–727.

Ma, Hao, Shu Lin, and Neng Liang. 2012. *Corporate political strategies of private Chinese firms*. New York: Routledge.

Martin, Nathalie. 2005. The role of history and culture in developing Bankruptcy and insolvency systems: The perils of legal transplantation. *Boston College International and Comparative Law Review* 28(1): 1. https://lawdigitalcommons.bc.edu/iclr/vol28/iss1/2. Accessed 25 September 2019.

Mattheis, Philipp. 2014. *Reich werden im Kommunismus*. Handelsblatt vom 26 April 2014. https://www.handelsblatt.com/unternehmen/mittelstand/special-existenzgruendung/start-ups-in-china-reich-werden-im-kommunismus/10363070.html?ticket=ST-2587437-TPta7OtI5cfcnydfE93d-ap2. Accessed 18 June 2019.

McGregor, Richard. 2010. *The party. The secret world of China's communist rules*. New York: Harper Perennial.

———. 2018. *Xi ignored private enterprise. Now he needs it*. https://www.bloomberg.com/opinion/articles/2018-11-20/the-chinese-economy-depends-on-the-private-sector-xi-ignores. Accessed 30 April 2019.

Molloy, Fran. 2016. *Enter the dragon: The new Australian technology park*. https://sciencemeets-business.com.au/tag/china-torch-program/. Accessed 18 June 2019.

Münzel, Frank. 2007. Einige Anmerkungen zum neuen Konkursgesetz der Volksrepublik China. *ZChinR* 1: 47–49.

National Bureau of Statistics of China. 2018. *China statistical yearbook 2017*. http://www.stats.gov.cn/tjsj/ndsj/2018/indexeh.htm. Accessed on 21.06.2019.

National Development and Reform Commission. 2016. *Zhōnghuá rénmín gònghéguó guómín jīngjì hé shèhuì fāzhǎn dì shísān gè wǔ nián guīhuà gāngyào 2016 nián*. http://www.gov.cn/xinwen/2016-03/17/content_5054992.htm. Accessed 23 June 2019.

Nee, Victor, and Sonja Opper. 2012. *Capitalism from below: Markets and institutional change in China*. Cambridge, MA/London: Harvard University Press.

Parry, Rebecca, Yongqian Xu, and Haizheng Zhang. 2010. *China's new enterprise bankruptcy law. Context, interpretation, and application*. Farnham/Surrey/Burlington: AShgate Pub (Markets and the law). http://site.ebrary.com/lib/academiccompletetitles/home.action. Accessed 25 September 2019.

PWC. 2015. *PwC private equity/venture capital 2014 review and 2015 outlook*. https://imaa-institute.org/docs/statistics/pwc_china_private-equity-2014-outlook-2015.pdf. Accessed 18 June 2019.

Reshetnikova, M.S. 2018. Innovation and entrepreneurship in China. *European Research Studies Journal* 2018 (11): 506–515.

Röhr, Nino. 2018. *Der Vertrag zwischen Venture Capital-Gebern und Start-ups*. Wiesbaden: Springer Fachmedien.

Schefczyk, Michael. 2006. *Finanzieren mit Venture Capital und Private Equity. Grundlagen für Investoren, Finanzintermediäre, Unternehmer und Wissenschaftler*, 2. Aufl. Stuttgart: Schäffer-Poeschel.

Sieren, Frank. 2018. *Zukunft? China! Wie die neue Supermacht unser Leben, unsere Politik, unsere Wirtschaft verändert*. München: Penguin.

Silicon Valley Bank. 2018. *China startup outlook 2018*. https://www.svb.com/globalassets/library/uploadedfiles/content/trends_and_insights/reports/startup_outlook_report/china/svb-suo-china-report.pdf. Accessed 18 June 2019.

Sommer, Theo. 2019. *China First. Die Welt auf dem Weg ins chinesische Jahrhundert*. München: C. H. Beck.

State Council of the People's Republic of China. 2018. *Guówùyuàn guānyú tuīdòng chuàngxīn chuàngyè gāo zhìliàng fāzhǎn dǎzào "shuāng chuàng" shēngjí bǎn de yìjiàn*. http://www.gov.cn/zhengce/content/2018-09/26/content_5325472.htm. Accessed 18 June 2019.

State Council of the PRC. 2017. *Guówùyuàn guānyú qiánghuà shíshī chuàngxīn qūdòng fāzhǎn zhànlüè jìnyībù tuījìn dàzhòng chuàngyè wànzhòng chuàngxīn shēnrù fāzhǎn de yìjiàn*. http://www.gov.cn/zhengce/content/2017-07/27/content_5213735.htm. Accessed 18 June 2019.

Tang, Alan, and Pauline Au. 2007. *China's enterprise bankruptcy law. The central government's new regulations on bankruptcy administrators*. A Plus, S. 48–50. http://app1.hkicpa.org.hk/APLUS/0711/p48_50.pdf. Accessed 25 September 2019.

Ten Brink, Tobias. 2014. Chinesischer Kapitalismus? Unternehmen und Unternehmertum in China. In *Länderbericht China*, ed. Doris Fischer and Christoph Müller-Hofestede, 681–738. Bonn: Bundeszentrale für politische Bildung.

Tse, Edward. 2015. *China's disruptors: How Alibaba, Xiaomi, Tencent, and other companies are changing the rules of business*. New York: Penguin.

———. 2016. *The rise of entrepreneurship in China*. https://www.forbes.com/sites/tseedward/2016/04/05/the-rise-of-entrepreneurship-in-china/#51479c063efc. Accessed 30 April 2019.

Unesco. 2019. *Tertiary graduates by level of education*. http://data.uis.unesco.org/index.aspx?queryid=162. Accessed 18 June 2019.

Wang, Jinmin, Jing Wang, Hua Ni, and Shaowei He. 2013. How government venture capital guiding funds work in financing high-tech start-ups in China: A 'strategic exchange' perspective. *Strategic Change* 22 (7–8): 417–429.

Weng, Charlie. 2011. To be, rather than to seem: Analysis of trustee fiduciary duty in reorganization and its implications on the new Chinese Bankruptcy Law. *The International Lawyer* 45 (2): 647–671.

Wübbeke, Jost, Mirjam Meissner, Max J. Zenglein, Jaqueline Ives, and Björn Conrad. 2016. *Made in China 2025. The making of a high-tech superpower and consequences for industrial countries*. Berlin: Mercator Institute for China Studies. https://www.merics.org/sites/default/files/2018-07/MPOC_No.2_MadeinChina2025_web_0.pdf. Accessed 18 June 2019.

Xi, Jinping. 2017. *Den entscheidenden Sieg bei der umfassenden Vollendung des Aufbaus einer Gesellschaft mit bescheidenem Wohlstand erringen und um große Siege des Sozialismus chinesischer Prägung im neuen Zeitalter kämpfen*. Report on the XIX. Chinese Communist Party Congress [German]. http://docs.dpaq.de/12860-rede_xi_jinping_19._parteitag_parteikongress_1_.pdf. Accessed 23 June 2019.

Xinhua. 2014. *Premier Li's speech at Summer Davos opening ceremony*. http://english.gov.cn/premier/speeches/2014/09/22/content_281474988575784.htm. Accessed 18 June 2019.

———. 2018. *Data shows strength of China's private enterprise*. http://www.xinhuanet.com/english/2018-11/05/c_137581763.htm. Accessed 21 June 2019.

———. 2019. *Chinese premier underlines mass entrepreneurship, innovation*. http://www.china.org.cn/business/2019-06/14/content_74885471.htm. Accessed 21 June 2019.

Xu, Hang. 2013. *Das chinesische Konkursrecht, Rechtshistorische und rechtsvergleichende Untersuchungen*. Berlin: LIT.

Yan, Fang. 2009. *China's east star air faces liquidation – Xinhua*. Xinhua vom 27.03.2009. https://www.reuters.com/article/china-eaststar-idUSSHA8512720090327. Accessed 25 September 2019.

Yan, Xiaojun, and Jie Huang. 2017. Navigating unknown waters: The Chinese communist party's new presence in the private sector. *The China Review* 17 (2): 37–63.

Yang, Lixin. 2008. Administrator in China's new enterprise Bankruptcy Law: Objective standards to limit discretion and expand market controls. *The American Bankruptcy Law Journal* 82: 533–534.

Yang, Yuan, and Nian Liu. 2019. China hushes up scheme to recruit overseas scientists. *Financial Times*, 10 January 2019. https://www.ft.com/content/a06f414c-0e6e-11e9-a3aa-118c761d2745. Accessed 12 June 2019.

Zhang, Jian. 2018. Public governance and corporate fraud: Evidence from the recent anti-corruption campaign in China. *Journal of Business Ethics* 148 (2): 375–396.

Zhu, Yuchao. 2011. Performance legitimacy and China's political adaptation strategy. *Journal of Chinese Political Science* 16 (2): 123–140.

Zweigert, Konrad, and Hein Kötz. 1996. *Einführung in die Rechtsvergleichung. Auf dem Gebiete des Privatrechts*, 3. Aufl. Tübingen: Mohr.

Luxury Consumption

8

Kevin S. Rowens

Economic growth and socio-economic conditions alone cannot explain the disproportionate luxury consumption of the Chinese. Rather, it is culturally rooted in traditional Southeast Asian gift-giving culture and Confucian feudal society, in which luxury made visible the status acquired through education and the rank of the respective family in imperial bureaucratic society. Chinese Millennials, the first generation of wealth inheritors of the reform era, connect this tradition with their consumption of luxury, often criticized as insensitive. In this increasingly self-confident society, the younger generations in particular are recalling their national traditions. As a result, the Western luxury brands that have dominated the Chinese market to date are increasingly facing competition from Chinese labels. Their reorientation requires not only new forms of marketing but also creative adaptation to traditional Chinese craftsmanship and design language.

The demand for luxury goods and services, which has been rising for years, is a global phenomenon "with local characteristics" (Wiedmann and Hennins 2013, p. 64). In 2015, for the first time, more than one trillion euros were turned over in the luxury goods market. In 2018, it was already 1.2 trillion euros. Annual growth rates of 3–5 percent are still expected up to 2025, albeit depending on current political and economic risks. In particular, the trade dispute between the US and China could be momentous for European luxury labels due to the "crucial" importance of the Chinese market. In addition to the great importance of the younger generations born since 1980, to which the growth in the high-end personal goods segment is largely attributable, emerging markets such as Russia, Brazil, India, but above all China are responsible for the rapid and sustained increase in demand for luxury goods. China now ranks first, accounting for 33 percent of global

K. S. Rowens (✉)
East Asia Institute of Ludwigshafen University of Business and Society,
Ludwigshafen am Rhein, Germany

B. Darimont (ed.), *Economic Policy of the People's Republic of China*,
https://doi.org/10.1007/978-3-658-38467-8_8

luxury consumption in 2018, ahead of the U.S. and Japan, and is considered a growth engine for the industry. By 2025, Chinese luxury consumption is expected to account for nearly half of the global luxury market at 46 percent (Bain and Company 2018, p. 1).

The unique increase in Chinese luxury consumption of over 33 percent in just under 35 years, which is forecast to continue in the future, cannot be explained solely by the size of the country and its nearly 1.4 billion people. Rather, despite all the progress and economic successes, China is a country in transition with profound social and economic upheavals.

The focus here is not on describing the success story of Western luxury brands in the Chinese market or exploring it from a market strategy point of view but on the social reasons and consequences of a booming market that seems to have become a "dance around the golden calf" for large and small manufacturers and marketing strategies.

The aim of this chapter is to justify excessive luxury consumption from China's hybrid socio-economic social system, which has created specific framework conditions for the development of luxury consumption, and from the country's materialist and collectivist tradition. To this end, the concepts of luxury and luxury goods are first clarified in general terms, and the Chinese luxury goods market is outlined in outline, then the social and economic conditions of luxury consumption are analyzed based on the Chinese social system, which, however, can only create the possibility of luxury consumption, since luxury is subject to ambivalent and relative evaluation. Historical and cultural factors play an important role in the actual consumption behavior of the Chinese so that marketing strategists assume a separate East Asian luxury segment on a global scale. Changes brought about by China's younger generation, the so-called Millennials, are the subject of a final consideration.

8.1 Luxury and Luxury Goods

> Historically, luxury was made for the few: It was the ordinary life of extraordinary people. (Kapferer 2015, p. 21)

8.1.1 Definition and Delimitation of the Concept of Luxury

The concept of luxury is derived from the Latin terms Luxus and Luxuria, which denote deviation from the straight and narrow, that which goes beyond the "usual measure", abundance, luxuriance, rich abundance, splendour, but also superfluity, waste, and dissipation (Mühlmann 1975, p. 20). It refers to the "expenditure that goes beyond what is necessary, i.e. what is generally recognized as necessary for the satisfaction of aspirations or the average standard of living" (Kambli 1892).

Luxury goods are not a category of goods in their own right, but the high-end market segment of industries producing everything from high-end watches and jewelry, designer fashion, and premium cars to hotels, yachts, and private jets. These are "relatively high-priced products or services by means of which, due to their relative scarcity, their owners wish to signal social distinction and belonging or experience a special intrinsic value. The luxury goods sector comprises the industries and companies that act as suppliers and competitors in this business segment with luxury goods" (Müller-Stewens 2013, p. 8).

According to the definition, products and services derive their value primarily not from their characteristics, but from the consumer's assessment, which depends essentially on the respective brand name and its brand essence (high price, high product quality, originality and creativity, aesthetics, uniqueness, and tradition). They satisfy needs located on the "higher" levels of Maslow's pyramid: Prestige, prestige, status, class (class) membership, and one's own values (Fig. 8.2). The idealistic benefits mean that, in addition to prices that are in no way based on production costs, they differ from normal consumer goods in that their price sensitivity is greatly reduced (the price is rather seen as an exclusivity factor), and that their consumption increases disproportionately with rising household income (Müller-Stewens 2013, p. 9).

8.1.2 The Chinese Luxury Goods Market

"Paradoxically, China has become the world's most important luxury market within 20 years but has so far hardly produced any luxury brands of its own" (Heine and Waldschmidt 2016, p. 1). Until a few years ago, luxury was exclusively associated with Western culture. The luxury brands from France, Italy, and also Germany are considered to be particularly valuable and prestigious because of their quality, craftsmanship, tradition, and design (Heine and Waldschmidt 2016, p. 1). The luxury fashion brands Chanel, Vuitton, and Gucci are still the most popular fashion brands, the top cosmetic brands are Lancôme, and Estée Lauder, and for cars, Audi, Porsche, BMW, and Mercedes are highly rated. Already in 2009, 66 percent of rich Chinese owned a Rolex watch at an average price of USD 2253 (Degen 2009, p. 5). Among others, Gucci and Chanel increased their sales of handbags and cosmetics, respectively, by 25 percent to USD 22.07 billion in the Chinese domestic market alone in 2017 (Bilanz Lifestyle 2018). Table 8.1 provides an overview of the market shares of individual product groups in 2017.

Until a few years ago, the following generalized characteristics applied to the Chinese luxury goods market:

1. Brands play a much bigger role than product quality or personal style. What is more important is that the brand's logo makes social status, a very important cultural value in China, visible. Successful Chinese also want to use luxury brands to signal their membership in a successful group.
2. Luxury goods are bought quickly and impulsively (Ma and Becker 2017, p. 193). The decision to buy a product is made quickly, as it is not based on the consumer's own decision, but depends on brand awareness, the personal recommendation of acquaintances or friends with the same status claims, and the intensity of advertising.
3. A large proportion of luxury goods in the goods segment are purchased for gifts. Their material value expresses appreciation for the recipient and their reputation. They play

Table 8.1 Main sectors of the Chinese luxury goods market

Luxury and premium cars (Audi, BMW, Porsche, jaguar	~53.50%	Cosmetics	28%
Portable luxury goods	~44.70%	Jewellery, jewels	27%
Luxury hotels	~1.80%	Lingerie	24%
		Shoes	21%
		Handbags, suitcases	18%
		Accessories	14%
		Watches	13%
		Men's underwear	8%

Source: Own representation according to Bain and Company (2017, p. 3) and World Luxury Index China (2013, p. 11)

a major role in maintaining social networks. The group highlighted in grey in Table 8.1 (luxury bags, cosmetics, and luxury goods) is generally popular as a gift and luxury jewelry. Luxury wristwatches and works of art especially among rich Chinese.

4. Luxury goods are considered an investment in China. "They are seen as a hedge for the future and against loss of value of money" (Ma and Becker 2017, p. 193). Since the abolition of gold backing, distrust of paper and fiat money has been a widespread motive for luxury investment. For China, with its huge dollar reserves, private luxury consumption also has to stabilize macroeconomic significance in this respect.
5. The Chinese make a large part of their luxury purchases abroad. The reason for this is not only the rapid increase in foreign travel, with some of the travel already representing a luxury good in itself. The Chinese have been considered the world's travel champions for several years now. Especially the "Golden Week" in October and the Chinese Spring Festival are used for "shopping tours". In addition, there are indirect purchases via acquaintances, relatives, friends, and even unknown compatriots abroad ("Daigou"). The travelers themselves are often contractors for the purchase of luxury goods.

The reasons for buying luxury goods abroad are:

- Fear of product counterfeiting in their own country.
- A gift from the Far West enhances its function of showing appreciation and respect for the recipient (Ma and Becker 2017, p. 196).
- The Chinese government's tax and duty policies. Import duty, consumption tax, and value-added tax add up to 65 percent extra of cost in China for domestically purchased luxury goods (Ma and Becker 2017, p. 195).
- The refund of the sales tax increases the lucrativeness of the foreign purchase.

The extent of foreign purchases depends on the government's tariff and tax policies and the exchange rate of the RMB, which had become increasingly favorable against the dollar and euro in recent years. Foreign purchases had risen to over 75 percent of luxury purchases. After lowering the import tax to strengthen domestic sales, the ratio of purchases is slowly changing in favor of domestic purchases.

Luxury goods are primarily purchased according to brand image. Searches and information are carried out via social networks (WeChat), and purchases are made in the ambiance of luxury shops. Service plays a major role here.

The buying behaviour of the younger generation, which has grown up alone in times of the market economy, is increasingly different from that of the previous generation. The demands on the luxury goods market are becoming more differentiated, individual, quality-conscious ("premiumization"), and tradition-conscious ("sinification", Berger 2013, p. 74). Younger Chinese are also increasingly purchasing high-priced luxury goods via internet portals and mobile payment systems (Bain and Company 2018).

8.1.3 The Relativity of Luxury

The example of the Chinese Millennials shows that consumer behaviour is not only shaped by socio-cultural values but that ideas and acceptance change depending on the times, e.g. luxury goods can become everyday consumption due to a loss of exclusivity. The moral-ethical attitude towards luxury is also ambivalent and dependent on the spirit of the times. It ranges from covetousness to wastefulness, from Veblen, who sees luxury as the ostentation, prestige, and extravagance of a rich idler class, to Sombart (luxury as a prerequisite and driving force of capitalism) and Reitzle, for whom "today's luxury is tomorrow's prosperity" and the "pacemaker of an economy". "Everyone earns from luxury – luxury creates work" (Reitz 2017, pp. 88 and 147).

According to Kühne and Boschart (2014, pp. 12–22) from the Gottlieb Duttweiler Institute in Zurich, ideas about luxury ideally pass through four phases in terms of type and intensity: Infantile, Adolescent, Maturity, and Seniority phases (Fig. 8.1).

8.2 Luxury in the Chinese Transformation Society

Figure 8.2 depicts China's transition from a socialist planned economy to a capitalist market economy against the background of the general satisfaction of needs. The economic reform policy initiated by Deng Xiaoping has been gradually changing Chinese society since 1978.

"The great chairman left […] a clean slate. The conditions for one of the most blatant social experiments in history seemed to have been created" (Schmid 2016). What is meant is the question of whether "the" Chinese in the "free market economy" first satisfy basic needs or hastily resort to luxury.

"It can […] be assumed that luxury – from antiquity to modern times – has always been very closely linked to the respective social system and as such was an important means by which the elite and upper class set themselves apart from the others socially and also

Infantile phase	consumer hunger young emerging market pent-up demand, exit demand "more is more
adolescence phase	Solvency Competitive pressure (peer pressure) Fear of social relegation Luxury with signal effect "more is a must
Maturity Phase	Luxury fatigue decrease in the marginal utility of the material luxury of product- au experience "more is less".
Seniority phase	Meaningful organization of time turning away from material luxury necessary maximum enjoyment
Saturated society millennials	stenciling demonstrative renunciation awareness of sustainability "show that you do not have to show

Fig. 8.1 Luxury phases. (Own representation according to Kühne and Boschart 2014, pp. 12–22)

asserted themselves" (Pöll 1980, p. 9). The change from the aristocratic understanding of luxury by the elites to the democratization of luxury consumption, which continues to this day, was brought about in the West by economic development and political democratization at the beginning of the twentieth century (Thieme 2017, p. 8), whereby the democratization of luxury initially meant the accessibility of luxury goods by broader circles of the population. These were primarily the bourgeois classes that had come to wealth and prosperity during the industrialization phase. In China, the democratization of luxury consumption only began at the beginning of the 1990s.

"After the Communist Party took power in the early 1950s, the government took everything away from its people, including social order and food" (Rudolf and Trester 2016, pp. 88–89). After the "opening and reform" (gaige kaifang) in 1978, it gradually allowed more and more needs to be satisfied. In doing so, Chinese society largely adhered to Maslow's hierarchy of need satisfaction, according to which the needs of one level must be satisfied before the needs of the next level can be met (Rudolf and Trester 2016, p. 90).

According to Kühne and Boschart's four-phase model of luxury (Fig. 8.1), Chinese society entered the infantile phase around 1980, which is characterized by a great need to catch up and the hunger for consumption typical of emerging markets. Initially, only a newly forming urban elite could afford this hunger for consumption. In the meantime, it has reached the adolescence phase, also with its descendants, the so-called Millennials, in

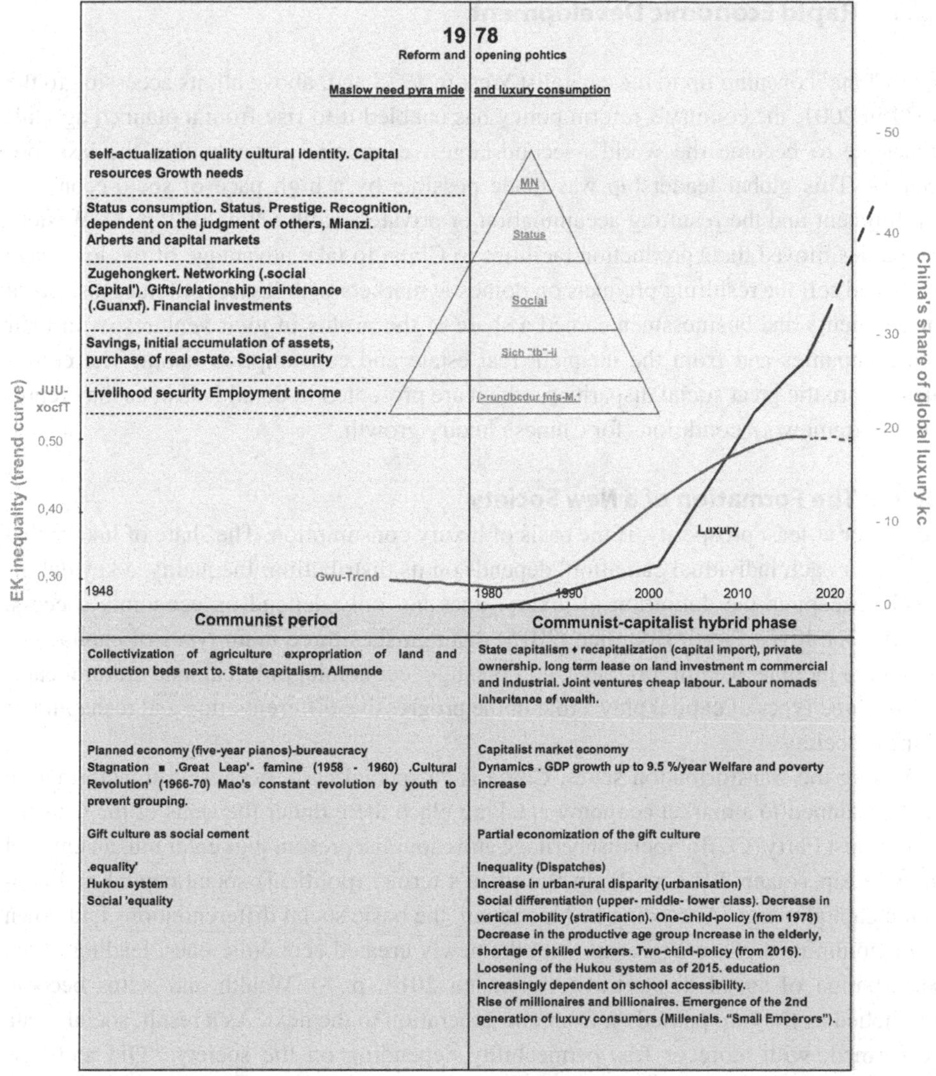

Fig. 8.2 Political economy development – need satisfaction – consumption level. (Own illustration based on Rudolf and Trester (2016, p. 89) and Bain and Comp (2018) as well as various volumes of the China Statistical Yearbook)

which it is primarily a matter of securing status and sending out a signal, while the hunger for consumption continues to spread to the rest of the growing middle class.

The general conditions and/or accompanying circumstances for the Chinese luxury boom, which began in 1990 with temporary slowdowns and has continued to this day, are the rapid economic development, the special nature of the Chinese transformation society, individual socio-economic factors, and the mentality of the Chinese.

8.2.1 Rapid Economic Development

With China's opening up to the capitalist West in 1978 and, above all, its accession to the WTO in 2001, the country's reform policy has enabled it to rise from a planned agricultural state to become the world's second-largest economic power within the last four decades. This global leadership was made possible by a high pace of socio-economic development and the resulting accumulation of private wealth (Zhu 2012, p. 1). Western companies moved their production facilities to China to take advantage of the low wage levels and sell the resulting products on domestic markets with high profit margins. Local entrepreneurs and businessmen earned a share of the profits in joint ventures, with their own companies and from the incipient real estate and consumption boom. The consequences are the great social disparities, which are presented in detail below, as they represent the framework conditions for Chinese luxury growth.

8.2.1.1 The Formation of a New Society

Wealth, or at least prosperity, is the basis of luxury consumption. The share of luxury that a group or each individual can afford depends on its distribution. Inequality, a key definitional concept in the definition of luxury, does not only depend on economic success. French sociologist Pierre Bourdieu (2015) distinguishes three main types of capital that can create income, wealth, privilege, and prestige: economic, political, and cultural capital. All three types of capital play a role in the progressive differentiation and reshaping of Chinese society.

Among the transformation states, China is "a special case, because a transformation from a planned to a market economy is taking place there under the aegis of the Chinese Communist Party (CCP). Socialist heritage and capitalist present thus enter into an unusual union" (Alpermann 2013, p. 2); in Bourdieu's terms, (political) social capital and economic capital merge. According to Alpermann, the basic social differentiations laid down in the communist phase aggregate with the newly created economic ones, leading to the exacerbation of social disparities (Alpermann 2013, p. 3). Wealth and status become entrenched as they are passed on from one generation to the next. As a result, social strata are formed, with more or less permeability depending on the society. "Old and new inequalities reinforce each other" (Alpermann 2013, p. 2).

The third type of capital, cultural capital (traditional status and education of traditionally wealthy families) has been preserved from pre-communist feudal times in the thinking and actions of party cadres and new elites. This is "incorporated cultural capital" (Bourdieu), acquired through education, transmitted through an educational tradition, or unconsciously adopted in behavior. This class of learned bureaucrats, whose origins date back to the fourth century (Zhou dynasty) and whose recruitment was institutionalized in the sixth century AD through a system of education and promotion (Ke Ju), still constitutes China's elite today. For 1300 years, this career path remained the only way for rural people to escape their fate in exploited village communities and achieve a better life. Male citizens achieved class membership by passing the civil service examination and demonstrating Confucian virtues. By raising his social status, the individual simultaneously raised that of his family (Lu 2008, p. 6).

This class was very fond of a luxurious lifestyle, which was reflected in the transformation of "incorporated" into "objectified cultural capital" (Bourdieu), which included not only writings, paintings, or machines that stored incorporated knowledge and skills, but also porcelain, gold and valuable furniture (Lu 2008, p. 5). "There are golden houses in the books and beautiful women like jade" is still a regularly quoted proverb that extols the glory of educational success in Chinese society (Dettmer 2016, p. 54). Although the ke-ju system was abolished in 1905 (Lu 2008, p. 2), as Chinese society internalized the tradition of education and upward mobility, it also retained a taste for luxury, works of art, and gold (Lu 2008, p. 4). Educational optimism and luxury as a motivation for upward mobility subliminally survived the Republic and People's Republic periods. With the reform and opening-up policy, education regained its old glory. In times of the one-child policy, parents spared no expense for their child's education, not infrequently to the point of financial ruin (Dettmer 2016, p. 55). However, good education and success increasingly depend on the wealth of the family rather than the abilities of the individual, confirmation of Alpermann's stratification theory described above. The combination of private debt, unrealized aspirations for advancement, and thus increasing social inequality may become a disruptive factor in forming Chinese society (Dettmer 2016, p. 55).

Unlike Alpermann, who sees luxury consumption as a consequence of wealth accumulation and concentration as well as class-specific social segregation, the China Weekly editor Zhu Xuedong (2012) assumes in an article published by the Goethe-Institut China that luxury is necessary as a driving force for the formation of a new social structure. In doing so, he draws on Sombart's "Love, Luxury and Capitalism", published at the beginning of the nineteenth century, and Bourdieu's concept of distinction (Bourdieu 2015).

After that, in the early days of capitalism in the nineteenth century, the emerging class of the wealthy bourgeoisie used money and the consumption of luxury goods to break up a social structure that had been handed down since the Middle Ages, dissolving old social differences and privileges and positioning itself as an upper class. An essential component of this positioning was taste, which not only reflected social differences but established new class identities and relationships. Similarly, according to Zhu (2012, p. 2), luxury consumption will also overcome the politically dominated class pattern in China. Symbols of success in the new constellation are luxury goods that signal wealth, status, and taste. Ultimately, it is a matter of maintaining differences, albeit not under the sign of political power, but between the economically successful and other strata of the population through the visible consumption of luxury goods (Zhu 2012, p. 3).

"If Bourdieu still assumed that the acquisition of cultural capital could mean social advancement, today's sociologists such as Oliver Nachtwey speak of descent societies" (Reitz 2017, p. 1). The "phenomenon" observed by Rudolf and Trester (2016, p. 91) of an office worker affording a luxury handbag at the price of 6 months' wages, thus experimenting with "relative wealth" in the pursuit of status and skipping a level of need, suggests acceptance and an unwavering desire for advancement – or pretense. In China, too, hybrid consumption phenomena can be observed as a result of limited social advancement or even social decline.

The traditional promise of social advancement through education and the disdain for physical labour has faltered in recent years. A university degree has long since ceased to be a guarantee of professional success and an elevated standard of living (Dettmer 2016, p. 60). Families' high investments in the education of their only child as a result of the one-child policy are paying off less and less. Due to the one-sided association of social recognition and career with an academic title and the disdain for physical labor, the country lacks the skilled workers it desperately needs. Consequently, the average starting salary of a university graduate fell below the average wage of a migrant worker for the first time in 2012 (Dettmer 2016, p. 60).

Despite the slowdown in economic growth and the structural problems in the labour market, the politically controlled construction of a "harmonious society" and the consumption of luxury goods continue unabated. Social disparities have little influence on luxury consumption anyway, as long as the general acceptance of luxury continues and the segment of the population eligible for luxury consumption is relatively price-insensitive (see above). Thorstein Veblen (2007) already noticed at the end of the nineteenth century as "conspicuous behaviour" of the "fine people", among other things, that the consumption of luxury goods increased proportionally with their price increases (Schipper 2014). The time factor in the wealth accumulation of traditionally thrifty middle-income Chinese (Zhu 2012) is also likely to lead to savings and support luxury consumption in China.

8.2.1.2 The Attractiveness of the Chinese Luxury Market

Due to its population wealth, the potential for luxury consumption is enormously high. However, buzzwords such as "world travel champion", "more billionaires than in Germany", "globally highest luxury consumption" lose considerable weight in relation to the size of the population. It is true that China's real GDP reached USD 13.4 trillion in 2018, ranking second behind the US with 20.5 trillion (Statista 2019). When converted to its citizens, China, despite its great dynamism, is still far behind in the global ranking along with other emerging economies such as Brazil, and Mexico in the middle of all countries. With an average income of 9608 USD, a Chinese reaches just under a quarter of the German income (40,862 USD/2017) and less than a sixth of the US income (62,606 USD/2017) (Statista 2019). Thus, the average Chinese per capita income does not explain the attractiveness of the Chinese luxury market. Its dynamism and attraction for Western luxury suppliers are based on the distribution of economic growth among individual population groups. In addition to socio-economic factors, demographic factors via age structure and urban-rural distribution of the population, income distribution, and propensity to consume influence access to luxury consumption.

8.2.2 Socio-Economic and Demographic Factors

> China's economic and demographic development is initially fuelling the boom, as are its cultural values. (Ma and Becker 2017, p. 196)

8.2.2.1 Purchasing Power

One internet platform ran the headline: "Fewer poor, more inequality" or, to put it another way: "In China there is a lot of money and a lot of inequality" (Ma and Becker 2017, p. 188). This aptly characterizes China's whole socioeconomic situation. To be sure, China has made great strides in poverty reduction. However, one-sixth of China's population is still considered poor. At the same time, the country has the most dollar millionaires after the USA.

Even though the price-adjusted gross domestic product of 16.807 USD in 2016 is almost twice as high as the nominal one, disposable income/capita 320.5 USD (Wirtschaftskammer Österreich 2019), and the savings rate is 12.7 percent of disposable income (GTAI 2018), these average figures, as well as the inequality parameter (Gini coefficient: 0.47) only, provide indications of the general consumer climate in the country: "Chinese consumers are among the most optimistic in the world … High consumer confidence is based on incomes that have been rising in real terms for decades" (GTAI 2018). For luxury consumption, they are not very meaningful. For luxury consumption, the lower limit of income is where it still allows for occasional, optional, or hybrid luxury consumption. Commonly, that is where the middle class begins.

However the relationship between wealth and poverty is seen in society, wealth and prosperity are the most essential prerequisites for luxury consumption, i.e. the purchasing power of the upper and middle classes are the most important parameters for it.

> In Chinese society [there is] a relatively large gap in income distribution, which initially benefits the luxury segment. (Ma and Becker 2017, p. 189)

8.2.2.2 China's Elite

According to the World Luxury Association China, there were 175 million customers "eligible" to buy luxury goods in China in 2013 (Ma and Becker 2017, pp. 188–189), with a middle class of about 400 million people and China's share of the global luxury market at 31 percent. Due to policy measures (anti-corruption campaign) and slowing economic growth, luxury consumption had weakened by 2016 and then increased to 33 percent of global consumption by 2018. As the government's anti-corruption intervention in the consistent growth of luxury consumption may have caused the upper segment of the elites, in particular, to merely exercise temporary restraint, the pool of potential luxury consumers may have expanded to over 200 million by the end of 2018.

China's elite, which is the stable backbone of China's luxury consumption, is divided by Lu (2008, p. 10) into three classes "upper-upper class, lower-upper class, and the upper-middle class" primarily according to their wealth. Ma and Becker (2017, pp. 189–190) classify by age. The complementary classifications are shown in Tables 8.2 and 8.3.

8.2.2.3 Middle Class

The definition of the lower income limit, up to which one can still speak of the middle class, is not uniform. In China, 350–450 million people are currently said to belong to the middle class.

Wages increased by 9.8 percent on average in recent years (Hua 2018). In 2018, purchasing power grew by 6.5 percent when adjusted for inflation (inflation at 2.1 percent higher than in previous years) (Xinhua 2019). Table 8.4 classifies annual incomes by level into income groups. Income by class provides initial indications of the delineation of a middle class.

The study by economic expert Wu Xiaobo is likely to provide realistic data for defining a class with an affinity for luxury. According to this, the net annual income of the middle class is between 100,000 and 500,000 RMB, i.e. between 14,000 and 70,000 euros, the fixed assets between 200,000 RMB (28,000 euros) and five million RMB (700,000 euros), but 82.5 percent already have additional further income (China SME Center Germany 2018). There are different figures on the size of sustainable luxury consumers. They vary between 10.7 percent of the total population, i.e. about 140 million and 175 million (Ma and Becker 2017, pp. 188–189). The growth rate for wealth is around 9 percent up to 2020.

The share of the middle class in the rapidly growing GDP has remained the same at 45 percent since the 1978 reform, i.e. it earns on average slightly more than the average per capita income of the country (Neinhaus 2015), while 10 percent of the highest income population increased its share from 27 to 41 percent. This means that the growth of the middle class, given its constant percentage share, is financed solely by the growth of GDP.

Table 8.2 Classification according to assets

Upper-Upper-Class (UUC)	Few super rich, mostly traditional or politically established.
	Wealth mostly inherited.
	Prestige and reputation also through sponsorship.
	Conspicuous public consumption (sponsoring events and charities).
Lower-Upper-Class (LUC)	Nouveau riche, not considered equal or elite by UUC.
	Successful businessmen and entrepreneurs.
	Conspicuous personal consumption to signal membership in the elite.
Middle-Upper-Class (MUC)	Bottom layer of the elite: Educated, career-oriented younger people, globally connected.
	Success is made visible.
	Conspicuous luxury, especially in the home, is meant to. Demonstrate success.

Source: Own representation according to Lu (2008)

Table 8.3 Classification from a target group perspective

Over 50-year-olds (Ü50)	Mostly high-ranking officials.
	Hardly any university degrees, low level of education,
	Little knowledge of foreign languages.
	Experience from cultural revolution.
	Traditionally oriented.
	Frugal and restrained in consumption.
	Not buyers, but consumers of luxury goods (gifts).
35–50 year olds (35/50)	Educated, knowledge of foreign languages, often experience abroad; already grew up in the time of the market economy.
	Traditional consumer restraint as a virtue but desire for higher standard of living.
	Luxury goods are signs of success, buy and consume luxury goods.
	The group coincides with MUC Lu's.
Under 35-year-olds (U35), "millennials" born after 1980	Generation of "little emperors", educated and drilled for success by parents; English is taken for granted.
	Result of one-child policy and family aspirations for advancement; alien to their parents' working world; spoiled.
	Confident and ambitious.
	Smart and hedonistic: Enjoyment of life, self-centered. Consumes uncritically (often recklessly beyond their means).
	Quality and tradition conscious, more refined and discreet luxury style.
	Represent 60 percent of Chinese luxury consumers (Europe: 38)
	Darlings of Gucci & Co; most important target group of the luxury industry

Source: Own representation according to Ma and Becker (2017) and Berger (2013)

Table 8.4 Income by class (in RMB)

EK class/household	2014	2017	Increase (%)
Upper class	50,968	64,934	27.4
Upper middle class	26,937.4	34,546.8	28.3
Middle middle class	17,631	22,496.3	27.6
Lower middle class	10,887.4	13,842.8	27.2
Underclass	4747.3	5958.4	25.5

Source: Own representation according to State Statistical Office, National Bureau of Statistics of China (2018)

8.2.2.4 Wealth and Inequality

Wealth is a sure indicator of luxury consumption. After China opened up to a market economy, according to research by the French economist Piketty (cited in Neinhaus 2015), the income of the top 1 percent of the population, i.e., the 1.2–1.4 million highest-income Chinese, grew 40-fold (3.817 percent), while that of the middle class grew by about 700 percent, slightly higher than that of the United States. During this period, the income of France's top promille increased by only 158 percent. Growth started in 1978 at a very low GDP. After a delay, luxury consumption followed in the late 1980s with its high-priced segment. So there is a direct link between economic growth and luxury consumption.

In 2006, there were 16 Chinese billionaires; by the end of 2017, there were 373 out of a total of 2158 worldwide. So one in five billionaires was Chinese. They had $1.12 trillion at their disposal, 39 percent more than in 2016. They are almost exclusively self-made billionaires. A quarter of their wealth growth in 2017 was based on increased property prices, as real estate makes up about 20 percent of their portfolios (Ritter and Ankenbrand 2018). Moreover, China is now home to the largest number of dollar millionaires in the world after the US. Their numbers continue to rise rapidly. While there were around 1.5 million in 2012, their number rose to 2.4 million by 2017, helped by the continued appreciation of the RMB against the dollar. For the luxury market in the higher price segment, these super-rich are important not only because of their purchasing power but also because of their function as trendsetters and upward mobility motivators, especially since in China wealth is ostentatiously displayed for everyone to see.

8.2.2.5 Demographic Development

For everyday consumption, China is seen as a market with a growing population of currently 1.395 billion/in 2018 (Statista 2018). For luxury consumption, this market is not only severely constrained by income and wealth inequality but is also increasingly influenced by the country's demographics. The age pyramid is becoming increasingly top-heavy, while the share of the productive middle age group is shrinking. This natural population trend, which is typical of industrialization, was accelerated by the one-child policy in effect since 1980, the consequences of which the relaxation since the mid-2000s is intended to reduce. While average life expectancy is rising due to improved medical care, the natural birth rate is declining at an accelerated rate compared to the historically slow development of Western industrialized countries due to political prescription.

8.2.2.6 Age Structure

The average age in China is 40 years (in Germany 45 years). It is expected to rise to 45 years by 2050 (Ma and Becker 2017, p. 190). The age group of 15- to 65-year-olds is of central importance for the economic performance of a state. This generation has been shrinking in China since 1980, and as the labor force declines, the per capita social burdens borne by the middle age group increase. The portions of income available for luxury consumption will decline if economic growth does not continue at least at the same rate. The extent to which productivity advances can stabilize or increase incomes remains uncertain.

In 2010, there were about 5.5 working people for every pensioner; by 2030, there will be only 2.5 (Ma and Becker 2017, p. 190). But this is not the only reason why disposable income for luxuries is declining. The unwritten intergenerational contract of Chinese society reinforces this effect: it states that parents should educate their children as well as possible and that the latter should then provide for their parents to the same extent in return.

The one-child policy has changed the level of aspiration. Status thinking and ambition for advancement led parents to want to make their only child's life and future as comfortable and successful as possible. According to the concept of extended family, parents, grandparents, and other relatives subordinate their demands to those of the child. They invest in upbringing, education, and comfortable living, which has significantly stimulated the purchase of luxury goods in the country (Ngai and Cho 2012, p. 258). Similar efforts are expected by their relatives from their children in old age. This removes significant purchasing power from the income of the young middle class to buy luxury goods and changes the structure of luxury consumption.

Luxury buyers are currently still very young. Eighty percent of luxury buyers are under 45 years of age, the majority of whom can be attributed to the Millennials, the "spoiled little emperors", who alone are responsible for 60–70 percent of Chinese luxury sales. The number of people aged 20–24 will have halved from 120 million to 60 million by 2024, whereas the number of people aged 50–54 will triple (Lu 2008, p. 20). Since the mobility behavior of younger people is much more flexible compared to older people, the urbanization that the government is aiming for is mainly carried out by the younger people. Luxury consumption will therefore be concentrated in cities even more than before. However, the hukou shift is being managed by the government in the context of the disparities described above, labour market policy, and the shortage of skilled workers in the production sector.

8.2.2.7 Spatial Disparities

In the 1980s and 1990s, the increase in the labour force was the most significant factor in GDP growth. Since then, the importance of this factor has been declining and urbanization, i.e. the migration of labor from the countryside to the cities, is becoming increasingly important. The Chinese government's urbanization plan assumes an urbanization rate of 0.9 percent (Botham 2016). The driving force behind this influx is the large income disparities between urban and rural areas.

Despite countermeasures by the Chinese government, significant income disparities have emerged over time between the booming regions in the east and the more rural regions in the interior and west of the country. Urban-rural incomes have remained persistently stagnant at a ratio of 3.21 to 1 (Stepan 2018). In 2018, urban income was RMB 39,251 and rural income was RMB 14,617, representing a real wage increase of 6.5 percent in urban areas compared to 5.6 percent in rural areas. However, the opening of the income gap slowed somewhat as per capita disposable income grew faster in rural areas than in urban areas in recent years (Xinhua 2019).

One of the most important causes of the great disparity between rural and urban regions is the hukou system. *Hukou* ("huji") classifies a Chinese person according to his registered main residence. The *hukou* is fixed to a person's whole life. Even if the person lives in the city for a longer period, as a registered resident of a rural region (nongye = citizen of an agricultural nature) he or she is excluded from state subsidies for social security, housing, school attendance, and other socio-economic entitlements. Migrant workers, like migrant workers, are referred to as the "fluid population" in China (Dettmer 2016, p. 62). With an internal visa, "suitable persons" with rural hukou can stay in the city. However, the visa must be constantly renewed. A citizen's social status is determined by their place of residence (hukou locality) and their *hukou* type (social entitlement). The result was the emergence of a spatial two-tier society with considerable disadvantages for the rural population beyond income:

- In the cities, 3.4 times as many children attend high school as in the countryside.
- There are 68.1 times as many students in cities as in rural areas.
- Medical care reaches only about 10 percent of rural areas.
- The current consumption level of the rural population corresponds to that of urban citizens in the nineties of the last century.
- The registered unemployment rate in urban areas is 5 percent, while in rural areas only 50 percent of the 40 million labor force (excluding migrant workers) can find work (Stepan 2018).

An often very flexible interpretation of the *hukou* system leads to the admission of only the elite in the big cities, i.e. rich, highly educated migrants who have family ties to urbanites, the required expertise, or the capital to buy housing (Dettmer 2016, p. 63).

The rural area, until today not a target area for luxury marketing strategists, offers a huge reservoir for increasing luxury consumption in two respects. Since 2006, the government has been trying to develop a "pleasant lifestyle" in rural areas as well and to strengthen purchasing power by promoting the construction of central utilities, training, and the establishment of businesses. At the same time, it is pushing ahead with urbanization. Partly under the impression of rising shortages of skilled workers in manufacturing and services, another 100 million farmers are expected to become urban residents by 2020 (as of 2014), and 70–80 percent of Chinese are expected to live in cities by 2030 (Botham 2016).

8.2.2.8 The Cultural Factor

Traditional values and traditions shape consumer behavior across all strata of the population. Chinese luxury buyers, who until now have been oriented exclusively towards Western brands, are becoming aware of their cultural heritage.

In China, luxury goods enjoy general acceptance and symbolize a person's status not only through their possession but also especially through gifts. The desirability and acceptance of luxury are based on a long feudal and materialistic tradition (Lu 2008). As long as economic growth keeps the hope of social advancement alive, wealth and luxury are unlikely to lose their motivational power.

Chinese culture is rooted in the East and Southeast Asian gift culture, or rather gift economy, as described above all by the French ethnologist Marcel Mauss in his treatise "The Gift. Form and Function of Exchange in Archaic Societies" in a "New Sequence" of the journal "L'Année Sociologique" in 1925 with his gift theory as a "total social phenomenon", i.e. as a forming force of the relations of individuals in a group and groups among themselves. The other root of Chinese behaviour is Confucianism with its hierarchical doctrine of virtue and relationships. The differences in Western buying behaviour are generalized in Fig. 8.3.

Confucian hierarchical thinking and Confucian education aimed at communal harmony are the basis of cultivating relationships through luxury gifts that express and confirm personal loyalties. "Title and seniority have a very great significance. Everyone is subordinate to everyone else in some way" (Schneider and Comberg 2013, p. 14). Relationships with others play an important role in the first place. Chinese consider a network of resilient relationships as "social capital" (Bourdieu). The guanxi system imposes an unspoken obligation that every favor and gift will be repaid equally whenever possible. The guarantor of

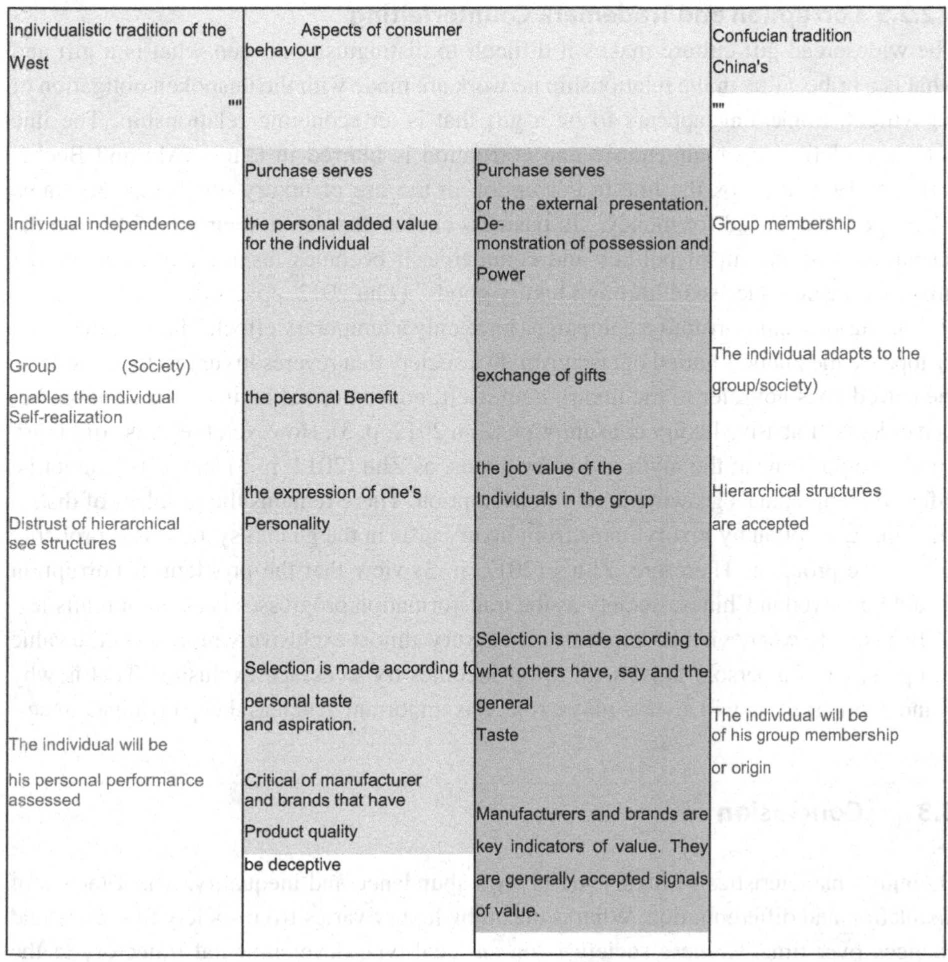

Individualistic tradition of the West ""	Aspects of consumer behaviour		Confucian tradition China's ""
Individual independence	Purchase serves the personal added value for the individual	Purchase serves of the external presentation. De-monstration of possession and Power	Group membership
Group (Society) enables the individual Self-realization	the personal Benefit	exchange of gifts	The individual adapts to the group/society)
Distrust of hierarchical see structures	the expression of one's Personality	the job value of the Individuals in the group	Hierarchical structures are accepted
The individual will be his personal performance assessed	Selection is made according to personal taste and aspiration.	Selection is made according to what others have, say and the general Taste	The individual will be of his group membership or origin
	Critical of manufacturer and brands that have Product quality be deceptive	Manufacturers and brands are key indicators of value. They are generally accepted signals of value.	

Fig. 8.3 Confucian tradition and purchasing behaviour. (Own representation after Lu 2008 and Wang and Ahuvia 1998)

this is the prestige and honor of the family for which the Chinese lives and work. The great Chinese holidays provide ample occasion for the purchase of luxury goods, especially in Southeast Asian foreign countries and Hong Kong.

Even more important is the preservation of face (mianzi), personally, and therefore the family. Social status is the most important thing in life. The consumption of prestige is to demonstrate social status. Luxury goods are an easy way to be respected in society. Their demonstrative use reflects success and affiliation and increases one's family's prestige in public (Lu 2008, p. 6). Brand, price, and their recognition value in public are therefore the most important criteria for a luxury purchase. Quality, service, or other knowledge about the product is less interesting. It is enough if the neighbour has it or recommends it and everyone knows what it is worth.

8.2.2.9 Corruption and Trademark Counterfeiting

The widespread gift culture makes it difficult to distinguish between what is a gift and what is a bribe. Gifts in the relationship network are made with the unspoken obligation of repayment. Something appears to be a gift that is an economic relationship. The line between relationship maintenance and corruption is blurred in China (Ma and Becker 2017, p. 191). Perhaps the line to corruption in the use of luxury goods can be drawn where power is traded for money. "In business circles that derive their windfall from the resources or cronyism of politics and commerce, it becomes fashionable to create the image of the new successful through luxury goods" (Zhu 2012, pp. 2–3).

The various anti-corruption campaigns have only a temporary effect. They are designed to appease the public's hatred of cronyism. In a society that reveres luxury as China's does, the hatred does not refer to the luxury item itself, nor to normal luxury consumption, but to incidents of abusive luxury consumption (Zhu 2012, p. 3). However, the abuse of luxury goods would gnaw at the myth of luxury goods, as Zhu (2012, p. 3) suggests, cannot be inferred from China's growing luxury consumption. There remains the problem of distinguishing corruption by luxury items from luxury gifts in the guanxi system. It is probably just a size problem. Therefore, Zhu's (2012, p. 3) view that the problem of corruption would be solved in Chinese society as the transformation progresses is also not realistic.

In a society where visible demonstrative luxury almost exclusively represents the value and prestige of a person, those with lower incomes try to escape exclusion. That is why brand counterfeiting will always play a role. It is important to at least keep up appearances.

8.3 Conclusion

Defining characteristics of luxury are always abundance and inequality. It is a means of distinction and differentiation. What is meant by luxury varies from society to society and changes over time. Chinese society is on an ideal-type developmental trajectory at the beginning of the adolescence stage. The peculiarity of China's transformation, developing a capitalist market economy under the umbrella of an authoritarian communist system, led to rapid economic growth, but with significant spatial and social disparities.

Among the most lasting consequences for the Chinese luxury market and the dominant Western luxury brands, has been the government's one-child policy since 1980. Based on the Chinese population's great acceptance of luxury, the importance of luxury gifts for relationship group cohesion and as a demonstrative status symbol, luxury consumption fully took off in 1990 and has since grown to 33 percent of global luxury consumption through the second generation of consumers, the luxury-savvy Millennials, who account for two-thirds of luxury consumption. By 2025, it is expected to be nearly half.

As Chinese society has solidified according to income classes and wealth distribution, and the percentage shares of the individual classes in GDP remain stable, the further growth of the luxury market depends above all on maintaining high economic growth. Current trends of renationalization and obstruction of free global trade pose great risks to the internal stability of the country. Nevertheless, policy drives luxury consumption. Prosperity and luxury, if possible for all, being educated and rich is the motto. To this end, it has recently even lowered the import tax on some luxury goods such as cosmetics. The question remains whether this can still be called luxury according to the above definition.

The "little emperors" born in 1980, nurtured as only children and pampered as their parents focused most of their resources on them, many of them the first generation of wealth inheritors, are responsible for the continued luxury consumption. Although only about 12 percent of the Chinese population, they account for 60–70 percent of China's luxury consumption.

The Chinese luxury consumers born after 1980, mostly educated, affluent, urban, spoiled, with an affinity for luxury and the Internet, spend their money more easily than their parents, who were brought up to be thrifty. The new "middle class" "are digitally savvy and very knowledgeable about luxury. They like fashion a casual, favor designer brands and start buying luxury goods at a young age with relatively high frequency" (Bain and Company 2017, p. 2). Although their decisions are still influenced by authority, criteria such as quality or personal benefit are playing an increasingly important role. They are increasingly informing themselves about the hottest luxury offers on internet portals. Purchases are still made in the ambiance of luxury shops, but increasingly also via online portals and payment systems. As by far the largest target group in the Western luxury industry, they are developing their tastes and are increasingly changing the Chinese luxury market and industry. "For a long time, the logic was that the industry would lead the way and buyers would follow … Meanwhile, formerly silent segments of the population have gained a voice and capital flows have changed. As a result, luxury customers now increasingly come from Asia, with cultural roots that are often alien to traditional European luxury manufacturers" (Albers 2014). The trends in the Chinese luxury market are: less ostentation/more value and sinification.

"In the big cities, a Western designer bag dangles from almost every hand, even workers save their salaries for foreign luxury products. Due to inflationary distribution, well-known designer names are no longer a unique selling point" (Seitz 2016). Luxury as a status symbol of the nouveau riche has now reached the middle income groups. Households with average annual incomes regularly buy luxury products.

The younger, saturated generation has other desires. They are redefining luxury. Increasingly, it matters what luxury brings them personally. Health and sophistication are gaining in importance for them, leisure activities, regional organic products, and poison-free fruit and vegetables, for example, are the new luxuries. Luxury is becoming more intangible and authentic with the younger generation. As they are more price-conscious and choosy in their purchases, overpriced products from abroad are having an increasingly difficult time.

This is one of the reasons why a new generation of luxury buyers is reflecting on its roots and craftsmanship tradition. Self-confidently, Chinese designers are developing their products and conquering the markets with them. In reference to the historical process of the spread of Han culture among non-Chinese peoples, this conquest of the market is called sinification. The Western luxury industry is reacting and adapting to the needs of the Chinese (Seitz 2016). "Strong Chinese players have emerged in the international luxury market, Chinese taste and style are becoming an export success" (Ma and Becker 2017, p. 198). With the increased sense of self and tradition, Chinese people are also interested in the history and stories behind a brand or luxury item to a much greater extent than in other countries. Consideration of local identity and storytelling are becoming an increasingly important part of luxury marketing in China.

China's increasing importance to the luxury industry is also reflected in the stakes held by Chinese investors in European luxury labels. For example, the Chinese consortium Fonsun has bought into the oldest French luxury couturier Lanvin, and Gangsu Gangtai Holding holds an 85 percent stake in the Italian luxury jeweler Buccellati (Seitz 2016).

Whether the future of the luxury market will become Asian, as some predict, will not be decided by taste and design, but by the growth of the Chinese economy. In this respect, the production-economy renationalization tendencies emanating from the US, with attempts to curb Chinese exports, do not bode well for Western luxury brands. For the fourth quarter of 2018, the Chinese central bank already corrected the economic growth to below 6.5 percent (Manager Magazin of 03.01. 2019).

The annual iPhone show of the "American flagship company" Apple went completely nowhere in China at the end of last year. The young Chinese target group, which accounted for about one-fifth of Apple's global sales, all but ignored the ritual. Revenue and profit warnings from the company and the loss of about 9–10 percent of its stock market value (over $50 billion) in a matter of hours were the consequences. With Apple CEO Cook's "China warning", the share prices of global luxury brands also came under pressure. The share values of Luis Vuitton, Burberry, or Prada fell by up to 6 percent (Spiegel Online from 04.01. 2019).

The reasons for the impending decline in luxury consumption are sought in the foreseeable slowdown in Chinese economic growth as a result of American tariff and foreclosure policies in the goods and services sector. However, Apple also sees the Chinese luxury consumer as a weighty reason. Although Apple was not affected by counter-sanctions of the government, according to the findings of Apple CEO Cook, the Chinese would have "consciously" (!) decided against the purchase of Apple products (Manager Magazin from 03.01. 2019). The process shows the importance of "soft factors" on economic processes in the age of boundless digital networking in the upper class phenomenon of luxury.

The luxury consumer, largely price insensitive and inert to negative economic developments, is nevertheless flexible, as the value of a luxury good depends on his projections of the luxury object. The Apple example is also an example of the limitations of superficial advertising strategies. The younger generation of Chinese luxury consumers, increasingly self-confident, informed, and unreservedly using digital media, discovers value, worth, and self-actualization not individualistically, but in shared cultural identity and tradition, a phenomenon that American marketing research describes with the terms "local identity" and "local mindset." Social data and Confucian stereotypes only make a limited contribution to understanding this. Even the most elaborate influencing techniques used by marketing departments have only a limited effect on such a deeply anchored mental disposition. China is not only a political-economic hybrid at the state level but also a political-social hybrid. Below the official state and economic organization with its institutions, Chinese society has preserved culture and informal form of organization rooted in the pre-state Southeast Asian cultural sphere, which will increasingly shape the tastes and consumption of the Chinese luxury buyer as political-economic strength and self-confidence increase.

References

Albers, Markus. 2014. China als Trendsetter. *Die Zukunft des Luxusmarktes ist asiatisch.* https://www.wiwo.de/erfolg/trends/china-als-trendsetter-die-zukunft-des-luxusmarktes-ist-asiatisch/10289024.html. Accessed 04 Sep 2019.

Alpermann, Björn. 2013. *Soziale Schichtung und Klassenbewusstsein in Chinas autoritärer Modernisierung.* https://zeithistorische-forschungen.de/2-2013/id%3D4388. Accessed 04 Sep 2019.

Bain and Company. 2017. *China luxury market study.* http://www.bain.com.cn/pdfs/201801180441238002.pdf. Accessed 04 Sep 2019.

———. 2018. *Luxury goods worldwide market.* https://www.bain.com/contentassets/8df501b9f8d6442eba00040246c6b4f9/bain_digest__luxury_goods_worldwide_market_study_fall_winter_2018.pdf. Accessed 12 June 2019.

Berger, Roland. 2013. *Luxus in China: Ein praktischer Ratgeber für die High-End-Industrie.* https://www.meisterkreis-deutschland.com/sites/default/files/meisterkreis-china-kompendium-de-2013.pdf. Accessed 04 Sep 2019.

Bilanz Lifestyle. 2018. *In China kehrt die Lust am Luxus zurück.* https://www.bilanz.ch/lifestyle/china-kehrt-die-lust-am-luxus-zuruck. Accessed 02 June 2019.

Botham, Craig. 2016. *Chinas Trendwachstum: Demografie.* https://www.schroders.com/de/at/privatanleger/insights/maerkte/chinas-trendwachstum-demografie/. Accessed 04 Sep 2019.

Bourdieu, Pierre. 2015. *Die verborgenen Mechanismen der Macht.* Hamburg: VSA.

China SME Center Germany. 2018. *Chinas neue Mittelschicht wächst – Trends und Interessen im Fokus – Studie des Wirtschaftsexperten Wu Xiaobo zeigt Entwicklung auf.* https://zhongdemetal.de/blog/blogdetailallgemein/detail/News/

chinas-neue-mittelschicht-waechst-trends-und-interessen-im-fokus.html. Accessed 03 June 2019.

Degen, Ronald Jean. 2009. *The success of luxury brands in Japan and their uncertain future.* International School of Management Paris, Glob Advantage. https://core.ac.uk/download/pdf/9306177.pdf. Accessed 04 Sep 2019.

Dettmer, Isabel. 2016. *HRM, Rekrutierung und Qualifizierung in China. Das Mismatch-Problem dargestellt am Beispiel der Hotellerie.* Würzburg: Würzburger University Press. https://opus.bibliothek.uni-wuerzburg.de/opus4-wuerzburg/frontdoor/deliver/index/docId/13895/file/978-3-95826-047-4_Dettmer_Isabel_OPUS_13895.pdf. Accessed 04 Sep 2019.

GTAI. 2018. *Kaufkraft und Konsum – China.* https://www.gtai.de/GTAI/Navigation/DE/Trade/Maerkte/Geschaeftspraxis/kaufkraft-und-konsumverhalten,t=kaufkraft-und-konsum%2D%2Dchina,did=2166244.html. Accessed 22 May 2019.

Heine, Klaus, and Vera Waldschmidt. 2016. *Luxusmarketing China.* https://upmarkit.com/de/neuigkeiten/luxusmarketing-china. Accessed 04 Sep 2019.

Hua, Sha. 2018. *Von wegen billig – die Löhne in China steigen rasant.* https://www.handelsblatt.com/unternehmen/mittelstand/familienunternehmer/produktionskosten-von-wegen-billig-die-loehne-in-china-steigen-rasant/22905244.html?ticket=ST-676921-tPwZ6g2QHnWYLSb1PeSG-ap4. Accessed 04 Sep 2019.

Kambli, Conrad Wilhelm. 1892. Der Luxus nach seiner sittlichen und socialen Bedeutung. *International Journal of Ethics* 2 (3): 398–399.

Kapferer, Jean Noel. 2015. *Kapferer on luxury. How luxury brands can grow yet remain rare.* London: Kogan Page.

Kühne, Martin, and David Boschart. 2014. *Der nächste Luxus. Was uns in Zukunft lieb und teuer wird.* Rüschlikon: GDI-Institut.

Lu, Pierre Xiaolu. 2008. *Elite China. Luxury consumer behavior in China.* Singapore, Wiley.

Ma, Xiaojuan, and Florian Becker. 2017. In *China: Kulturelle Werte und Konsumentengewohnheiten im größten Luxusmarkt der Welt*, ed. W.M. Thieme . Wiesbaden, Springer.und Luxusmanagement

Manager Magazin. 2019. *Apple-Chef Cook schockt die Märkte.* https://www.manager-magazin.de/digitales/it/apple-tim-cook-kassiert-umsatzprognose-a-1246204.html. Accessed 04 Sep 2019.

Mühlmann, Horst. 1975. *Luxus und Komfort – Wortgeschichte und Wortvergleich.* Bonn: Rheinische Friedrich-Wilhelms-Universität.

Müller-Stewens, Günter. 2013. *Das Geschäft mit Luxusgütern. Geschichte, Märkte, Management.* https://www.alexandria.unisg.ch/226030/1/Brosch%C3%BCre_Das%20Gesch%C3%A4ft%20mit%20Luxusg%C3%BCtern_final_1.2.pdf. Accessed 04 Sep 2019.

National Bureau of Statistics of China. 2018. *China statistical yearbook.* http://www.stats.gov.cn/tjsj/ndsj/2018/indexeh.htm. Accessed 22 May 2019.

Neinhaus, Andreas. 2015. *Ungleichheit in China nimmt rapide zu.* https://blog.tagesanzeiger.ch/nevermindthemarkets/index.php/41804/ungleichheit-in-china-nimmt-rapide-zu/. Accessed 04 Sep 2019.

Ngai, Joann, and Erin Cho. 2012. The young luxury consumer in China. *Young Consumers* 13 (3): 255–266.

Pöll, Günther. 1980. *Luxus: Eine wirtschaftstheoretische Analyse.* Berlin: Duncker & Humbolt.

Reitz, Michael. 2017. *Noch feinere Unterschiede?* https://www.deutschlandfunk.de/das-denken-pierre-bourdieus-im-21-jahrhundert-noch-feinere.1184.de.html?dram:article_id=398990. Accessed 04 Sep 2019.

Ritter, Johannes, and Hendrik Ankenbrand. 2018. *Nahezu jeder fünfte Milliardär ist ein Chinese.* https://www.faz.net/aktuell/finanzen/finanzmarkt/chinas-reiche-werden-immer-reicher-woher-kommt-das-ganze-vermoegen-15856795.html. Accessed 04 Sep 2019.

Rudolf, Joachim, and Elizabeth Trester. 2016. *China. Der nächste Horizont. Ein Kompass für Anleger und Unternehmer*. Frankfurt am Main: Faziz-Communication GmbH.

Schipper, Lena. 2014. Thorstein Veblen – Spott auf die feinen Leute. *Frankfurter Allgemeine Zeitung*. https://www.faz.net/aktuell/wirtschaft/wirtschaftswissen/die-weltverbesserer/thorstein-veblen-spott-auf-die-feinen-leute-12978070-p2.html. Accessed 04 Sep 2019.

Schmid, Simon. 2016. *Wie Maslows Pyramide in China getestet wurde*. https://www.handelszeitung.ch/blogs/free-lunch/wie-maslows-pyramide-china-getestet-wurde-1110774. Accessed 04 Sep 2019.

Schneider, Gerd, and Jufan Comberg. 2013. *Geschäftskultur China Kompakt*. 3rd ed. Neuss: Conbook.

Seitz, Janine. 2016. *Neuer Luxus, created in China*. https://www.zukunftsinstitut.de/artikel/neuer-luxus-created-in-china/. Accessed 04 Sep 2019.

Spiegel Online. 2019. *Apples Umsatzwarnung verstört auch Luxus-Investoren*. https://www.spiegel.de/wirtschaft/unternehmen/china-warum-apples-umsatzwarnung-luxusmarken-wie-louis-vuitton-schockt-a-1246426.html. Accessed 04 Sep 2014.

Statista. 2018. *China: Gesamtbevölkerung von 1980 bis 2020 und Prognosen bis 2027*. https://de.statista.com/statistik/daten/studie/19323/umfrage/gesamtbevoelkerung-in-china/. Accessed 01 Aug 2022.

———. 2019. *China: Bruttoinlandsprodukt (BIP) in jeweiligen Preisen von 1980 bis 2021 und Prognosen bis 2027*. https://de.statista.com/statistik/daten/studie/19365/umfrage/bruttoinlandsprodukt-in-china/. Accessed 01 Aug 2022.

Stepan, Matthias. 2018. China altert und erlebt die Grenzen der Planbarkeit. *Neue Zürcher Zeitung*. https://www.nzz.ch/meinung/china-altert-und-erlebt-die-grenzen-der-planbarkeit-ld.1406076. Accessed 04 Sep 2019.

Thieme, Werner M. 2017. *Luxusmanagement*. Wiesbaden: Springer.

Veblen, Thorstein. 2007. *Theorie der Feinen Leute: Eine ökonomische Untersuchung der Institutionen*. 3rd ed. Frankfurt/Main: Fischer.

Wang, N.Y., and A.C. Ahuvia. 1998. Personal taste and family face: Luxury consumption in Confucian and western societies, University of Hawaii. *Psychology and Marketing* 15: 423–441.

Wiedmann, Klaus Peter, and Nadine Hennins. 2013. *Luxury marketing: A challenge for theory and practice*. Wiesbaden: Springer Gabler.

Wirtschaftskammer Österreich. 2019. *Statistik: Länderprofil China*. https://wko.at/statistik/laenderprofile/lp-china.pdf. Accessed 04 Sep 2019.

World Luxury Index. 2013. *China 2013*. http://www.digital-luxury.com/reports/World_Luxury_Index_China_2013_by_Digital_Luxury_Group.pdf. Accessed 04 Sep 2019.

Xinhua. 2019. *Einkommen der chinesischen Bewohner steigt 2018 um 6,5%*. http://german.xinhuanet.com/2019-01/22/c_137765458.htm. Accessed 04 Sep 2019.

Zhu, Xuedong. 2012. *Die geheime Logik des Luxus in China*. https://www.goethe.de/ins/cn/de/kul/mag/20693705.html. Accessed 04 Sep 2019.

Labour Market

9

Alexandra Ens, Louis Margraf, and Paul Gebel

For a long time, China's most important resource for the economic boom of the past decades was cheap labour. Low labour costs attracted foreign investment and helped China to become the world's second-largest economy (Grzanna 2010). The Chinese labour market has only developed in recent years, as under the planned economy there was no such concept, but jobs were allocated.

In the following, the developments of the labour market and the current challenges in China are presented. Then the so-called migrant workers will be discussed, followed by the discussion about the achievement of the Lewis Turning Point in the PRC, which describes the end of the surplus of unskilled personnel: A shortage of personnel then results in salaries rising and working conditions improving. The problem is that wages are rising faster than the country can transform economically. China could remain stuck in the so-called "middle-income trap" and not continue to rise to become an advanced industrial nation.

9.1 Development of the Labour Market

The statistical determination of the situation of the labour market is difficult due to the unreliability of data in China. According to Taubmann (2004, p. 21), the official data on unemployment differ greatly from the estimated real unemployment of various politicians

A. Ens
Friedrich-Alexander-University of Erlangen-Nürnberg, Nürnberg, Germany

L. Margraf (✉)
Renmin University, Beijing, PR China

P. Gebel
SAP, Munich, Germany

B. Darimont (ed.), *Economic Policy of the People's Republic of China*,
https://doi.org/10.1007/978-3-658-38467-8_9

and scholars. Several indicators clearly show that the official figures cannot be correct. For example, between 2002 and 2017, the official unemployment rate always hovered near the 4 percent mark, reaching a maximum of 4.3 percent during the 2009 financial crisis (Trading Economics 2019). As already described in Chap. 2, all official data from the Chinese government should be viewed critically.

9.1.1 Development from 1949 to 1976

Jutta Hebel and Günter Schucher (1992, pp. 7–9, 11) described that with the introduction of a centrally planned economy, the state took over tasks such as identifying, balancing, and planning all resources in the country. This included manpower which was allocated to different regions, industries, and enterprises. This was legitimized by raising the standard of living in material and cultural terms. The goal was full employment for all people of working age. Employment policies of this period were particularly characterized by full-time employment, lifetime provision by the enterprise, low mobility, and centralized assignment to so-called work units (dānwèi). A symbol of this policy became the "iron rice bowl" (tiě fànwǎn), which expresses lifetime tenure (Zhu 2005, p. 17). It means a secure job, income, and lifelong social security for the employees, as well as accommodation in the company housing (Lüthje 2010, p. 474).

All males aged 16–59 and females aged 16–54 were considered to be able to work. In addition, all those who were not in the age group but were working were included (Hebel and Schucher 1992, pp. 29–30). The employees were divided into cadres and workers, with cadres assuming leading roles in the enterprises and workers performing manual work. Performance evaluation of cadres included annual evaluations of their performance, the opinion of peer cadres, and the opinion of superiors. Workers' performance was evaluated by superiors only according to criteria such as political and ideological thinking (Zhu 2005, pp. 20–21).

The nationalization of enterprises and the centralized distribution of labour during this period led to inefficient operation of workers and cadres, as there were no incentives to work economically (Wu 2003b, p. 41). Thus, on the one hand, there was an excessive demand for labour from enterprises to meet the given numbers. On the other hand, structural unemployment existed due to the misallocation of labour, as there was not enough work for workers. There was no incentive to work efficiently because qualifications and performance were not taken into account in wages (Hebel and Schucher 1992, pp. 9–10).

In 1958, the so-called *hukou* system was introduced to control the population. It allowed control over the residence of individual households and until today over migration within China. Changing residence is only possible with government permission, which is sometimes difficult. It also affects job opportunities and access to public schools, which is why owning an urban *hukou* is still considered advantageous today. Holders of a rural *hukou*, for example, cannot work for the government (Bao et al. 2011, p. 564).

However, not all labor policies and reforms of this period were discriminatory against the rural population. For example, one of the government's biggest human resource development programs in the 1950s and 1960s was to reduce the proportion of illiterate people in China by making primary education compulsory. Thus, people in rural areas gained access to education, even if it was only primary education (Yang et al. 2004, p. 301).

Much of the progress made in the early stages of the Communist Party was undone during the Cultural Revolution. Moreover, during the Cultural Revolution, about 15 million pupils and students were sent to the countryside for re-education, where they were supposed to learn from the peasants. For them, entering the labour market after its emergence was particularly difficult, as they had hardly any education to show for it (Mühlhahn 2017, p. 79).

9.1.2 Development from 1977 to 2000

After Mao's death, Deng Xiaoping announced in 1978 at the third plenary session of the 11th Party Congress that China's economy must be reformed and opened up (Wu 2003b, p. 38). Deng Xiaoping's opening-up policy was to reform the country's economy in slow controlled steps. His primary concern was to stimulate the economy. To do this, farmers first had to be motivated to produce more efficiently. In Anhui Province, the household responsibility system was introduced in 1978. This system allowed the free sale of the surplus produced goods on the market and at the same time lowered the quotas to be paid to the state. In 1983, the household responsibility system was finally officially recognized by the CCP and manifested as an important component of the Chinese economy. By December 1983, about 99 percent of all production groups had adopted this system (Thiel 1998, pp. 22–23). Thus, between 1978 and 1983, the yield of the agricultural sector grew by about 4.4 percent compared to 1970–1978 (Zhang 2003, p. 3). To make SOEs more profitable, the contract responsibility system was introduced in 1987, which was adapted by about 80 percent of all large and medium-sized SOEs in the same year. While SOEs were given more autonomy to make decisions, it was also contracted with each individual to determine what the relationship between the enterprise and the state was and how profits were subsequently shared (Chen 1993, p. 14).

Although the opening policy had already brought about many changes, an ineffective distribution of work prevailed until 1989. Shortly before the event of the student protest at Tiananmen Square on 4 June 1989, reports on the existing employment problems were published for the first time. According to the planned economy, these should not have existed at all. In the reports, it was announced, for example, that about 50 million Chinese were unemployed and about 20 percent of the employees in the state enterprises were superfluous. Nevertheless, the state had to pay for wages, pensions, and social benefits, so state enterprises operated ineffectively (Hebel and Schucher 1992, p. 1). Many workers who joined the student protests in 1989 were state employees who had lost their jobs due to the reforms and could barely afford food due to rising prices (Bundeszentrale für politische Bildung 2014). Illegal workers who did not have work permits due to a change of residence without registration in the hukou system joined the protests (Kristof 1989). After these events, reforms stalled. In September 1992, the 14th Party Congress finally announced that it endorsed the socialist market system, laying the groundwork for the privatization of state-owned enterprises. In addition, the same year saw the enactment of the Trade Union

Law, which, with the adoption of the Labour Law in 1994, represented great progress for labour market development. Now, for example, it was allowed to create individual contracts between employees and employers as well as collective agreements. In addition, every company that employed more than 25 employees was required to establish a trade union within the company (Zhu et al. 2011, pp. 129, 133).

Throughout the period from 1978 to 2000, the proportion of workers employed in a state-owned enterprise declined rapidly, leading the Chinese government to establish re-employment centers in 1996 (Lu et al. 2002, pp. 27–28). By the end of 1996, up to 70 percent of all SOEs had been privatized, which meant layoffs for several million SOE workers. In 1997 alone, 11 million people in state-owned enterprises were laid off (Qian 2000, pp. 171–172).

However, the development of the market-oriented economic system enabled the free choice of occupation (Chi et al. 2012, p. 6). Now private enterprises became a real employment engines. Between 1989 and 1998 alone, the number of people employed in private enterprises grew from 1.7 million to about 171 million. Since private enterprises had limited financial resources and technology, they initially preferred labor-intensive production (Lu et al. 2002, pp. 28–30). In general, the education level of the working-age population in China at the time was not high, so human capital was inexpensive. In 1996, 48.3 percent of the working-age population did not have a high school degree higher than primary school. However, as the standard of living increased, the level of education increased noticeably. By 1999, this number had dropped by 4 percent, leaving only 45.3 percent without a high school diploma (Chi et al. 2012, p. 2). As a result, the age of entry into the workforce increased as people invested more time in education (Lu et al. 2002, pp. 28–30).

Other important changes in the labour market during this period concern the distribution of the workforce among the different sectors and the associated job opportunities for the female part of the population. By the end of the 1980s, the secondary sector reached the highest share of employment in the population at almost 16 percent. After 1989, this figure fell to 11 percent by 2000. The share of the primary sector tended to decline. Instead, the proportion of the population working in the tertiary sector grew (Lu et al. 2002, pp. 26–28). As the tertiary sector grew, more work opportunities for women emerged because it did not require heavy physical labor. The percentage of the female population that worked in the service sector increased by about 10 percent from 1982 to 2000 (Ma 2004, p. 43). The percentage of women in the total employed population grew from 32.9 percent in 1978 to 37.9 percent in 1998 (Lu et al. 2002, p. 31).

9.1.3 Development After 2001

By joining the WTO in 2001, China agreed to several commitments, such as adopting international labor standards, respecting human rights, and promoting trade unions (Zhu and Warner 2004, pp. 311–312). Even though China is still criticized today for non-compliance with some WTO guidelines, such as discrimination, lack of transparency, and rule of law, its integration into world trade boosted the Chinese labor market. This is because with the accession to the WTO, more foreign companies invested in China, creating jobs. In doing so, foreign companies increasingly invested in the tertiary sector as government restrictions in this area were lifted (Davies 2013, p. 13).

However, with the increasing number of foreign companies, problems in the Chinese labor market have intensified. One of these problems, which is currently being discussed, is the lack of qualified personnel. With the increasing demand from foreign companies, the situation has intensified to the point that there is now often talk of a "war for talent" (Wenderoth 2018). Chinese enterprises had high losses of qualified personnel at the beginning of the millennium because most local enterprises did not attach much importance to human resource development compared to foreign enterprises. The outdated human resource practices of Chinese companies did not provide enough incentives to retain their employees who wanted to benefit from the better working conditions of foreign companies (Wu 2003a, p. 50).

For the first time, the Chinese government gave human resources development its own chapter in the 12th Five-Year Plan from 2001 to 2005, according to which more emphasis was to be placed on education to counteract the shortage of skilled workers – especially in the high-tech sector and in science (Yang et al. 2004, p. 302).

The number of students at Chinese universities has risen steadily in recent years. In 2000, only 2.2 million students were enrolled in a university. As the population grew wealthier, investment in children's education increased, so that by 2009 there were already about 6.4 million enrolled (Dian 2014, p. 6). In 2018, a record 8.3 million graduated (Leng 2019). As the number of graduates grew, so did the number of unemployed graduates. In China, a degree is now no longer a guarantee of a secure job, so higher education degrees, such as masters and doctorates, are becoming more attractive. With the growing average length of study, the age of the working population continues to rise (Dian 2014, p. 6). The number of students studying abroad has been increasing in recent years. Most of them study in the US and Europe. In total, about 5.2 million Chinese have studied abroad in the first 40 years of opening up, with about 1.45 million Chinese enrolled abroad in 2018 alone (Ministry of Education of the PRC 2018).

According to Mark Robbins (2016, p. 5, p. 12), the Chinese government has been trying to improve the quality of education since 2008 with programs such as the "Thousand Talents Program" by financially supporting research by Chinese and foreign scientists. This is intended to promote and recruit international experts. Funding is provided to doctoral students under the age of 55 who have received part of their education abroad and are working as professors. Initial successes of this program are credited to the increasing number of scientific research results and registered patents.

Chinese companies, such as Huawei, focus on participation in the education of talented Chinese to attract qualified personnel. For this purpose, the company founded Huawei University in 2005, where courses are offered for employees, but also customers. With 1–6 month courses on corporate culture, product portfolio, and sales strategies, future leaders for Huawei are trained (Scullion and Collings 2011, p. 143). However, the consequences of the shortage of qualified personnel have not yet been resolved. The shortage resulted in a high turnover rate of qualified personnel because they can develop their career faster with company changes than waiting for promotion in the same company (Woo and Zhang 2019). Because of the demand for qualified personnel, skilled individuals can enter new employment within a short period without much risk. The motivation for this often stems from monetary reasons as well as insufficient future prospects in the company (Woo and Zhang 2019).

Entry-level workers report having increased their starting salary from about 5000 renminbi within 6 years and five job changes to about 22,000 renminbi through job hopping (Jing and Chen 2019). In a survey conducted by a local newspaper in Hangzhou, one of China's largest high-tech cities, about 76 percent of respondents said they were currently looking for a new job (Woo and Zhang 2019). Due to the current trade conflict between China and the US, the demand for skilled workers is expected to decrease and this will affect the duration of an average employment. With early signs of a weak economy, the desire for social and financial security is increasing. Accordingly, changing jobs is less attractive (Jing and Chen 2019).

This development means a difficult entry into the job market for career starters and the annual 8 million university graduates. Compared to the previous year, job offers for career starters have fallen by around 13 percent overall in 2019, and by as much as 19 percent in state-owned enterprises. These are currently gaining in attractiveness as employers because they promise more social security (Leng 2019). In 2013 alone, 1.2 million Chinese had applied for 19,000 open civil service positions (Dettmer 2016, p. 41). University graduates are increasingly applying to state-owned enterprises. Last year, the application rate increased by about 5 percent from the previous year, even though there were 19 percent fewer positions available. Large companies with more than 10,000 employees have 11 percent more applicants, although they are currently cutting an average of about 37 percent of their positions. In contrast, small private companies have seen a decline in applications, although they offer comparatively more positions for entry-level workers (Leng 2019).

Xi Jinping has set himself the goal of eradicating poverty in China by 2020. According to official data, in 2017 about 43 million Chinese were still living in absolute poverty with an annual income below 2800 renminbi. Therefore, the minimum wage is increased approximately every year, which is determined regionally by the individual provincial governments. Accordingly, in the economically developed regions in the east of the country, the minimum wage is higher than in the west (Cai 2017). In 2019, the average minimum wage in China was 2480 renminbi per month, up about 2.5 percent from the previous year. Compared to the minimum wage of 10 years ago, there is an increase of over 250 percent. In 2009, the average minimum wage was 960 renminbi per month (Trading Economics 2019). Due to rising production costs and imposed tariffs by the U.S., many companies are moving their production bases to other countries, making President Xi's goal difficult to achieve (Hao 2019). Due to rising production costs, many companies in China have had to cut jobs. The government has announced that it will take countermeasures. For example, tax breaks for companies are to be waived for small job cuts (Wang 2019). Thus, according to official figures, the Chinese government created 13 million new jobs in 2018 and even exceeded its own targets by 2 million jobs (Balding 2019). Nevertheless, it seems unrealistic that the unemployment rate fell to 3.78 percent in the first quarter of 2019 in the context of the trade war (Trading Economics 2019).

9.2 Labour Migration

Since the 1980s, the PRC has experienced the largest population migration in world history. This massive migration and the resulting social challenges are consequences of the previous economic reforms. At the beginning of Deng Xiaoping's economic reforms, the number of migrant workers in rural areas was about one to 2 million, which was less than 1 percent of the total rural labor force. By the end of 2018, according to official figures, 244 million people had migrated, accounting for 17.5 percent of the total population. The inflow of migrants has been declining again since 2014, as can be seen from Table 9.1, at which time the number of domestic migrants had peaked at 253 million (National Bureau of Statistics 2019). Although the urbanization rate had surpassed the 50 percent level of other middle-income countries in 2016 at 57.36 percent, it will take another 20–25 years to reach the 80 percent urbanization rate of developed nations (Zheng 2018, pp. 187–188).

This development raises questions, with migrants facing discrimination in different ways. On the one hand, the Chinese economy needs cheap labour; on the other, income inequality continues to grow, which in turn is a reason for growing social unrest involving more and more migrant workers. While there were a total of 49 major protests in 2014, each involving more than 1000 workers, about 95 percent of the total 712 protests in the first half of 2019 involved fewer than 100 workers. This can be attributed to the fact that, on the one hand, more acceptable working conditions exist in large factories and, on the other hand, social harmony is prioritized under Xi Jinping. Furthermore, as a result of changing working conditions and a restructuring of the Chinese economy as a whole,

Table 9.1 Internal migration in the PRC

Year	Migrants (million)	Population (million)	Percentage (%)
2000	121	1267.43	9.55
2005	147	1307.56	11.24
2010	221	1340.91	16.48
2011	230	1347.35	17.07
2012	236	1354.04	17.43
2013	245	1360.72	18.01
2014	253	1367.82	18.5
2015	247	1374.62	17.97
2016	245	1382.71	17.72
2017	244	1390.08	17.55

Source: Own calculation based on China Statistical Yearbook 2018. http://www.stats.gov.cn/tjsj/ndsj/2018/indexeh.htm

factory workers are no longer the group of workers with the greatest propensity to strike; instead, most strikes are observed in the service sector: in addition to taxi drivers, suppliers and other comparatively simply structured service providers, some of the better-paid service providers, such as management consultants or workers in the health sector, are on strike. Although salary increases are the primary trigger for these, strikes are generally over non-payment or delayed payment of wages; for example, a full 80 percent of worker strikes in the first half of 2019 were over delayed payment of wages (China Labour Bulletin 2019). However, most migrant workers do not have a high school diploma and are employed in the low-wage sector. Cai et al. (2005) studied the occupational distribution and education levels of migrant and local urban workers, finding that 92 percent of migrant workers were unskilled and 95 percent were employed in the low-wage sector.

9.2.1 Internal Migration Policy

According to Yang (2018, pp. 53–55), a professor of society and demography at the Chinese People's University, Chinese statistics distinguish different groups. The most common terms for migrant workers include "floating residents", "agricultural workers", "temporary resident population", etc. As the various terms are used inconsistently in different statistics, it is difficult to classify them precisely. For example, statistics from the State Health and Family Planning Commission include interurban migrants, while statistics from the State Council only include rural residents migrating to cities; statistics from the Bureau of Statistics include migrants of all ages, whereas statistics from the State Health and Family Planning Commission only count working-age migrants. In addition, China is currently undergoing a huge urbanization process, and many migrants are not correctly registered in terms of residence, which can lead to incorrect statistical figures. Moreover, people who have not worked in their home province for more than 1 month are classified as "migrant workers", while other statistics only show such workers as "migrant workers" after 3 months (Cai 2006, p. 299).

The term migrant in the context relevant here means those people who move from rural areas to cities to work there for a certain period. Occasionally, migrants also move from one city to another. Individuals belonging to this group usually work in urban areas but have their registered residence in rural areas. Academics consider the terms discriminatory as it titles migrant workers as different from ordinary workers (Cai 2006, p. 299).

Domestic migration played a significant role in Deng Xiaoping's economic reform policies, as it was the rural labor force that made the economic boom of eastern China possible. It was the rural workers who, on the one hand, built the necessary infrastructure on the east coast of China and, on the other hand, formed the reservoir of cheap labor for the industrial sector that made Chinese enterprises internationally competitive. The east coast developed into a prosperous region. The western regions remain relatively poor and underdeveloped, especially compared to the developed east coast. However, migrants contribute to the transfer of money, education, and information to western China. This contribution manifests itself, among other things, by sending money to their families or bringing education and information home. However, there is a danger that migrants will forever remain in a kind of underclass, as they are unlikely to free themselves from their status as agricultural labourers. So far, China's migration policy has not been a tool to improve rural development, reduce poverty or promote labour migration. The precarious situation of migrants caused a stir, especially from 2005 onwards, when an investigation by the State Council (2006) revealed shortcomings: wage payments below the minimum wage, late or unpaid wages, exceeding legal working hours, lack of safety at work, lack of access to social security schemes, frequent occupational diseases and accidents at work, and a lack of access to education for migrants' children. Subsequently, a statement on "Solving the Problems of Migrant Workers" (State Council of the PRC 2006) was published. Local governments and relevant authorities were urged to improve occupational safety, strictly enforce the prohibition of child labour, implement the legal requirements for accident insurance, and initiate projects regarding the establishment of health insurance schemes for migrant workers that cover the risks of serious diseases. Migrants should be supported by workers' representatives in enterprises and trade unions. In addition, legal assistance for migrant workers should be improved, especially with regard to accident compensation. Many of these measures have been implemented, at least on paper. However, the most important equalization would be achieved by reforming the *hukou* system. Further improvement should be ensured by unifying the labour market (Wong et al. 2007, p. 37).

9.2.2 Reform of the Hukou System

Resource allocation was necessary for the implementation of the planned economy. For this purpose, the so-called *hukou* system was introduced (Mühlhahn 2017, p. 19). This is the state population registration system in China, with the help of which food vouchers and benefits for social security were distributed and in some cases still are. Changing from a rural *hukou* to an urban *hukou* was only possible with difficulty. The *hukou* was usually transferred by the parents at birth, so if someone was born in the countryside, he or she had little opportunity to move to the city. This system persists to this day and is criticized for reinforcing the rural-urban divide (Naughton 2018, pp. 122–123).

The *hukou* system was introduced in 1958. Although the restrictions on migration were not directly anchored in the *hukou* system, other regulations inhibited migration to the cities. It was not until 1984–1988 that the *hukou* system was deregulated and rural residents were allowed to migrate to urban areas. This was related to the fact that they were surplus labor in the countryside and were needed as workers in the developing coastal regions. The influx of labor increased in the years from 1989 to 1991, so the *hukou* system was further deregulated from 1992 to 1994. Due to the reforms of state-owned enterprises in the 1990s, the unemployment rate of urban residents increased significantly, making migrant workers competitors of urban residents in the labor market. Therefore, from 1995 to 2000, immigration to urban regions was strictly regulated by the state, especially migrants who could only be employed in certain occupations specified by local governments. After WTO accession in 2001, the demand for labour increased again, so reforms of the *hukou* system were promoted. However, to this day, there is a difference in housing allocation, education system, and access to social security (Ma 2018, pp. 72–74).

Since 2001, the complete abolition of the *hukou* system has been discussed and experimented with in various regions. Since then, any person who works legally in small towns or cities and has accommodation there can officially register as a resident in these places. Some medium-sized cities and provincial capitals have also lifted their restrictions in this regard. However, there are regional differences. In megacities such as Beijing or Shanghai, there is an "open door policy", although the number of migrant workers allowed to register is still limited (Huang and Pieke 2003, p. 21).

The "National Plan for New Urbanization (2014–2020)" aims to abolish the *hukou* system, although individual local governments can still determine the criteria for the influx. The cities of Beijing and Shanghai in particular need restrictions to ensure the supply and safety of the population (Chen and Fan 2016, pp. 13–14). However, after this plan was published, it turns out that many people, with rural hukou, did not want to exchange it because they would have had to give up their leased land (Chen and Fan 2016, p. 17).

9.2.3 Standardisation of Labour Market Regulations

The dualism between urban and rural areas results in a divided labor market, which is reinforced by the *hukou* system. The aim is to unify labour market regulations while using redundant labour from rural areas as cheap labour for urban development and urban industry. Under the 1994 Labor Law and June 29, 2007, Labor Contract Law, companies in urban areas are required to enter into labor contracts with each worker, including their migrant workers. Through these contracts, companies are required to pay their share of statutory social security contributions for their workers. However, these provisions can be easily circumvented. According to official data, in 2009 only 42.8 percent of migrant workers had employment contracts with their respective employers. According to the most recent data, in 2016 the proportion of migrant workers who had employment contracts was 35.1 percent, down 1.1 percentage points from the previous year (National Bureau of Statistics 2017). Accordingly, a decrease in the number of signed employment contracts can be observed.

Although informal work has existed in the People's Republic since 1949, the concept per se was initially not officially applicable due to the strict ideology of the Chinese state. It was only as a result of the reform and opening-up policy from 1978 onwards, with the increase in private ownership, that informal employment was able to consolidate itself as a concept. From the mid-1990s, the reform of state-owned enterprises led to widespread layoffs, which is why the communist government became interested in alternative enterprise models. In fact, informal work is applied by enterprises to avoid costs, such as tax or insurance payments. In 1990, there were an estimated 23 million workers; this number increased rapidly in subsequent years to about 175 million in 2005 (Chen and Hamori 2014, pp. 80–81). Informal work is particularly attractive to migrant workers, as it can generate far higher wage payments than rural work, due to the elimination of taxes and social security contributions (Zhang 2016, p. 22).

Overall, the situation of migrant workers has improved. Inequalities in social benefits still exist (Chan and O'Brien 2019, pp. 103–122), but the Labor Contract Law provides for equal treatment. At least, in theory, migrant workers and urban workers are equal in terms of their social and political rights according to the "Provisional Regulations of Residence Permit" enacted by the State Council on 1 January 2016 (Yao 2018, p. 19). However, the reality is quite different, as many departments at the provincial level or below still use bureaucratic procedures to obtain additional fees (Nielsen and Smyth 2008, p. 3).

Whether cheap labor will remain relevant for the Chinese labor market in the future is analyzed in the following section using the Lewis Turning Point.

9.3 China at Lewis Turning Point

The decades-long rapid growth of the Chinese economy could only be achieved through a surplus of cheap labour from the rural areas, which supplied the industrial sector in the economically flourishing cities with their labour. This labor force, which consisted mostly of migrant workers, fostered the growth of export-oriented China through cheap wages (Cai and Wang 2005a, pp. 34–36). In 2004, however, the first reports of a shortage of migrant workers in China emerged. The Pearl River Delta was initially affected, from which the shortage spread to other regions of the southeast coast shortly thereafter. For a time, employers found it difficult to recruit new workers. Thus, the first debates began among scholars as to whether the surplus of labor from the rural regions had been used up and China had thus reached the so-called Lewis Turning Point (Cai and Wang 2005b, pp. 5–6). This marks the point in a country's development where labour from the traditional agricultural sector can no longer supply the modern sector with cheap labour (Lewis 1972, pp. 75–78). The consequences include a shortage of labour and rising wages.

Some scholars claimed that the shortage of migrant workers in the PRC in 2004 was only a temporary phenomenon that would occur occasionally. The reason, they argued, was labour market segmentation or the overwhelming disadvantages of the hukou system, as the shortage only occurred in coastal regions and there was still a surplus further inland (Knight et al. 2011, pp. 25–26). Others suggested that this was the result of demographic change and a logical consequence of recent decades due to the rapid growth of the Chinese economy (Cai 2010a, p. 107). Labour market developments at the beginning of the current decade were reminiscent of this situation. After a brief period of dormancy between 2005 and 2009, wages started to rise more strongly again, including in some of the domestic provinces (Mitali and N'Diaye 2013, p. 5). Researchers disagree on the current situation and try to explain it in different ways to find out whether China has reached the Lewis

Turning Point or not. Reaching the Lewis Turning Point will have an impact on the Chinese economy and force the government to deal with the possible consequences.

First, a look at the empirical framework is presented, after which the debate on whether China has reached the Lewis Turning Point is outlined. Finally, the implications of reaching the Lewis Turning Point for the PRC and the government's actions are analyzed.

9.3.1 Lewis Model

Sir William Arthur Lewis was a British economist who was primarily concerned with developing countries. He was awarded the Nobel Prize in 1979 for his model of the dual economy, which he first described in 1954 (Lewis 1954, pp. 139–141).

According to this model, also called the Lewis model, an economy is dualistic at the beginning of its development (Lewis 1954, pp. 139–140). This means that it is divided into two sectors. Once into the traditional agricultural sector and once into a modern industrial sector (Fields 2004, p. 727). According to Lewis, the traditional sector has a surplus of labor and the marginal productivity of labor is much lower than that of the modern sector. Marginal productivity describes the change in the output of a firm, given a change in the input of a factor of production by one unit, in this case, labour (Mankiw 2001, p. 419). Surplus labour, for Lewis, is labour with very low or almost no marginal productivity. According to Lewis, there is a one-way transfer of labour between the two sectors. The transfer of this surplus labour from the agricultural to the industrial sector does not affect the output of the traditional sector while it allows the growth of the modern sector without affecting wages (Gosh 1985, p. 95). Lewis (1954, pp. 139–140) describes that wages are generally higher in the modern sector than in the traditional. This encourages surplus labour to move from the traditional to the modern sector.

The modern sector grows with the help of the surplus labour from the traditional sector. This process continues until the point is reached where the surplus is used up. This point is called the Lewis Turning Point, which arose from a development and refinement of the original Lewis model by John C. H. Fei and Gustav Ranis (1961, pp. 533–535) and the original author Sir William Arthur Lewis (1972, pp. 75–77). Although the surplus of labor has been used up, labor continues to flow into the modern sector because of the higher marginal productivity of labor in it and the still higher wages compared to the traditional sector. This increases the previously constant wages in the modern sector. According to the law of diminishing marginal returns (Mankiw 2003, p. 61), this onward flow of labor will cause the marginal productivity of labor in the traditional sector to increase to the point where it is equal in both sectors. Thus, after the Lewis Turning Point is reached, the consumption of the surplus labor in the agricultural sector should be followed by an equalization of the marginal productivity of both sectors. In the long run, this should lead to the disappearance of inequalities in both sectors, and with them the dual economic structure (Zhu and Cai 2012, p. 4).

9.3.2 Applicability to the People's Republic of China

In what framework are the Lewis model and the concept of the Lewis Turning Point applicable to the PRC? The Chinese economy has exhibited a dualistic characteristic since the time of Mao Zedong. This dual system distinguishes between rural and urban households (Naughton 2007, p. 114). Rural households were granted the right to a piece of land and urban households were given the right to a job in the city (Gransow 2014). For the most part, only urban dwellers benefited, as agricultural cultivation was not very profitable due to the government-set price of grain, which was very low (Naughton 2007, p. 115). This led to the first wave of rural exodus between 1949 and 1957, in which about 26.3 million people migrated from the countryside to the cities (Windrow and Guha 2005, p. 2). The first parallels can be drawn here with the sectors described by Lewis in his model. Urban households represent the modern industrial sector and rural households the traditional agricultural sector.

This rural exodus was not initially intended, as it was feared that the social systems of the cities would be overburdened and slums could form. In 1958, therefore, the *hukou* system was introduced to stop the rampant rural exodus. With the economic reform under Deng Xiaoping in 1978, the household registration system of the PRC was relaxed again. The transformation from a planned to a market economy led to a surplus of 150 million people in the countryside. The relaxation of the *hukou* system was intended to enable these people to migrate to the cities to meet the demand for labour there and to promote the growth of the coastal provinces. Again, parallels can be drawn with Lewis's model of the dualistic economic system. The surplus labour from the countryside (traditional sector) migrates to the cities and coastal provinces (modern sector) to promote their growth. However, according to one study, the migration of these migrant workers had not affected agricultural production (Richardson 2008, p. 9). This finding is also consistent with the description in Lewis' model, as the transfer of surplus labour does not affect the output of the traditional sector and does not increase wages in the modern sector. Similarly, non-rising wages have been the case for years in the modern sector among migrant workers (West and Zhao 2000, p. 137).

It can be noted that to look at the economic development of the last 30 years in China, the Lewis model and the concept of Lewis Turning Point can be used. Much of Ranis', Fei's, and Lewis' thinking can be applied to it. Building on this, in recent years many scholars have tried to clarify whether China has reached the Lewis Turning Point or not (Cai and Wang 2005b; Cai 2007; Garnaut 2010; Knight et al. 2011; Mitali and N'Diaye 2013; Zhang et al. 2011).

However, Arthur Lewis's model and the Lewis Turning Point derived from it define factors such as excess labor only vaguely and not in enough detail to make a calculation. Similarly, marginal productivity of labor is difficult to calculate. It is especially difficult to apply the concept of Lewis Turning Point to a country like China, which has undergone unprecedented development. The size of the country and the fact that its individual provinces could be states in their own right, which are also very different from each other, make it difficult to analyze the relevant factors. That is why various opinions regarding this issue circulate among scholars. The opinions can be broadly divided into two directions. One side assumes that China has already reached or passed the Lewis Turning Point (Cai 2010c, p. 136), while the other claims that it is still too early to speak of the concept of the Lewis Turning Point in China's economic development (Minami and Ma 2010, pp. 163–165).

In the following, the surplus of labor in the countryside, the development of wages, supply and demand in the labor market, and the inequality of the income gap between urban and rural areas are considered the most important indicators for assessing the situation in China.

9.3.2.1 Surplus Labour and the "China Paradox"

Kam Wing Chan (2010, pp. 513–514) described the labour situation in China's rural regions as the "China Paradox". Here, it was described that, on the one hand, reports of a shortage of migrant workers were increasing, but at the same time a significant proportion of workers in the primary agricultural sector were surplus.

Since surplus labor is used up when the Lewis Turning Point is reached, looking at the volume of surplus labor is an important factor in determining the timing of the Lewis Turning Point (Zhu and Cai 2012, p. 4). Looking at the average wages of migrant workers, it was found that the surplus labor force did not decrease in the last years of the 2000s decade (Zhao 2010, p. 79).

Xianzhou Zhao (2010, pp. 78–79), in a 2010 Economist article, explained the "China Paradox" in terms of rising mobility costs for migrant workers. He argued that the high cost of traveling from rural provinces to modern coastal provinces combined with the high turnover in the labour market there would deter many potential migrant workers. He additionally calculated that there was a surplus of 190 million rural workers from 2003 to 2005.

Similar findings could be drawn from looking at the Chinese Household Income Project Survey in 2007, conducted by the China Institute of Income Distribution. Many potential migrant workers chose to stay in the countryside because they were too old, believed they could not find work in the city or had to take care of the elderly and children in their families. However, these potential migrant workers would be willing to work in the cities if joining their families and permanent migration to the city was possible through the *hukou* system (Knight et al. 2011, pp. 14–17, 26).

Scholars Ryoshin Minami and Xinxin Ma (2010, pp. 163–168) calculated for the years 2001–2005, a rural labor surplus between 159 million and 297 million people. They compared the marginal productivity of labor with the per capita net income of rural households and their per capita consumption expenditure. However, as mentioned earlier, marginal productivity of labor is difficult to calculate and Lewis's definition of surplus labor is vague. Thus, other scholars conclude that the rural labor surplus is much smaller (Cai and Wang 2005b; Cai 2007; Ma and Ma 2007).

The best known of the scholars mentioned above is Fang Cai. He argues in his consideration of the labour surplus with the demographic change in China and claims that the existing figures regarding the agricultural labour force do not reflect the changed situation. In his view, the other researchers misjudge these figures (Cai 2010a, pp. 109–110). Cai (2007, pp. 2–10) further explains that the number of people of working age in rural areas is 485 million and that by 2005 200 million of them had already migrated to the cities. Of the 285 million left behind, 170 million would have been needed in the agricultural sector, while 50 percent of the 115 million remaining workers would have been over 40 years old.

However, the labour force in the industry mainly concerns the age groups from 20 to 39 (Mitali and N'Diaye 2013, p. 7). Consequently, according to Cai, in 2007 the rural labour surplus was at most 58 million and would be negligible compared to the rapid development of China's economy (Cai 2007, pp. 2–10). In a working paper of the International Monetary Fund, using the so-called "disequilibrium approach" (Quandt and Rosen 1986, pp. 235–238), a surplus of 151 million workers in the country was calculated for the year 2010, which should fall to 57 million by the year 2015 and only amount to 33 million in 2020. According to this study, the Lewis Turning Point should be reached between the years 2020 and 2025. In addition, further progressions in the development of the labour surplus in China were calculated for different scenarios (Mitali and N'Diaye 2013, p. 14). In these calculations, it is striking that only an increase in labor force participation has a significant impact on delaying the Lewis Turning Point. This could be achieved, for example, by increasing the retirement age or improving training measures, since many migrant workers do not have the necessary qualifications for many areas of industry in which there is a labour shortage. Other scholars, however, are closer to Cai compared to these scenarios. Xiahe Ma and Jianlei Ma (2007, pp. 4–7) assume that the surplus of labor in the countryside was about 55 million in 2007. For this purpose, they considered the difference between the number of workers in the agricultural sector and the number of workers needed for agricultural production. Therefore, Ma and Ma, as well as Fang Cai (2007, pp. 5–8), believe that the surplus of labor in the traditional agricultural sector cannot meet the needs of the rapid development of the modern industrial sector (Cai 2007, pp. 5–8). A working paper by Tsinghua University (Zhu and Cai 2012, p. 6) provides an overview of the rural labor surplus in China calculated in various studies. This includes the 46 million surplus labor force calculated by Jiangui Wang and Shouhai Ding (2005, pp. 33–36) in 2003 using classical and neoclassical estimation methods. The different approaches and methodologies used to calculate the surplus rural labour force lead to significantly different results and thus to conflicting views on whether and when the Lewis Turning Point is or has been reached (Zhu and Cai 2012, p. 6).

9.3.2.2 Development of Wages

According to Arthur Lewis' model, once the Lewis Turning Point is reached, a sharp increase in wages should occur in the traditional agricultural sector and the modern industrial sector (Ranis and Fei 1961, pp. 533–536). Some scholars, such as Xiaobo Zhang (Zhang et al. 2011, p. 13) and Fang Cai (Cai 2010b, pp. 4–6), believe that there is a clear trend regarding rising real wages since 2003. According to their interpretations, it is therefore becoming apparent that the era of surplus labour is over. Jane Golley and Xin Meng

(2011, p. 27) disagree, claiming that there is no evidence that the increase in migrant workers' nominal wages resulted from a shortage of migrant workers.

The surge in wages in 2005 and 2006 is striking at first glance, as the growth in migrant workers' wages is almost as high as that of urban workers. However, this was related to the increase in the minimum wage by the government and not to the effects of labour shortages (Zhou 2010, pp. 17–19). Still, others believe that the rising wages would be a result of labour market segmentation. They argue that there are enough workers available, but that they do not always meet the requirements of the individual segments of the labour market. The necessary qualifications would often be lacking and would lead to locally occurring wage cost increases (Knight et al. 2011, pp. 25–26).

The International Monetary Fund Working Paper argues that despite local evidence of tensions in the labor market, overall wage growth has been steady at 12–15 percent over the past decade (Mitali and N'Diaye 2013, p. 5). When the Lewis Turning Point is reached, however, wages should jump. A general labor shortage in coastal regions – representative of the modern industrial sector – would be reflected in a divergence in the growth rates of wages in these and inland regions (representative of the traditional agricultural sector). According to Mitali and N'Diaye (2013, p. 5), this is not the case. What is increasingly observed is that industrial firms are moving from the coastal provinces to the inland provinces. There, wages are often even lower, and more rural workers can be found willing to work under the conditions of the industrial firms (Mitali and N'Diaye 2013, p. 5). This could be an indication that the Lewis Turning Point has already been reached in the coastal provinces and is yet to be reached in the inland provinces. Scholarly opinions differ regarding the evolution of wages as a sign of reaching the Lewis Turning Point.

9.3.2.3 Supply and Demand in the Labour Market

According to the Lewis model, the closer an economy moves towards the Lewis Turning Point, the greater the demand in the labour market. The rapid growth of the industrial sector, this leads to a tight labor market. Conversely, this means that the unemployment rate should fall after the Lewis Turning Point. Minami and Ma (2010, p. 164) estimated the unemployment rate at 12 percent and argued that this is a counterargument to a shortage of labor in the countryside.

Nevertheless, evaluations of the statistics of the State Statistical Office of China indicate a worsening of the overall situation in the labour market. According to these, since 2001 the difference between urban demand and the supply of labour has been steadily narrowing until it no longer exists at all. However, it is questionable whether all firms kept records of their demand and whether informal workers were included in the labour supply measures (Mitali and N'Diaye 2013, pp. 5–6).

9.3.2.4 Income Disparities Between Urban and Rural Areas

When the Lewis Turning Point is reached, the surplus of labour is used up; this leads, among other things, to wages rising in both the traditional and the modern sector and to the marginal productivity of labour converging in both sectors. Thus, the income gap between the two sectors should narrow significantly. One indicator to look at the inequality of income distribution is the Gini coefficient. This is given with a value of 1 to 100 or 0.1 to 1. The closer the value approaches 100, the greater the inequality. Here a Gini coefficient of over 40 or 0.4 is considered critical by the United Nations (Amiel and Cowell 1999, p. 136). Between the years 1988 and 2002, the value increased from 0.382 to 0.445 in the PRC. From the years 2005 to 2017, the value always remained so high that due to the inequality of income distribution, it could be assumed that there should be a large amount of surplus labour. Figure 9.1 shows the development of the Gini coefficient in China from 1997 to 2017.

The development after 2008 clearly shows that the Gini coefficient has fallen continuously in recent years, which could be a sign of the end of the era of labour surpluses. However, the Gini coefficient is still above the critical level. Observation of future values could provide new insights. Critically, the Gini coefficient looks at labour and capital income but says nothing about the inequality of wealth positions.

The existing income gap between urban and rural areas in China was explained by Fang Cai and Meiyan Wang (2007, pp. 4–6) using the Kuznets curve. This represents the empirical relationship between economic growth and income distribution in the development of an economy (Kuznets 1955, pp. 1–4). The Kuznets curve states that as a country develops,

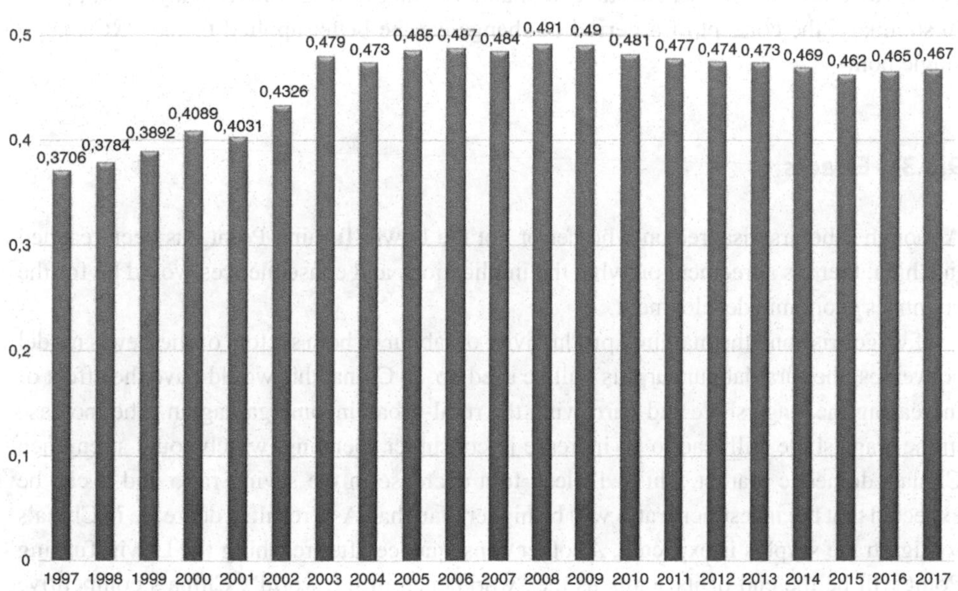

Fig. 9.1 Gini coefficient from 1997 to 2017 (Source: Various years of China Statistical Yearbook from the State Statistics Bureau)

economic inequality first rises and then falls. According to Cai and Wang, the opinion of other scholars that the income gap is greater today than it was at the beginning of the economic reforms in the 1980s would be wrong. In calculating the income gap, the difference in the price index of urban and rural areas would not have been included. If this difference were included in the calculations, the ratio of rural workers' to urban workers' incomes in 2006, for example, would be 1–2.57, not 1–3.28. Thus, the income gap in that year was as high as it was in 1978 and would contradict an income gap that continues to grow (Cai and Wang 2007, pp. 4–8).

The development of the Chinese labour market allows for different opinions regarding the surplus of labour in the country. Similarly, interpretations of the development of wages go in different directions. The assessment of the unemployment rate does not lead to a clear result and the widening income gap between urban and rural areas is equally controversial among scholars. Most scholars consider the Lewis Turning Point as a specific, precisely calculable point in China's economic development. Fang Cai (2010c, p. 136), for example, dates it to 2004, while Ryoshin Minami (1968, pp. 380–384), in his consideration of Japan's economy in 1968, sees the Lewis Turning Point less as a precise point in time than as a period of change in a country's economic development. This opinion was also held by Ross Garnaut (2010, p. 187). Due to the size of the country and its labour market, it is difficult to speak of a specific Lewis Turning Point for the whole PRC. This is supported by the aforementioned relocations of industrial enterprises to the inner provinces. Moreover, the propensity for industrial action is greater in the coastal provinces than in the inland provinces, which can be seen as an indicator that different Lewis Turning Points can be considered for coastal and inland provinces (Mitali and N'Diaye 2013, p. 5). In summary, the concept of a period of change can be better applied to the PRC than a single point.

9.3.3 Effects

Although scholars disagree on whether or not the Lewis Turning Point has been reached in China, there is agreement on what the implications and consequences would be for the country's economic development.

If wages rise and the marginal productivity of labour in both sectors of the Lewis model converges, the rural labour surplus will be used up. In China, this would have the effect of increasing the wage share and narrowing the rural-urban income gap again. The increase in the wage share will lead to an increase in consumer spending, which could strengthen China's domestic market. This will lead to a decrease in the saving ratio and it can be expected that the investment ratio will be higher than that. As a result, a decrease in China's foreign trade surplus is expected. Another consequence after reaching the Lewis Turning Point will be the end of its status as the "workbench of the world". China's competitive advantage in international trade through mass production of labor-intensive products by cheap labor will no longer be sustained. A shift towards capital-intensive and technologically sophisticated products can be expected (Garnaut 2010, p. 189). This will increase the pressure on international trade in certain developed countries, but at the same time lead to a transfer of labour-intensive manufacturing industries to countries with cheap labour (Zhu and Cai 2012, p. 10). Moreover, with the increase in real wages, could come a period of high inflation. China's incredible growth rates in recent decades came in large part from the advantage of cheap and huge masses of workers from the countryside. A shift away from this "workbench of the world" model will inevitably lead to lower growth rates (Huang and Jiang 2010, p. 203).

China's economic development is closely linked to the Lewis Turning Point. After years of immense growth, the country is moving towards normal development. In the process, the step from an emerging market to an industrialized country is to be completed once and for all. One danger that could threaten the country on this path is the so-called "middle income trap". This describes a situation in which a country's economy loses its competitiveness in industries with low value added, but at the same time fails to establish itself in industrial sectors with high value added potential. As a result, the country fails to rise above a median per capita income as a share of GDP. Having reached the Lewis Turning Point, China faces the challenges of the "middle income trap" (Huang and Jiang 2010, pp. 206–210).

9.3.4 Government Measures

How the PRC will manage the transformation of its economy is closely linked to the government's actions. The latter must draw the consequences from the effects of the Lewis Turning Point. First of all, measures have already been taken to delay the achievement of the Lewis Turning Point. For example, the one-child policy that had been in place for years has been repealed to counteract the ageing of society. However, it is questionable whether this measure will lead to a higher birth rate in modern China. Moreover, it would take at least 15–16 years for children to even reach working age. A "Higher Fertility" scenario would have little impact on the timing of the depletion of the surplus labor force due to the dissolution of the one-child policy (Herrmann 2018, p. 32).

Migrant workers should be given better access to education so that they can be trained as skilled workers. In this way, the change from an industry with low value added by cheap unskilled labour to an industry with a focus on capital-intensive and high-quality products is to be completed. The Chinese government has recognized the dangers of ending low-wage production in the context of reaching the Lewis Turning Point. In the 13th Five-Year Plan, scientific and technological advances play an important role in the future growth of the nation. This is sometimes reflected in the "Made in China 2025" strategy, a plan to modernize Chinese industry. The Chinese government is attempting to counter the dangers associated with the end of cheap wages with a flight to the future. Modernizing the industrial and economic sectors and strengthening the domestic market through urbanization and greater prosperity for all is an alternative compared to the measures of delaying the Lewis Turning Point.

9.4 Conclusion

China's future economic development is closely linked to the end of the rural labor surplus and the concept of the Lewis Turning Point. Whether the point has been passed or is yet to come is a matter of dispute among researchers. Researchers disagree on key points regarding the end of the surplus labor force. It is difficult to determine a specific Lewis Turning Point for a country of this size and with such a huge labor market. Provinces on the coast may have already reached the Lewis Turning Point, while inland it is not yet in sight. Therefore, the concept of a period of change may be more applicable.

What scholars do agree on, however, is the impact and consequences of the Lewis Turning Point on China's economy. The growth rates of the past can most likely no longer be sustained and there will be a shift away from being the "workbench of the world" to more capital-intensive and sophisticated production. The looming danger in the transformation of the Chinese economy is the "middle income trap". The country's government has recognized this and is trying to counter it with measures aimed at modernization and a stronger domestic market. But the end of cheap wages and surplus labor is not all negative. It has the potential to narrow the income gap again and increase the overall prosperity of the population. Formerly unskilled workers, most notably migrant workers, can benefit from the fact that the coming shortage of skilled workers will be addressed by providing them with opportunities for further training. In addition, they can get permanent rights to stay in the cities due to the intended urbanization.

References

Amiel, Yoram, and Frank Cowell. 1999. *Thinking about inequality: Personal judgment and income distributions*. Cambridge: Cambridge University Press.

Balding, Christopher. 2019. Bad jobs data could bite China. *Bloomberg*, 19. https://www.bloomberg.com/opinion/articles/2019-02-19/dodgy-unemployment-data-could-be-china-s-achilles-heel. Accessed 03 June 2019.

Bao, Shuming, Örn B. Bodvarsson, Jack W. Hou, and Yaohui Zhao. 2011. The regulation of migration in a transition economy. China's Hukou system. *Contemporary Economic Policy* 29 (4): 564–579.

Bundeszentrale für politische Bildung. 2014. *Verordnetes Schweigen am Platz des Himmlischen Friedens*. http://www.bpb.de/politik/hintergrund-aktuell/185616/25-jahre-tiananmen-massaker-03-06-2014. Accessed 23 May 2019.

Cai, Heping. 2006. Ländliche Wanderarbeitnehmer in der Volksrepublik China – Probleme und Lösungsansätze. In *Zeitschrift für ausländisches und internationales Sozialrecht 20*, 297–319. Heidelberg: C.F. Müller.

Cai, Fang. 2007. The myth of surplus labor force in rural China. *Chinese Journal of Population Science* 2: 2–10.

———. 2010a. Demographic transition, demographic dividend, and Lewis turning point. *China Economic Journal* 3 (2): 107–119.

———. 2010b. On the challenges of and the path for China's development of the Lewis turning point of large country's economy. *Journal of Guangdong University of Business Studies* 1: 4–12.

———. 2010c. The Lewis turning point and the reorientation of public policies: Some stylized facts of social protection in China. *Social Science in China* 6: 125–137.

Cai, Jane. 2017. Minimum wages on the march in China as labour pool shrinks. *South China Morning Post*, 13. https://www.scmp.com/news/china/economy/article/2115121/minimum-wages-march-china-labour-pool-shrinks. Accessed 17 June 2019.

Cai, Fang, and Dewen Wang. 2005a. China's demographic transition: Implications for growth. In *The China boom and its discontents*, ed. Ross Garnaut and Ligang Song, 34–52. Canberra: Asia Pacific Press.

Cai, Fang, and Meiyan Wang. 2005b. An economic analysis on shortage of migrant workers. *Guandong Social Sciences* 2: 5–10.

———. 2007. Re-assessment of rural surplus labor and some correlated facts. *Chinese Rural Economy* 10: 4–12.

Cai, Fang, Du Yang, and Meiyan Wang. 2005. *How close is China to a labor market? (zhongguola-odongli shichang zhuanxing yu fayu?)*. Beijing: The Commercial Press.

Chan, Kam Wing. 2010. A China paradox: Migrant labor shortage amidst rural labor supply abundance. *Eurasian Geography and Economics* 51 (4): 513–530.

Chan, Alexsia T., and Kevin J. O'Brien. 2019. Phantom services: Deflecting migrant workers in China. *The China Journal* 81 (1): 103–122.

Chen, Derong. 1993. *The contract management responsibility system in China*. Basingstoke: University of Aston.

Chen, Chuanbo, and Cindy Fan. 2016. China's hukou puzzle: Why don't rural migrants want urban hukou. *The China Review* 16 (3): 9–39.

Chen, Guifu, and Shigeyuki Hamori. 2014. Formal and informal employment in urban China: Income differentials. In *Rural labor migration, discrimination, and the new dual labor market in China*. Berlin/Heidelberg: Springer.

Chi, Wei, Richard Freeman, and Hongbin Li. 2012. Adjusting to really big changes: The labor market in China, 1989–2009. In *The Chinese economy*, ed. Masahiko Aoki and Wu Jinglian, 93–113. London: Palgrave Macmillan.

China Labour Bulletin. 2019. *The shifting patterns of labour protests in China present a challenge to the union*. https://clb.org.hk/content/shifting-patterns-labour-protests-china-present-challenge-union. Accessed 23 Aug 2019.

Davies, Ken. 2013. *China investment policy. An update*. OECD Publishing (OECD working papers on international investment). https://www.oecd.org/china/WP-2013_1.pdf. Accessed 03 June 2019.

Dettmer, Isabel. 2016. *HRM, Qualifizierung und Rekrutierung in China*. Dissertation. Würzburg: Würzburg University Press.

Dian, Liu. 2014. Graduate employment in China: Current trends and issues. *Chinese Education & Society* 47 (6): 3–11.

Fields, Gary S. 2004. Dualism in the labor market: A perspective on the Lewis model after half a century. *The Manchester School* 72 (6): 724–735.

Garnaut, Ross. 2010. Macro-economic implications of the turning point. *China Economic Journal* 3 (2): 181–190.

Golley, Jane, and Xin Meng. 2011. Has China run out of surplus labour? *China Economic Review* 22 (4): 555–572.

Gosh, Dipak. 1985. A Lewisian model of dual economy with rural-urban migration. *The Scottish Journal of Political Economy* 32 (1): 95–106.

Gransow, Bettina. 2014. *Hintergrund und Problemaufriss: Stadt-Land-Gefälle und Meldesystem (hukou).* http://www.bpb.de/gesellschaft/migration/kurzdossiers/151283/stadt-land-gefaelle-und-meldesystem. Accessed 23 Aug 2019.

Grzanna, Marcel. 2010. Aus für Billiglöhne. *Süddeutsche Zeitung*, 19. https://www.sueddeutsche.de/wirtschaft/china-aus-fuer-billigloehne-1.989810. Accessed 03 June 2019.

Hao, Nicole. 2019. Around 200 US companies considering moving production from China to India. *The Epoche Times*, 29. https://www.theepochtimes.com/around-200-us-companies-considering-moving-production-from-china-to-india_2897876.html. Accessed 17 June 2019.

Hebel, Jutta, and Günter Schucher, Ed. 1992. *Zwischen Arbeitsplan und Arbeitsmarkt. Strukturen des Arbeitssystems in der VR China.* Hamburg: Institut für Asienkunde (Mitteilungen des Instituts für Asienkunde Hamburg, 204).

Herrmann, Birgit. 2018. Individuelles Paarglück statt "Kinder vom Fließband": Gewollte Kinderlosigkeit im städtischen China. In *Aspekte des sozialen Wandels in China*, ed. Björn Alpermann, Birgit Herrmann, and Eva Wieland, 32–70. Wiesbaden: Springer.

Huang, Yiping, and Tingsong Jiang. 2010. What does the Lewis turning point mean for China? A computable general equilibrium analysis. *China Economic Journal* 3 (2): 191–207.

Huang, Ping, and Frank Pieke. 2003. China migration country study. In *Conference on migration, development and pro-poor policy choices in Asia.* Dhaka, Bangladesh.

Jing, Meng, and Celia Chen. 2019. Job-hopping days are over in China's tech sector as start-ups feel chill winds of slowing economy. *South China Morning Post*, 24. https://www.scmp.com/tech/start-ups/article/2183320/job-hopping-days-are-over-chinas-tech-sector-start-ups-feel-chill. Accessed 03 June 2019.

Knight, John, Quheng Deng, and Shi Li. 2011. The puzzle of migrant labour shortage and rural labour surplus in China. *China Economic Review* 22 (4): 585–600.

Kristof, Nicholas D. 1989. China erupts … the reasons why. *The New York Times*, 04. https://www.nytimes.com/1989/06/04/magazine/china-erupts-the-reasons-why.html. Accessed 23 May 2019.

Kuznets, Simon. 1955. Economic growth and income inequality. *The American Economic Review* 45 (1): 1–28.

Leng, Sidney. 2019. China's 8.34 million graduates fighting for fewer jobs as vacancies dwindle and gap years increase, survey shows. *South China Morning Post*, 31. https://www.scmp.com/economy/china-economy/article/3012635/chinas-834-million-graduates-fighting-fewer-jobs-vacancies. Accessed 03 June 2019.

Lewis, Arthur William. 1954. Economic development with unlimited supplies of labour. *The Manchester School* 22 (2): 139–191.

———. 1972. Reflections on unlimited labour. In *International economics and development*, ed. L.E. di Marco, 75–96. Cambridge: Academic.

Lu, Ming, Jianyong Fan, Shejian Liu, and Yan Yan. 2002. Employment restructuring during China's economic transition. *Monthly Labor Review* 125 (8): 25–31.

Lüthje, Boy. 2010. Arbeitsbeziehungen in China: 'Tripartismus mit vier Parteien'? *WSI-Mitteilungen* 63 (9): 473–479.

Ma, Jingkui. 2004. *Zhongguo she hui zhong de nü ren he nan ren: Shi shi he shu ju.* Beijing: Zhongguo tong ji chu ban she.

Ma, Xinxin. 2018. *Economic transition and labour market reform in China.* London: Palgrave Macmillan.

Ma, Xiaohe, and Jianlei Ma. 2007. How much surplus labor in rural China on earth? *Chinese Rural Economy* 12: 4–9.

Mankiw, Gregory Nicholas. 2001. *Grundzüge der Volkswirtschaftslehre*. 2nd ed. Stuttgart: Schäffer-Poeschel.

———. 2003. *Makroökonomik*. 5th ed. Stuttgart: Schäffer-Poeschel.

Minami, Ryoshin. 1968. The turning point in the Japanese economy. *The Quarterly Journal of Economics* 82 (3): 380–402.

Minami, Ryoshin, and Xinxin Ma. 2010. The Lewis turning point of Chinese economy: Comparison with Japanese experience. *China Economic Journal* 3 (2): 163–179.

Ministry of Education of the PRC. 2018. *2017 sees increase in number of Chinese students studying abroad and returning after overseas studies*. http://en.moe.gov.cn/News/Top_News/201804/t20180404_332354.html. Accessed 18 June 2019.

Mitali, Das, and Papa N'Diaye. 2013. *Chronicle of a decline foretold: Has China reached the Lewis turning point?* Prepared by Mitali Das and Papa N'Diaye. IMF working paper, WP/13/26. http://www.imf.org/external/pubs/ft/wp/2013/wp1326.pdf. Accessed 23 Aug 2019.

Mühlhahn, Klaus. 2017. *Die Volksrepublik China*. Berlin: de Gruyter.

National Bureau of Statistics. 2017. *2016 Nián nóngmín gōng jiāncè diàochá bàogào*. http://www.stats.gov.cn/tjsj/zxfb/201704/t20170428_1489334.html. Accessed 01 Sep 2019.

———. 2019. *China Statistical Yearbook 2018*. http://www.stats.gov.cn/tjsj/ndsj/2018/indexeh.htm. Accessed 23 Jan 2020.

Naughton, Barry. 2007. *The Chinese economy – Transitions and growth*. Cambridge: MIT Press.

———. 2018. *The Chinese economy – Adaptation and growth*. 2nd ed. Cambridge: MIT Press.

Nielsen, Ingrid, and Russel Smyth. 2008a. The rhetoric and the reality of social protection for China's migrant workers. In *Migration and social protection in China*, ed. Ingrid Nielsen and Russel Smyth, 3–13. Singapore: World Scientific Publishing Co.

Qian, Yingyi. 2000. The process of China's market transition (1978–1998). *Journal of Institutional and Theoretical Economics* 156 (1): 151–171.

Quandt, Richard E., and Harvey Rosen. 1986. Unemployment, disequilibrium and the short run Phillips curve. *Journal of Applied Econometrics* 1 (3): 235–253.

Ranis, Gustav, and John Fei. 1961. A theory of economic development. *The American Economic Review* 51 (2): 533–565.

Richardson, Sophie. 2008. *One year of my blood: Exploitation of migrant construction workers in Beijing*. https://www.hrw.org/report/2008/03/11/one-year-my-blood/exploitation-migrant-construction-workers-beijing. Accessed 23 Aug 2019.

Robbins, Mark. 2016. *The thousand talents program: lesson from China about faculty recruitment and retention*. https://www.researchgate.net/publication/321172646_The_Thousand_Talents_Program_Lesson_From_China_About_Faculty_Recruitment_and_Retention. Accessed 23 Aug 2019.

Scullion, Hugh, and David G. Collings. 2011. *Global talent management*. New York: Routledge. (Routledge global human resource management series).

State Council of the PRC. 2006. *Guānyú jiějué nóngmín gōng wèntí de ruògān yìjiàn*. http://www.gov.cn/jrzg/2006-03/27/content_237644.htm. Accessed 27 July 2019.

Taubmann, Wolfgang. 2004. Arbeitslosigkeit und Armut im städtischen China. In *Sozialer Sprengstoff in China?. Dimensionen sozialer Herausforderungen in der Volksrepublik*, ed. Kristin Shi-Kupfer. Essen: Asienstiftung (Focus Asien, 17, S. 21–32). https://www.asienhaus.de/publikationen/detail/sozialer-sprengstoff-in-china-dimensionen-sozialer-probleme-in-der-volksrepublik/. Accessed 23 Aug 2019.

Thiel, Ingo. 1998. *Der dörfliche Bodenübernahmevertrag (nongcun-tudi-chengbao-hetong) in der VR China. Ökonomische Funktion und rechtliche Gestalt im Wandel (1985–1995)*. Marburg: Tectum. (Wirtschaftspolitische Forschungsarbeiten der Universität zu Köln, 23).

Trading Economics. 2019. China minimum monthly wages. *Trading Economics.* https://tradingeconomics.com/china/minimum-wages. Accessed 17 June 2019.

Wang, Yue. 2019. *Job cuts and no more snacks: China's Internet companies brace for slowest growth in years.* Forbes Media LLC. https://www.forbes.com/sites/ywang/2019/03/13/job-cuts-and-no-more-snacks-chinas-internet-companies-brace-for-slowest-growth-in-years/#43534c696729. Accessed 17 June 2019.

Wenderoth, Michael C. 2018. *China's growth is slowing but the war for talent is not.* Forbes Media LLC. https://www.forbes.com/sites/michaelcwenderoth/2018/11/30/chinas-growth-is-slowing-but-the-war-for-talent-is-not/#22e3c144479. Accessed 03 June 2019.

West, Loraine A., and Yaohui Zhao. 2000. *Rural labor flows in China.* Berkeley: Institute of East Asian Studies, University of California.

Windrow, Hayden, and Anik Guha. 2005. The hukou system, migrant workers, & state power in the People's Republic of China. *Northwestern Journal of International Human Rights.* http://scholarlycommons.law.northwestern.edu/cgi/viewcontent.cgi?article=1014&context=njihr. Accessed 23 Aug 2019.

Wong, Daniel Fu, Changying Li Keung, and Hexue Song. 2007. Rural migrant workers in urban China: Living a marginalised life. *International Journal of Social Welfare* 16 (1): 32–40.

Woo, Ryan, and Lusha Zhang. 2019. China's tech hub Hangzhou sees surge in job-hopping: Newspaper. *Reuters.* https://www.reuters.com/article/us-china-economy-hangzhou-jobs/chinas-tech-hub-hangzhou-sees-surge-in-job-hopping-newspaper-idUSKCN1R20T1. Accessed 03 June 2019.

Wu, Fuxin. 2003a. Jiaru WTO yu shenzhen renli ziyuan quanli wuyixin. *Southern Forum* 2: 49–52.

Wu, Li. 2003b. Gaige kaifang qianhou renli ziyuan peizhi ji xiaolü bijiao yanjiu. *China Academic Journal* 2: 37–45.

Yang, Juhua. 2018. *In search of the urban dream among internal migrants of China.* Beijing: Economic Science Press.

Yang, Baiyin, De Zhang, and Mian Zhang. 2004. National human resource development in the People's Republic of China. *Advances in Developing Human Resources* 6 (3): 297–306.

Yao, Xianguo. 2018. *Research on equalization of employment rights of urban-rural laborers.* Beijing: Economic Science Press.

Zhang, Linxiu. 2003. *Agricultural development and the opportunities for aquatic resources research in China.* Penang: WorldFish Center (WorldFish Center contribution, no. 1668). http://pubs.iclarm.net/Pubs/china/pdf/china_agricultural.pdf. Accessed 22 May 2019.

Zhang, Y. 2016. *Urbanization, inequality, and poverty in the People's Republic of China.* ADBI working paper series no 584. Tokyo: Asian Development Bank Institute. https://www.adb.org/sites/default/files/publication/189132/adbi-wp584.pdf. Accessed 23 Aug 2019.

Zhang, Xiaobo, Yang Jin, and Shenglin Wang. 2011. China has reached the Lewis turning point. *China Economic Review* 22 (4): 542–554.

Zhao, Xianzhou. 2010. Some theoretical issues on Lewis turning point. *The Economist* 5: 75–80.

Zheng, Z.J. 2018. *Middle income trap. An analysis based on economic transformations and social governance.* Beijing: Tsinghua Publishing Group.

Zhou, Tianyong. 2010. Whether the labor surplus in China. *Shanghai Economy* 11: 17–19.

Zhu, Cherrie Jiuhua. 2005. *Human resource management in China. Past, current and future HR practices in the industrial sector.* London: Routledge Curzon.

Zhu, Andong, and Wanhuan Cai. 2012. *The Lewis turning point in China and its impacts on the world economy.* AUGUR working paper (WP#1). Beijing: School of Marxism, Tsinghua University. http://ss.rrojasdatabank.info/chinalewis11.pdf. Accessed 17 Nov 2019.

Zhu, Ying, and Malcolm Warner. 2004. Changing patterns of human resource management in contemporary China: WTO accession and enterprise responses. *Industrial Relations Journal* 35 (4): 311–328.

Zhu, Ying, Malcolm Warner, and Tongqing Feng. 2011. Employment relations "with Chinese characteristics": The role of trade unions in China. *International Labour Review* 150 (1–2): 127–143.

E-commerce

10

Barbara Darimont, Marianne Friedrich, and Jonas Henselmann

The cooperation between the state and private companies allows a symbiotic relationship to develop, with the result that investments by e-commerce companies such as Alibaba and Taobao are often made in the interests of the state. This not only allows the state to allow private companies to promote state interests but also to control and regulate the market. In addition, the state uses the digital trend to digitally monitor the population (social credit system). The establishment of so-called e-commerce villages in rural areas represents a concise example of where private companies, such as Alibaba and Taobao, invest in the infrastructure of rural regions to both promote their own e-commerce business and grow the weak regional economy. Different models of e-commerce villages have already been implemented as pilot projects in selected rural regions of China.

The Chinese population makes most of its purchases via e-commerce, making the PRC a global leader in this industry due to its population (Clark 2016, p. 10). This fact raises some questions. Why are Chinese consumers internet savvy to such a high degree? What business concepts are behind the success of Chinese Internet companies?

According to the China Internet Network Information Center (CNNIC), the number of internet users in China was around 772 million in 2017, increasing by another 40.74 million compared to the previous year. Of these, 753 million users used the internet through

B. Darimont (✉)
East Asia Institute of Ludwigshafen University of Business and Society,
Ludwigshafen am Rhein, Germany
e-mail: darimont@oai.de

M. Friedrich
Universtity Würzburg, Würzburg, Germany

J. Henselmann
Arrow ECS, Colorado, USA

B. Darimont (ed.), *Economic Policy of the People's Republic of China*,
https://doi.org/10.1007/978-3-658-38467-8_10

their mobile phones (CNNIC 2018). Thus, 98 percent of internet users were mobile internet users, which is partly due to the network quality in China and partly shows how much time Chinese people spend on their mobile phones browsing the internet. With over 157 million purchases made daily, this clearly shows the potential and weight of this market. Alibaba turned over 250.27 billion renminbi (RMB) in 2018 with circa 636 million shoppers and currently holds a market share in the e-commerce space of just under 75 percent. According to estimates by the consulting firm McKinsey, China's share of global online trade is over 40 percent, making it larger than the United States, the United Kingdom, Germany, France, and Japan combined (Ma 2018, p. 12).

Electronic payment systems, such as Alipay and WeChatpay, simplify internet purchases, as these programs are usually linked directly to the mobile phone and do not require further verification. Another point is that in China, both young people and the older class of society own mobile phones. This means that the target group for e-commerce in China is the entire population. Compared to Germany, network coverage is also significantly higher. In the rural regions of China, a large proportion of people have their first contact with the internet via smartphone, as a PC or laptop is comparatively expensive and only of limited use. An internet broadband line is not required to use the smartphone, only a good mobile network and a large number of locally produced and cheap mobile phones (Ma 2018, p. 12).

Compared to other sectors, such as telecommunications, oil, electricity, or finance, the Internet sector is predominantly dominated by private companies. Characteristically, some companies have emerged that rule with a near monopolistic position in their territory. In the e-commerce sector, this company is Alibaba, whose sales platform is called Taobao. Other companies include both Tencent, which is best known for its messenger service WeChat, and Baidu, which is comparable to Wikipedia and Google. Therefore, the abbreviation BAT has become common for these companies.

With the internet coming to China in 1994, the e-commerce business developed only marginally in the 1990s. It was not until 2008 that rapid growth began, as the government created a positive environment with legal rules. Overall, the development of e-commerce is divided into four phases: From 1996 to 2000, the first companies started e-commerce business, including Alibaba. The accelerated development phase from 2001 to 2007 was initially characterized by competition between eBay and Alibaba and other companies. In June 2007, the National Development and Reform Commission adopted the first five-year plan for e-commerce. This marked the standardization of e-commerce business at the national level. On December 17, 2007, the Ministry of Commerce released the "Views on the Improvement for the Regulated Development of E-commerce" (Wu and Weng 2018, p. 252). From 2008 to 2014, the e-commerce market was standardized. During this period, various plans and standards were adopted by policymakers to promote this industry. In the beginning, e-commerce was defined as "consumption based on information technology" (Ma 2018, p. 10), but then expanded to include "high-value information products for consumption" (Ma 2018, p. 10), such as motion pictures and online videos. E-commerce consequently refers to a transaction on the Internet that involves the flow of information, money, or goods (Li 2017, p. 57). In the year 2014, the whole Chinese e-commerce business becomes international. The characteristic of this phase is that Alibaba went public on the New York Stock Exchange and the international breakthrough of Chinese companies in this field began (Wu and Weng 2018, p. 254).

10.1 General Conditions

The success of e-commerce and internet giants is based on several factors. First, the development of the Internet coincided with the economic growth in the PRC. Due to the former planned economy, before the start of e-commerce, there was no retail industry, there was no logistical infrastructure, and the marketing industry was also only just emerging (Zeng 2019, p. 222). Moreover, these developments met a society that was at the beginning of the consumption pyramid. A similar emergence would be hard to imagine in societies whose consumption needs were already saturated. Furthermore, one of the favored leisure activities of the Chinese population is shopping.

Digital media and technologies are drivers of the Chinese economy. One advantage is the open-mindedness of the Chinese population to embrace this new technology; there is also a high networking mentality. There are regional differences, however, because within the cities 70 percent of the population has access to the internet, and in rural regions only 30 percent. Accordingly, there are still opportunities in rural areas to further expand and spread the internet. Some companies, such as Alibaba, recognized the opportunities in the Chinese domestic market early on and are supporting the expansion of the internet infrastructure to induce an increase in their customer numbers (Alibaba Group 2015).

The mobile network and mobile phones have leapfrogged a stage of development, which is more conducive to innovation. This is due to the so-called "leapfrogging". The English term "leapfrogging" means "jumping over" and describes in an economic context the phenomenon that stages are skipped within a development process (Schumpeter 1934). A distinction is made between demand-oriented leapfrogging, which starts with the consumer, and supply-oriented leapfrogging by firms. Demand-driven leapfrogging can be observed especially in the technology industry, because technology innovations occur in a very short time. Accordingly, from the consumer's perspective, it may be useful to delay the purchase of a product to wait for a new innovation if it is already foreseeable that it will occur within a short period. Supply-side leapfrogging, on the other hand, can be illustrated by retail behavior. Farmers and retailers in rural China could only market their products in small regional markets for a long time because there were no structures to bring farmers' products into a business relationship with wholesalers. The development of the internet, smartphones, and digital marketplaces, such as Taobao, created the opportunity for farmers and other small producers or retailers to market their products directly to end customers, rather than exclusively regionally. The "brick-and-mortar" stage within the retail development process was leapfrogged with these innovations (Boos and Peters 2016, p. 133).

In Western markets, such as Germany, it has long been highly common to use landlines and mobile phones simultaneously. In China, the majority of people never owned a landline phone (Statista 2019) and the mobile phone was the first phone for many (CNNIC 2018). Therefore, the majority of Chinese have unconsciously leapfrogged by skipping both landline telephony and using a computer to access the internet. On the positive side, the Chinese have not had to spend any resources on mending old telecommunications structures, but have been able to invest directly in new and innovative mobile networks and relevant data lines (Boos and Peters 2016, p. 133).

Other characteristics of the development are rapid network coordination and digitalization or data intelligence. Ming Zeng (2019, pp. 41, 66), a former strategic planner of Alibaba, understands data intelligence as enabling companies to "rapidly and automatically improve their processes using machine learning technology." Alibaba has built an extensive network, so many productions and sales steps can be done within the network, such as marketing, services of all kinds, and logistics. Compared to other countries, it can be argued that a company that has to compete under an authoritarian system is much more pragmatic with networks and less affected by personal animosities. In particular, the establishment of Alipay and convenient payment functions, in general, have facilitated a tremendous development of e-commerce. The trust merchant function that Alibaba has adopted helps to provide a trust advantage. A customer does not have to fear not receiving money, nor does he have to file a lawsuit or similar complaint in case of quality defects, which can be costly (Zeng 2019, pp. 247–248). In terms of data intelligence, the PRC has an advantage simply because of the mass of internet users, as the amount of data encourages further development in e-commerce.

The Chinese state has recognized the potential of e-commerce. In rural areas, the government is trying to combat poverty with e-commerce. Poverty reduction is part of the CCP's program, which is why, for example, taxes were waived for the rural population, and investments were made in medical care and education (Ahlers et al. 2016, pp. 58–59). As part of the 11th Five-Year Plan in 2006, it was decided to develop rural areas much more than before. The program dealing with these tasks was titled by the government "building a new socialist countryside" (shèhuì zhǔyì xīnnóngcūn jiànshè – BNSC) (Ahlers 2014, p. 3). The goal is to industrialize rural regions and make them competitive. Problematically, agricultural products are unstable in price and overpriced because they have no direct sales market, but middlemen are the profiteers (Wang and Shen 2018, p. 70). Furthermore, unemployment is prevalent in the countryside, which the new e-commerce companies aim to remedy as both young and older residents are involved in the businesses. It is estimated that the rural population will decrease by more than one-third by 2030 and more than 300 million people are expected to move to cities (Sun et al. 2017, p. 945). Given this trend, the question arises whether e-commerce has the potential to counteract this development and stop the migration of the rural population to the cities. Should this possibility be available in rural regions in the future, it will become clear whether the cities still offer the same attractiveness to the Chinese population as they do today. With the help of e-commerce, many Chinese could potentially find a job, receive an education, and have medical coverage in their home country.

10.2 Platform Economy

A special form of e-commerce is the platform economy. The platform economy model achieved its breakthrough with the help of the internet and the dawn of the so-called new economy. Shares of platform companies such as Apple, Amazon, Alphabet Inc. and Tencent Holdings, along with Microsoft, represent the five companies with the largest market capitalization in 2019. Although the idea of the platform has existed in the form of marketplaces, flea markets or shopping malls, entirely new opportunities are emerging due to digitization and the previously unmatched reach of the internet. Corporate models, such as those of Airbnb, Uber, Alibaba's Taobao, or WeChat, which merely provide a platform for providers and customers (so-called matching), are, on closer inspection, very similar to a marketplace or shopping mall where rent is charged for a location (Herda et al. 2018, p. 15). Furthermore, Amazon shows that it is equally possible to offer its products on a supplier-customer merging platform and thereby profit from a specially created platform. However, the monopoly-like market positions and expansion strategies of these platform companies are increasingly meeting with resistance and concerns at the political (Herndorn 2019) and academic levels because monopoly-like structures are emerging that reduce competition and against which little antitrust action has been taken to date (Srnicek 2016, pp. 45–47, 62–68; Herda et al. 2018, pp. 15–16; Evans et al. 2011, pp. 279–280).

Platform economies can maximize their benefits through network and scale effects (cf. Economies of Scale). The advantage of networking stems from the fact that a platform is of greater interest to companies the more potential customers access it. This applies vice versa from the customer's point of view. The more providers offering their services or products on a platform, the more competition and choice there is. In addition, cost items can lead to higher revenues and, respectively, larger profit margins when there is a larger number of users on a platform. For example, the development and maintenance of software are of the same or similar expense regardless of whether it is used by tens of thousands or hundreds of thousands of users (Evans et al. 2011, pp. 14–17). Companies using the platform economy business model do not own any of the traditional factors of production but act as intermediaries between customers and producers (Herda et al. 2018).

In the PRC, completely new business concepts have emerged within the framework of the platform economy. For example, in the clothing industry, there are so-called wang hongs (wǎng hóng), a Chinese form of social media influencer who only sell their products at a certain time – usually for one day – as part of on-demand marketing. They only sell a certain number of garments; when the day is over, the products can no longer be obtained. Demand usually exceeds supply. This is called starvation marketing, which seems to work particularly well in a country where there was limited supply under a planned economy a few decades ago (Zeng 2019, pp. 25, 45–46, 106). This form of selling requires an on-demand supply chain, which in turn works best when all parties are excellently connected. This business model thrives on identifying fashion trends early on or setting new trends. For example, Instagram is analyzed with special software for this purpose. Ming Zeng

(2019 p. 120) assumes that the future lies in the consumer-to-business (C2B) market. In old-fashioned terms, this is custom manufacturing. This would mean, for example, that customers set the new trends and the company then translates these trends into fashion.

10.2.1 Policy Objective

The PRC has been focusing on innovation since the 1990s and has been supporting research and development in this area since 2006 with the 13th Five-Year Plan and the "Made in China 2025" program, the innovation program that aims to promote the automation and digitization of industry within ten industries, and the "Internet Plus" plan that aims to drive digital innovation (Tse 2015). The "Internet Plus" plan, presented by Chinese Premier Li Keqiang in March 2015, supports the interconnection of mobile internet, cloud computing, Big Data, and the Internet of Things, as well as the growth of online retail, Industry 4.0, and online banking. It also aims to promote Chinese internet companies internationally. The focus is on manufacturing, the financial sector, the health sector, and government administration (Boos and Peters 2016, p. 135).

In the PRC, there is censorship covering the internet. The control of the internet is carried out using various instruments, firstly the great firewall, which blocks websites and VPN protocols, then using keyword filters, which filter the content of the internet specifically according to certain keywords, as well as manual control, which is intended to prevent collective actions (Tai 2015). Such censorship leads to competition from the outside becoming almost impossible. As a result, Western internet companies Google and eBay failed to enter the Chinese market. However, a lack of knowledge about the Chinese market and customers, as well as a lack of support from the Chinese government compared to Chinese companies, also contributed to the failure of Google and eBay (Boos and Peters 2016, p. 136).

10.2.2 Social Credit System

The social credit system, which rates individuals, businesses, social organizations, and government agencies based on their trustworthiness, will radically change the governance of society and the economy (Kostka 2018). The Chinese government is considered a global pioneer in the use of data and internet-based technologies as tools to govern the population. The political leadership is investing in new IT systems and surveillance technologies to realize a new IT-based authoritarian state (Lee 2019, p. 208). In doing so, the CCP focuses on ensuring social stability and security.

The Chinese government plans to introduce the social credit system nationwide in 2020. So far, only pilot projects exist. Although there are already commercial scoring platforms, such as Sesame Credit from the Alibaba subsidiary Ant Financial, this is not mandatory, unlike the social credit system (Campbell 2019). A prerequisite for the social credit system is an infrastructure that can store and process large amounts of data. Most importantly, it requires data sets from various institutions, both private and public. However, Chinese firms are reluctant to disclose their data (Kharpal 2019). There is little incentive for companies to fully disclose their data to the government, as this data has a tremendous competitive advantage and economic value. According to the January 2017 Cybersecurity Law, companies operating in the IT industry in China, for example, have to disclose their software solutions, including their source codes (Pattloch 2018). This applies equally to government entities at central and local levels, as data sets are associated with valuable political power and influence. Regarding the storage of data, studies show that the majority of the Chinese population is positive about projects such as the social credit system. They can understand the government's intention and even support it, as this can be used to track down criminals, for example, and create a harmonious society (Kostka 2018). The lack of basic data sharing is currently one of the biggest hurdles to the expansion of a social credit system (Chorzempa et al. 2018, p. 8).

10.2.3 Alibaba

One of the largest internet companies in the world, Alibaba Holding Limited, was founded by Jack Ma in Hangzhou in 1999 and has been led by Daniel Yong Zhang since 2014. Its core business is in e-commerce and includes the platforms Alibaba.com (B2B international), 1688.com (B2B China), Taobao (C2C), and Tmall (B2C). Alibaba Group diversifies into six different industries, such as e-commerce, internet-related businesses, financial services, social networking services, digital healthcare, and cultural and entertainment sectors. Within this corporate structure, an ecosystem of approximately 55 subsidiaries and affiliates has emerged (Greeven and Wei 2018, pp. 20–29).

Initially, Alibaba generated no profit despite a rising profile and increasing user numbers. In the mid-2000s, Ma decided to migrate the English website Alibaba.com from China to the American Silicon Valley in California. The language barriers, long distances, and time difference led to increasing costs and high losses, so the "Back to China" strategy was initiated by transferring Alibaba's English website to Hangzhou (Erisman 2015, pp. 40–53). Initially, the marketplace platform remained free of charge, eventually offering services for fees. This includes the establishment of Alipay as a secure payment option (Erisman 2015, pp. 53–63; Tse 2015, p. 34). Alipay's success made the online payment system lead the market with 61.5 percent in 2018. With an active user base of more than 600 million, a transaction volume of USD 1.7 billion, availability of the system on mobile devices of any kind, and support for 18 currencies in 110 countries, Alipay took the place of the most successful online payment system in the world in 2018 (Nitsche 2018).

In just two decades, Ma has transformed his company into one of China's largest business ecosystems, whose combined online and offline businesses encompassed an estimated value of over $500 billion in market capitalization by the end of 2016 (Greeven and Wei 2018, p. 3). Alibaba's success can be attributed to numerous factors. Alibaba realized early on that the expansion of internet infrastructure, would significantly impact e-commerce commerce, and invested. Another success factor for Alibaba is the closure of the Chinese market due to internet censorship (Tai 2015).

Due to the insufficient logistics infrastructure in China, Alibaba has decided to organize a merger of China's five major logistics companies (Sitongyida, Shunfeng, Zhongtong, Yuantong, and Shentong) and establish a company called Cainiao Logistics. It is Alibaba's goal with Cainiao to build its own logistics network in China to be able to transport goods from e-commerce platforms. Compared to Alibaba's competitor, Jingdong, Alibaba does not rely on its warehousing and delivery of goods from merchants and middlemen but focuses on joining forces with well-known delivery companies (Kim 2018, pp. 230–231).

The corporate culture at Alibaba includes a permanent "trial and error process". There is no lengthy strategy planning, but rather trial and error (Zeng 2019, p. 204). This principle, which is also applied in Chinese politics, leads to success in the Chinese market environment. Furthermore, entrepreneurship, venture capital, internet boom, growing middle class, and dynamic regulatory frameworks are among the success factors. Chinese market players also prefer the speed and adaptability of businesses and diversity in customer segments. In addition, Alibaba has been able to cope with market and legal uncertainty, as well as absorb capital that is difficult to invest profitably in China (Greeven and Wei 2018, p. 9).

Shopping is one of the biggest leisure activities in China and Alibaba can profit from this: In 2017, on November 11, the so-called "Singles' Day", goods worth USD 1.5 billion were ordered online in the first three minutes of the day alone. On this day, there is heavy price cutting in China; the day was created by Alibaba. The total trade volume on this day in 2017 was over 25 billion USD. On Singles' Day 2017, the proportion of sales generated through Alipay was 90 percent of total external sales. In previous years, Alibaba held a gala celebration called "Tmall's Double-11 Night Carnival" in the hours before Singles' Day began. For this, stars from abroad are invited every year; the aim of the event is to keep viewers and customers awake until midnight, and then start the day right with a purchase. Spectators are also given discount vouchers which are randomly drawn to their smartphones. This gala is broadcast by Youku, China's YouTube, and Tudou, another Chinese streaming service acquired by Alibaba in 2016. By presenting products during the gala, mostly by famous personalities and always provided with a direct product link, Alibaba links shopping events with entertainment content and puts viewers and customers in a continuous buying mood. The percentage of online transactions is particularly high in Tibet, as the infrastructure of the autonomous region makes retail shopping difficult and residents resort to online mail-order products (Ma 2018, pp. 11–15).

10.3 E-commerce in the Countryside

E-commerce is a powerful means to connect the unconnected to global trade. (González 2016)

The development of rural e-commerce began in the mid-2000s and became known as "Taobao villages". The emergence and proliferation of these villages fueled expectations for e-commerce: e-commerce was expected to drive China's stagnant agriculture and address rural poverty. The government launched a program in 2006 to inform and educate villages about e-commerce and digitization: the Village Informationisation Programme VIP. However, the program was only partially successful as there was only one program for all provinces, but it was not tailored to the needs of each province and therefore did not meet the different conditions. However, the number of rural people involved in e-commerce increased from 15.4 percent in 2006 to 47.3 percent of the total rural population in 2015. In 2014, Taobao announced the project to invest several billion RMB to build service centers in rural areas, subsequently, the number of those in rural areas engaged in e-commerce increased. The number of total rural internet users fluctuated slightly and remained at around 28 percent (Li 2017, pp. 57–58).

In 2016, the Chinese Ministry of Commerce (MofCom) reported that, for the first time, e-commerce in rural areas had grown faster than in cities. Rural e-commerce is not only gaining importance in China; in countries such as India and Indonesia, sales are predominantly generated in rural areas. In India, for example, circa 60 percent of Jabong.com's sales were generated in smaller villages, and in Indonesia, the website BliBli generated over a third of its sales in rural areas (Kshetri 2018, p. 91). In China, in 2015, the government demanded that telecom companies lower the price of broadband internet while increasing speeds. In 2016, $300 million was invested by the Ministry of Commerce in 200 rural areas to be used primarily for department stores, training, and other e-commerce related activities (Kshetri 2018, p. 3). The government has limited ability to help with education, which is why they often work with e-commerce experts and companies, predominantly Alibaba, as these experts have been in e-commerce for a long time and have the expertise. By operating villages in e-commerce, the money generated can be used to invest in new infrastructure. At the same time, the construction of schools or hospitals improves local education and medical care (Cui et al. 2016, p. 5). In 2017, Alibaba stated that Taobao had established branches in 29 provinces, 600 counties, and 30,000 villages (Li 2017, p. 57).

10.3.1 State Objectives

Reasons for the state to spread e-commerce in the country can be divided into three categories: First, the state wants to use the money generated from e-commerce to invest in infrastructure in the countryside and further the BNSC project. At the same time, the state will be relieved as it will have to invest less money. Moreover, better digital infrastructure will enable the collection of data, which is essential for e-governance. Furthermore, agriculture will be further expanded and developed, and rural poverty will be tackled (Wang and Shen 2018, pp. 70–72).

On 31 October 2015, the General Office of the State Council published "Guiding Thoughts on Advancing the Development of Rural E-commerce" (State Council of the PRC 2015). In the beginning, it is emphasized that e-commerce is seen as an important means to become less dependent on agriculture and to fight poverty. The role of e-commerce is supposed to be to promote mass entrepreneurship and innovation and to combine internet shops with "physical shops". Other goals mentioned are the promotion of rural consumption, the increase of domestic demand, and the promotion of agricultural renewal.

The CCP wants to merge logistics, finance, and marketing to generate a system that is accessible to inexperienced e-commerce operators and makes it easier to get started. Furthermore, the party itself wants to provide a shopping platform. A market for agricultural and local products is to be created and rural tourism is to be expanded. It is emphasized that products can be delivered to the cities so that sales in the villages do not have priority. In addition to the expansion of the internet, infrastructure, such as railway stations and roads, is to be further expanded and improved. In terms of policies and guidelines, the party is planning a set of guidelines to control e-commerce and set rules for trade. Classification of agricultural products and inclusion of e-commerce in the catalogue of poverty reduction and development aid are planned (State Council of the PRC 2015).

University graduates who go to the countryside to start an e-commerce business are to be supported. This is to be done primarily by granting loans, which are to be approved more easily for younger entrepreneurs. To do so, the procedure for granting loans is to be simplified. The CCP wants to do more propaganda and, for example, send successful e-commerce entrepreneurs back to villages so that they can present their success there. Furthermore, successful e-commerce platforms, such as Alibaba, will be encouraged to set up e-commerce centers in the countryside to train new entrepreneurs and teach farmers how to use smartphones. Quality and safety are to be increased, and illegal production is to be tracked and punished. This is to be implemented by governments at the provincial level (State Council of the PRC 2015). Furthermore, Party members should be motivated to open shops, as the status as a CCP member would promise a good quality of the product. This type of e-commerce is referred to as Red Taobao (Li 2017, p. 60).

In the Chinese government's guiding principles published in 2015, a section is devoted to recruiting experts. By these experts, the government understands above all students after the gaokao, university graduates, but also young people returning home or simple workers. They are supposed to offer free training to entrepreneurs who run their shops in the villages. The main aim is to increase financial support for these ventures: Thus, according to the government, a simpler procedure for granting loans to young people and entrepreneurs in rural areas will be enacted so that there are more attractive acquisition opportunities in the countryside for this target group (State Council of the PRC 2015).

During the BNSC campaign, villagers had to bear parts of the costs for projects to be implemented in the villages (Ahlers et al. 2016, p. 65). Of course, it is difficult for the rural poor to raise such sums, as farmers here are mostly still self-supporters. E-commerce offers itself as a solution to this: The government has little risk in investing, as costs are borne by the residents themselves. In return, in addition to the profits from e-commerce, the residents simultaneously have the opportunity to invest in the development of their community and gain certain independence from the state. In bearing these costs, they act as creditors; however, in the case of successful implementation, the money is reimbursed by the state (Ahlers et al. 2016, p. 63). Provincial governments mandate municipalities to monitor and evaluate village activities. The municipalities then pass on which villages are most likely to be eligible for funding. This takes into account the economic situation, the villages' implementation of tasks, i.e. how effectively they have implemented policies in the past, and how much control the municipality has and can exercise over the village (Ahlers et al. 2016, p. 62). This creates competition between municipalities and villages; with a good performance in e-commerce, more funding support can be obtained from the state at the same time, which can bring additional infrastructure and benefits to the business.

The state requires the establishment of e-commerce companies to receive benefits (State Council of the PRC 2015). According to the State Council, these are composed of farmers, members of Alibaba and Taobao, respectively, and state administration. The Chinese government states that local governments themselves should write implementation plans regarding e-commerce plans (Generalbüro des Staatsrates 2015). Grassroots organizations or "grass root level organizations" (jīcéng zǔzhī) are supposed to take a leading role in implementation. They have the task of communicating with other villages, suppliers, or business partners, thus representing the interests of the village, and also act at the local political level as representatives of the village's interests (Leong et al. 2016, p. 477).

With the help of Taobao's investment and with the revenue generated by e-commerce, local governments have gained more resources as they are no longer dependent on funding support from the central government. At the same time, however, local governments have become more dependent on e-commerce companies for economic performance and approval. Consequently, local governments are increasingly dependent on the companies in their area and need their support. Therefore, these societies can exert influence on local governments and their decisions (Unger and Chan 2015, p. 10). As private companies or individual entrepreneurs increasingly have resources at the local level and participate directly in infrastructure projects, they gain influence as they are indispensable for the development of the village. This in turn can lead to influence at the political level (Ahlers et al. 2016, p. 57). Cooperation and interaction between private companies and local governments have thus become crucial for the successful implementation of policies in rural areas.

The goal of the state is to solve rural problems by using e-commerce to perform governmental tasks. The first task is to increase the efficiency of agriculture. This will be done by conducting a market analysis based on e-commerce and electronic commerce. Current agricultural production is based on the previous year's data and therefore cannot respond to requests in a timely manner; the result is price fluctuations and uncertainty for farmers (Wang and Shen 2018, p. 69). With the help of the spread of e-commerce for agricultural products, this danger can be contained, as data is constantly collected and can be evaluated and analyzed in real time. These are in turn transmitted to the farmers: Demand can thus be adjusted and production risk reduced. In addition, there is a general reduction in costs for farmers and consumers, as new distribution channels and opportunities are created. In current agriculture, products go through many intermediate stages before reaching the consumer, which generates higher prices. With the help of e-commerce, agriculture should therefore be changed from 'business-to-business' (B2B) to 'business-to-consumer' (B2C). Thus, consumers can have direct contact with farmers; moreover, costs for middlemen can be saved and the pricing of agricultural products becomes more transparent (Wang and Shen 2018, pp. 69–70).

10.3.2 Taobao Villages

Alibaba awards the title "Taobao Village" to villages that are particularly successful in e-commerce. To get into this category, some requirements must be met. These are as follows:

1. Annual e-commerce transactions exceed RMB 10 million.
2. At least 10 percent of households are engaged in e-commerce or more than 100 shops are operated by villagers.
3. Villagers use Taobao as their primary e-commerce platform.

Over 90 percent of these villages are located in coastal regions and the developed part of China, of which over 70 percent are in southern China. Often these villages form near already developed and economically important cities. Logistical proximity and good connectivity benefit these villages. The requirements to join the ranks of the Taobao village category show that Alibaba has an economic interest in expanding e-commerce in the countryside. The sheer volume and previously unknown products of the rural population represent a profit opportunity economically and are therefore attractive to e-commerce companies, such as Alibaba (Li 2017, pp. 57–58).

In 2013, there were 20 Taobao villages across China, and by 2014 there were 211. This was due to a project announced by Taobao in 2014 to establish service centers at the provincial and village levels, thereby connecting Taobao and the people living there. In 2016, there were 1331 Taobao villages across China (Li 2017, p. 59). These service centers also provide medical service, education, and training on online retailing and e-commerce. Alibaba works closely with the government on these projects (Li 2017, p. 58). Furthermore, Alibaba has invested in research and development in recent years, establishing AliResearch and Alibaba Damo Academy. The information collected in the service and training centers can be used for research purposes. The pilot projects and model villages are strategically interesting, as they had not been tapped at all before (Ahlers 2014, p. 157).

10.3.3 Examples of E-commerce Villages

Since e-commerce in rural areas is new to all stakeholders, there is a need to launch trial villages and pilot projects to minimize risks. Various models in diverse provinces and villages provide information on how the different village cultures react to the government's measures (Ahlers 2014, p. 156). On the one hand, these model villages serve as models for other local governments, and on the other hand, they are propagated as showcase projects (Ahlers 2014, p. 157). This promotes competition between the individual villages. In some cases, e-commerce villages emerged without government intervention. In the following, three examples will be analyzed that emerged due to individual entrepreneurs or because of the special location and opportunities or special products. The state often only took on a supporting role after the entrepreneurs had already taken the first step.

10.3.3.1 Qingyanliu/Yiwu

Qingyanliu is the first official e-commerce village in China. In Yiwu, about seven kilometers away from Qingyanliu, more than 1.7 million products are sold by about 70,000 shops. These products include art, metal fittings, everyday goods, electronic products, and toys. 75 percent of all products sold online come directly or indirectly from Yiwu (Cui et al. 2016, pp. 7, 9).

This shows the strategic importance of the city for e-commerce and the areas around the city, as there are often favorable opportunities to set up as e-commerce entrepreneurs. The emergence of e-commerce villages around Yiwu, thus including Qingyanliu, is interpreted as industry-driven development. Furthermore, the region around Yiwu was ranked first among areas for e-commerce development and potential in a 2016 ranking by the Ali Research Institute. Yiwu is the world's largest distribution and sourcing center for small goods. Many entrepreneurs have recognized and successfully capitalized on the fact that this offers opportunities not only for the city itself but for the villages and communities surrounding the city. Thus, an e-commerce ecosystem has formed with Yiwu at its center, and it is constantly growing. This is because villages are often at the beginning of the value chain, and products can be produced more cheaply by villages that jointly perceive services such as product design, etc. Taking advantage of Yiwu's reputation and opportunities in the field of e-commerce, villages can successfully offer their products. In addition to the size and scope of the market in Yiwu, there is an attempt to link physical markets with virtual ones (Wang and Shen 2018, p. 90).

In 2005, entrepreneur Wengao Liu saw the problem that many apartments in Qingyanliu village were empty because the young population had left for the cities due to job opportunities. His plan was to use the buildings as work spaces for e-commerce entrepreneurs, giving villagers a chance to rent out their empty apartments and thereby improve their monthly income. With this move towards e-commerce, the number of villagers increased. While this was just 1486 people in 2005, it had already increased to 8000 by 2010. In 2013, there were approximately 2000 online shops in the village, which had a turnover of about 330 million USD (Cui et al. 2016, pp. 7–8).

Liu teamed up with other entrepreneurs to form an e-commerce community, where services such as product photography, design, and product description were taken from the village's most successful businesses, saving newcomers money while providing expertise. Logistics companies set up shop and transported approximately 80,000 shipments daily. This brought additional cost savings for local entrepreneurs, as they no longer had to hire outside logistics contractors. Other businesses such as photo studios, design firms, and customer service companies can now also be found in the village. Strikingly, many entrepreneurs use the village as a springboard: They start their business in the village, steadily increase their sales, and after a few years reach the point where they have become too big for the village's warehouses. Then they usually have no choice but to move their business to another location, making room for a new business. So the village can be seen as a kind of e-commerce school: An entrepreneur can start running his business there with just a laptop. Free training and all the services he needs to successfully run a shop are available to him. If the entrepreneur is successful, he can expand his business further. If his business has become too big for the village, he has to move his location back to the city (Cui et al. 2016, p. 9).

10.3.3.2 Suichang

Suichang is a county in the Lishui county-level city and the first county to operate an online store at the county level (Cui et al. 2016, p. 7). Suichang started to engage in e-commerce in 2005. Then, in 2013, the first shop for local products was opened at the county level. The county is located in the southwest of Zhejiang province and is best known for its high-quality agricultural products, such as bamboo charcoal, potatoes, chamomile oil, and chrysanthemum rice (Cui et al. 2016, pp. 7–9).

The Suichang model is generally understood to be the linking of e-commerce with agricultural products. The farmers in the region could not sell their products outside the village because of the lack of infrastructure and contact with logistics service providers. Therefore, a system was established in which a service station was set up in each village so that contact could be made directly with the farmers and the products could be collected locally (Wang and Shen 2018, p. 88).

This development was initiated by the entrepreneur, Dongming Pan, who previously lived and worked in Shanghai. He pushed through the idea of going back to his home village to promote development in the village with the help of e-commerce. He co-founded the "Suichang Online Store Association" and established the link between sellers and online stores to distribute products while educating farmers. In 2013, the association had circa 1473 members, including suppliers, online stores, and other service providers, such as logistics entrepreneurs and designers (Cui et al. 2016, p. 10). The development in Suichang is considered to be driven by service providers (Wang and Shen 2018, p. 87). This implies that the spread of e-commerce comes from service providers; in this example, there was support, especially financial, from the local government.

After joining the society, some farmers specialized in one product and then registered their brand. However, because the demand was so great, a farmer often could not meet it alone and consequently joined forces with other farmers. Furthermore, the society established its own company, roughly equivalent to a supermarket, with its d (Maitelong MTL). This company provides services such as photography, web design, and product shipping to the farmers, and the farmers pay a share to MTL after selling their products. This is especially advantageous if the products do not have a brand, so the reputation and respectability of a brand can be used in sales (Cui et al. 2016, p. 10).

Furthermore, this model involves the government, which supports dissemination mainly because of its developmental impact. The government helps finance projects and build infrastructure, advises on planning, and supports training and education for workers as well as businesses (Wang and Shen 2018, p. 89).

Other investments included opening up transport routes and internet connectivity. In addition, the government introduced a quality test for foodstuffs to increase food safety. A total of approximately USD 480,000 was invested in this system. Product standards were set and a system was developed to track the origin and route of shipments. The government itself thus took over the assurance of quality and safety, which was identified as the main concern of Chinese buyers (Cui et al. 2016, p. 10).

10.3.3.3 Jinyun

Jinyun is a county in Zhejiang Province, which had the largest number of poor villages in the whole province – half of the inhabitants lived on an income equivalent to about $400 a year. 92 percent of the population is engaged in agriculture. In 2013, there were approximately 1300 online businesses in the entire county, generating USD 72.4 million in sales (Leong et al. 2016, p. 477).

Beishan is a village in Jinyun that particularly benefits from e-commerce, mainly because villagers sell outdoor equipment such as tents, backpacks, sleeping bags, and barbecue equipment (Leong et al. 2016, p. 477). The village is now seen as one of the most successful e-commerce villages and circa a quarter of the outdoor products sold on Taobao come from Beishan (Cui et al. 2016, p. 11). An entrepreneur, hereafter referred to as Mr. Wang, started to buy and then resell products, in this case, outdoor equipment, cheaply (Cui et al. 2016, p. 7). At this point, these products were still from Yiwu, where Mr. Wang, a former bread baker, had already gained experience in e-commerce. So, compared to the villages already analyzed, these were not products that were typical and therefore well-known for the region. Wang decided in 2006 to start his brand together with his brother instead of continuing to buy outdoor products from outside: BSWolf is the name of the brand (Leong et al. 2016, p. 477). The reason Wang decided to produce his products and no longer just resell them was mainly the problem of the poor quality of the products supplied by Yiwu (Cui et al. 2016, p. 12).

Word of his success spread through the village, leading villagers to want to learn e-commerce and eventually ask the entrepreneur for help. They were particularly fascinated by the fact that he had been able to rise from a simple bread baker to a BMW driver.

Moreover, Wang supported villagers' new e-commerce activities: for example, if they were worried about having money left over to buy and distribute products after investing in computers, he provided his brand's products at no extra cost (Leong et al. 2016, pp. 479–480). Although he himself lost about 20 percent of his sales, he saw the problems of the villagers who were plagued by unemployment or poor wages and wanted to help the village to recover (Cui et al. 2016, p. 12). He thus became the villagers' sales agent. In Jinyun, as in many other provinces, skilled workers left for the larger cities because of better job opportunities. However, this changed with the success of the BSWolf brand: for example, a 26-year-old graphic designer returned to Beishan because, although she did not earn as much money as before, she was ultimately able to live more cheaply because of the cost of living in the village was much lower than in the city (Leong et al. 2016, p. 480).

The local government in Jinyun trains e-commerce entrepreneurs and financially supports the establishment of their brands. Entrepreneur Wang's successful model was copied throughout the county; thus, villagers were not limited to BSWolf products, but began to sell, for example, car accessories throughout Jinyun province, in addition to other outdoor products (Leong et al. 2016, pp. 481–482).

The development in Beishan is due in no small part to entrepreneur Wang, who both opened up e-commerce as a new business opportunity and helped teach villagers, foregoing his profit to enable or facilitate villagers to get started (Cui et al. 2016, p. 14). On the one hand, this motivation can be interpreted as benevolent towards the other villagers, but on the other hand, the entrepreneur needed further expertise from Alibaba, which he could only achieve by adopting the village as a Taobao village (Cui et al. 2016, p. 14). In this way, more professionals could be attracted and the brand could be further expanded. Ultimately, both e-commerce entrepreneurs, who could share services, and ordinary villagers, who gained access to new infrastructure, benefited.

10.4 Conclusion

Overall, e-commerce in the PRC is characterized by a very high rate of transactions, the trend of online purchasing, rapid implementation of O2O (online-to-offline) models, and online sales of agricultural products (Wu and Weng 2018, pp. 255–256). Success factors of Chinese e-commerce enterprises are network coordination, high flexibility, willingness to experiment, and data intelligence. Electronic payment systems are another factor, as credit cards were hardly used in the PRC before. Due to the fiduciary function of companies like Alibaba, fraud has been prevented. The Chinese platform economy is constantly developing new business concepts, so there is a high potential for innovation.

The state can act much more extensively in China than in non-authoritarian states, for example by implementing a social credit system with which citizens and companies can be almost completely controlled in their everyday lives. Often the Chinese state enters into a symbiotic relationship here with companies that are officially private but cannot be described as private under these conditions.

By merging private companies like Alibaba into communities with farmers and entrepreneurs, it is easier for the government to oversee and monitor the actions of e-commerce actors. What is clear is that village governments have become dependent on e-commerce companies for their actions, as they can evaluate how the local government implements policies on the one hand, and are involved in funding projects, especially infrastructure projects, on the other. Local governments are measured as well as evaluated on such projects and can only rise if they successfully implement many projects and policies. This dependency consequently means a say on the part of e-commerce companies and Taobao, as they are funders and supporters of local projects.

References

Ahlers, Anna L. 2014. *Rural policy implementation in contemporary China – New socialist countryside*. London: Routledge.

Ahlers, Anna L., Thomas Heberer, and Gunter Schubert. 2016. Whither local governance in contemporary China? Reconfiguration for more effective policy implementation. *Journal of Chinese Governance* 1: 55–77.

Alibaba Group. 2015. Alibaba Group announces March quarter 2015 and full fiscal year 2015 results. Company also appoints new CEO and new member of the board of Hangzhou. https://www.alibabagroup.com/en/news/press_pdf/p150507.pdf. Accessed 01 July 2019.

Boos, Patrick, and Christina Peters. 2016. Digitales Wachstum in China am Beispiel von Alibaba. In *Digitale Transformation oder digitale Disruption im Handel*, ed. Gerrit Heinemann, H. Mathias Gehrckens, and Uly J. Wolters, 127–147. Wiesbaden: Springer Fachmedien.

Campbell, Charlie. 2019. How China is using big data to create a social credit score. https://time.com/collection/davos-2019/5502592/china-social-credit-score/. Accessed 19 June 2019.

China Internet Network Information Center – CNNIC. 2018. *Anzahl der Nutzer des mobilen Internets in China in den Jahren 2007 bis 2017 (in Millionen)*. Hg. v. Statista 2019. https://de.statista.com/statistik/daten/studie/39488/umfrage/mobiles-internet-nutzer-des-mobilen-internets-in-china/. Accessed 03 September 2019.

Chorzempa, Martin, Paul Triolo, and Samm Sacks. 2018. *China's social credit system: A mark of progress or a threat to privacy?*, 1–11. Peterson Institute for International Economics. https://www.piie.com/publications/policy-briefs/chinas-social-credit-system-mark-progress-or-threat-privacy. Accessed 03 September 2019.

Clark, Duncan. 2016. Alibaba. *The house that Jack Ma built*. New York: HarperCollins Publisher.

Cui, Miao, Shan L. Pan, Sue Newell, and Lili Cui. 2016. Strategy, resource orchestration and e-commerce enabled social innovation in rural China. *Journal of Strategic Information Systems* 26 (1): 3–21.

Erisman, Porter. 2015. *Alibaba's world. How a remarkable Chinese company is changing the face of global business*, International. Aufl. London: Palgrave Macmillan.

Evans, David S., Richard Schmalensee, Michael D. Noel, Howard H. Chang, and Daniel D. Garcia-Swartz. 2011. *Platform economics: Essays on multi-sided businesses*. Competition Policy International. https://papers.ssrn.com/sol3/papers.cfm?abstract_id=1974020. Accessed 03 September 2019.

González, A. 2016. Innovation drives trade and investment promotion at global conference in Morocco. http://www.intracen.org/news/Innovation-drives-trade-and-investment-promotion-at-global-conference-in-Morocco/. Accessed 31 July 2019.

Greeven, Mark, and Wei Wei. 2018. *Business ecosystems in China. Alibaba and competing Baidu, Tencent, Xiaomi and LeEco*. London/New York: Routledge.

Herda, Nils, Kerstin Friedrich, and Stefan Ruf. 2018. Plattformökonomie als game-changer. *Strategie Journal* 03: 2–18.

Herndorn, Astead W. 2019. Elizabeth Warren proposes breaking up tech giants like Amazon and Facebook. *The New York Times*. https://www.nytimes.com/2019/03/08/us/politics/elizabeth-warren-amazon.html. Accessed 04 June 2019.

Kharpal, Arjun. 2019. Huawei would have to give data to China government if asked: Experts. https://www.cnbc.com/2019/03/05/huawei-would-have-to-give-data-to-china-government-if-asked-experts.html. Accessed 19 June 2019.

Kim, Young-Chan. 2018. Alibaba: Jack Ma's unique growth strategy and the future of its global development in the Chinese digital business industry. In *The digitization of business in China*, ed. Young-Chan Kim and Pi-Chi Chen, 219–248. Cham: Springer International Publishing.

Kostka, Genia. 2018. Chinas soziale Bonitätssysteme Sind – Noch – Beliebt. https://www.merics.org/de/blog/chinas-soziale-bonitaetssysteme-sind-noch-beliebt-englisch. Accessed 19 June 2019.

Kshetri, Nir. 2018. Rural e-commerce in developing countries. *IT Professional* 20 (2): 91–95.

Lee, Claire. 2019. Datafication, dataveillance, and the social credit system as China's new normal. *Online Information Review*. https://doi.org/10.1108/OIR-08-2018-0231. Accessed 31 July 2019.

Leong, Carmen Mei Ling, Shan-Ling Pan, Sue Newell, and Lili Cui. 2016. The emerge of self-organizing e-commerce ecosystems in remote villages of China: A tale of digital empowerment for rural development. *MIS Quarterly* 40 (2): 475–484.

Li, Anthony H.F. 2017. E-commerce and Taobao villages. *China Perspectives* 3: 57–62.

Ma, Winston. 2018. *Die digitale Seidenstraße: Chinas neue Wachstumsstory*. Berlin: Nicolai Publishing & Intelligence.

Nitsche, Nicole. 2018. *Alipay, WeChat & UnionPay – Chinas big three. Ein Vergleich der tech-Giganten Chinas*. Hg. v. Payment & banking. https://paymentandbanking.com/alipay-wechat-unionpay-chinas-big-three/. Accessed 31 July 2019.

Pattloch, Thomas. 2018. Update zum cyber security law in China. https://www.plattform-innovation.de/files/Update%20zum%20CSL%20in%20China%20Layout-Pattloch.pdf. Accessed 19 June 2019.

Schumpeter, Joseph. 1934. *The theory of economic development*. Cambridge, UK: Harvard University Press.

Srnicek, Nick. 2016. *Platform capitalism*. Cambridge, UK: Polity Press.

State Council of the PRC. 2015. State Council Guideline to Accelerate Rural E-Commerce Development. http://www.gov.cn/zhengce/content/2015-11/09/content_10279.htm. Accessed 31 July 2019.

Statista. 2019. Number of fixed telephone lines in China from January 2018 to February 2019 (in millions). https://www.statista.com/statistics/278202/china-number-of-fixed-telephone-lines-by-month/. Accessed 05 September 2019.

Sun, Dongqi, Liang Zhou, Li Yu, Haimeng Liu, Xiaoyan Shen, Zedong Wang, and Xixi Wang. 2017. New-type urbanization in China: Predicted trends and investment demand for 2015–2030. *Journal of Geographical Sciences* 27 (8): 943–966.

Tai, Katharin. 2015. Made in China/Zensur ist nicht gleich Zensur. https://www.gq-magazin.de/auto-technik/articles/wie-die-neue-zensur-chinas-funktioniert. Accessed 31 July 2019.

Tse, Edward. 2015. *China's disruptors. How Alibaba, Xiaomi, Tencent and other companies are changing the rules of business*. New York: Penguin Random House.

Unger, Jonathan, and Anita Chan. 2015. State corporatism and business associations in China: A comparison with earlier emerging economies of East Asia. *International Journal of Emerging Markets* 80: 1–16.

Wang, Tianyu, and Xueqing Shen. 2018. *Nóngcūn diàn shāng píngtái jiànshè yǔ yánjiū*. China Machine Press.

Wu, Chung-Shen, and Chih-Yuan Weng. 2018. Conclusion: The Sino digital economy: Development history, current status, and challenges going forward. In *The digitization of business in China*, ed. Young-Chan Kim and Pi-Chi Chen, 249–265. Cham: Springer International Publishing.

Zeng, Ming. 2019. Smart Business. *Alibabas Strategie-Geheimnis*. Frankfurt: Campus.

Fiscal and Financial Policy

11

Barbara Darimont, Paul Gebel, and Alyssia Reißler

Economic growth in recent years has been financed in many cases by ever higher debt. As the central body, the Chinese state implements national fiscal and monetary policy. Revenues are generated through taxes. It is difficult to understand how the tax revenues are distributed among the individual provinces. There are no legal regulations for this, but it is negotiated. Since the opening policy in 1978, various tax reforms have been implemented; the latest reform was the income tax reform in 2018. The PRC's debt is a topical issue where a wide variety of data can be found, depending on the source. The current debt burden is weighing on the stability of the Chinese economy. Total debt in the PRC is growing faster than economic output. A crisis in the Chinese financial sector would have consequences for the entire world (Dieter 2019, pp. 12–13).

Overall, the Chinese financial market is considered unstable, which is because the economy is seen as a vehicle for maintaining the CCP's grip on power and the financial sector thus does not enjoy much attention. High financing support in the financial sector is accepted for the maintenance of power (Hess 2014, pp. 775–778).

Due to the Chinese government's digitalization strategy, the FinTech market is particularly virulent and is recording high growth rates, so many transactions are conducted via this market. However, the FinTech market is susceptible to illegal transactions, so the

B. Darimont (✉)
East Asia Institute of Ludwigshafen University of Business and Society,
Ludwigshafen am Rhein, Germany
e-mail: darimont@oai.de

P. Gebel
München, Germany

A. Reißler
Ludwigshafen, Germany

B. Darimont (ed.), *Economic Policy of the People's Republic of China*,
https://doi.org/10.1007/978-3-658-38467-8_11

Chinese government is striving to control this market more closely (Tobin and Volz 2019, pp. 17, 31–32).

Lending is done through state-owned banks and is very restrictive, so other forms of lending and investment have become established, such as a shadow banking sector and crowdfunding. In 2012, the informal credit market covered about 60 percent of loans to small and medium-sized enterprises (Tobin and Volz 2019, p. 15). Since 2018, this development has been curbed by government restrictions to be prepared for a possible crisis in the financial market (Dieter 2019, p. 17).

11.1 General Fiscal and Financial Policy

China's financial system consists of banks, insurance institutions, the non-banking sector, such as venture capital institutions, microcredit associations, etc., the stock market, and the bond market in Shenzhen and Shanghai (Tobin and Volz 2019, p. 18). China's financial sector is highly controlled, unlike any other major economy. The banks and insurance companies are among the largest in the world, as some of them generate well over USD 100 billion in revenues annually (Kroeber 2016, p. 148).

11.1.1 Tax Policy

The distribution of public funds gives rise to conflicts in all systems of government. In the PRC, the central government has claimed sole competence in the area of taxation since the tax reform in 1994. Financial management is exercised by the provincial governments or the respective regional administration. The 1994 tax reform has resulted in a form of fiscal federalism, as it is a mixed system of separate and composite taxes. The central government receives 75 percent of value-added tax, customs duties collected by customs offices, and corporate income tax from state-owned enterprises under the control of the central government. For local governments, the remaining 25 percent is VAT, income tax, land use tax, and corporate income tax of local businesses (Heilmann 2016, p. 91). The tax reform in 1994 led to distribution between local governments and central government, with local governments receiving less revenue (Gordon and Li 2018, p. 180). For local governments, the tasks to be financed with local revenues have permanently increased. Currently, the tasks to be financed by local governments include education, health care, promotion of agricultural technology, family planning, and salaries of village cadres (Heilmann 2016, pp. 89–93).

Companies – including those with foreign capital – are subject to corporate income tax, value added tax, and business tax. Individuals have to pay income tax on their salaries. At the local level, companies have to pay stamp duty, deeds tax, land valuation tax, vehicle and ship use tax, and property tax, among other taxes.

The financial revenue in 2018 was RMB 15.38 billion (Ministry of Finance of the PRC 2019, p. 12). Of the total government revenue, the central government receives about half. Government expenditure was RMB 13.5 billion in 2018. In terms of expenditure, the central government only accounts for about 30 percent of total government expenditure, while about 70 percent is funded by local governments. The central government creates inter-regional fiscal equalization between rich coastal areas and poor western regions, which is negotiated between provinces and the central government. Disputes between the central government and the provinces often end in favour of the provincial governments, as the central government does not have direct access to provincial governments' tax revenues (Heilmann 2016, p. 91). At the central level, the Ministry of Finance is responsible for tax revenues. The individual provinces have their ministries of finance. In addition to corporate and personal income tax, there are luxury tax, business tax, and taxable services, such as real estate sales at 5 percent of the value of the property. The tax system was reformed in 1991, 1994, 2005, 2007, 2011, and 2018.

11.1.1.1 Corporate Taxation

The corporate income tax was legally revised in 2007. The primary objective of the Corporate Income Tax Act was to put domestic and foreign companies on an equal footing and to promote sustainable economic development. The government deliberately pursued the goal of promoting technological development and privileging environmentally friendly and resource-saving techniques. In previous years, foreign companies were often given better tax treatment than domestic companies. Regional investment support for foreign companies – especially in special development zones – was prohibited, and more relevant tax support was implemented instead. Until the law was passed, domestically owned companies were taxed at 33 percent of revenues, while wholly or partially foreign-owned companies received significant tax incentives at 15 or 24 percent, especially in the "special economic zones." The concessions only applied to certain industries that China just deemed important, such as environmental protection. With the new law, a uniform tax of 25 percent of business income is now payable if there is no special benefit. Taxable income is total income fewer deductible expenses and losses. A deductible is e.g. maintenance costs, and depreciation for inventory (Münzel 2007, note 1).

11.1.1.2 Income Tax

The income tax has been revised several times since 1994, in 2005, 2007, 2011, and 2018. The National People's Congress passed the seventh amendment to the Individual Income Tax Law on 31 August 2018 (National People's Congress of the PRC 2018), which applies to both Chinese citizens and foreigners. Those who have resided in China for more than 5 years are subject to Chinese tax liability. A double taxation agreement has existed between Germany and China since 14 February 1986, which was revised in 2016 (Bundesministerium für Finanzen 2015).

The tax reform in 2011 was a tax break for the lower income groups, this is also true for the reform in 2018, which took effect in 2019. The state must at least symbolically, tax the rich more. The income tax rate is progressive, starting at 3 percent if someone earns one RMB. The highest rate is 45 percent for an annual income of more than RMB 96,000 a year (National People's Congress of the PRC 2018).

11.1.2 Public Debt

The debt of a government is the difference between government revenue and government expenditure. Government expenditure is, for example, expenditure on government employees and government investment, e.g. infrastructure projects, as well as interest expenditure. Government revenues are mainly tax revenues and other revenues, e.g. profit distribution from state-owned enterprises. Since 2007, the debt of all countries has worsened; Germany, for example, has a debt of 60.9 percent of GDP in 2018 (Eurostat 2019). The PRC has an official debt of 50.5 percent of GDP in 2018 (Ministry of Finance of the PRC 2019). The actual debt of the PRC can only be estimated. Various estimates of total government, corporate, and individual debt range from 200 to 300 percent of GDP, with 80–90 percent attributed to the government alone (Hua 2019; Reuters 2019b; Dieter 2019, p. 5).

At the end of 2013, China's National Audit Office published a report on public debt. According to this report, local governments have the highest debt and were indebted to the tune of RMB 17,891 billion in 2013, while the central government owed RMB 12,384 billion (Wang and Färber 2016, p. 11). Debt varied widely among provinces in June 2013. Guizhou province was the worst, with 94 percent of provincial GDP. However, this province is not one of the economically prosperous areas, so the actual figures were not particularly high. The city of Shanghai had a debt of 43 percent of its GDP at the time, which in actual terms was then much higher than Guizhou's debt. In percentage terms, Shandong was the province with the lowest debt, at 11 percent of GDP (Wang and Färber 2016, p. 11). Zhou Xiaochuan, the former president of the central bank, has criticized the local government for the debt but has not presented a proposed solution (Huang 2018, p. 209).

The causes of the debts lie in the financial crises. The Asian crisis in 1998 prompted the Chinese government to launch economic stimulus packages. These debts had not yet been paid off when the global economic crisis of 2008 necessitated further stimulus packages. Half of these stimulus packages had to be borne by the local governments, which had to take out loans and repay the interest. Mainly infrastructure projects such as roads, railways, and airports were implemented. With such large investments, the likelihood of malinvestment is high. Overinvestment is a phenomenon of capitalist economies, so the PRC is catching up with Western economies in this area (Dieter 2019, p. 7).

There was no incentive for the provincial governments to keep the debt low. The assessment of the top provincial officials was based on the GDP of the respective province, which was increased by the construction of infrastructure. This was the most important indicator, while the level of debt played no role. As a result, there is an unhealthy regional competition among regions, where long-term goals are not set because the respective provincial governor is in another province after a few years (Huang 2018, pp. 207–210). Many local governments have pledged land use rights as collateral for their loans. In some cases, land has been encumbered as collateral multiple times due to poor management. Administrative reform for land could result in local governments being unable to repay their debts and would no longer be liquid (Liu and Li 2013). Local governments have not been allowed to take on new debt since 01 January 2015. Provinces are allowed to directly issue bonds themselves for the first time. A large part is used to replace old debt. Especially as no limit has been set on the provincial governments' borrowing via bonds (Wang and Färber 2016, p. 6).

Another factor leading to uncertainties in the debt situation in the PRC is the shadow banks and their debt. However, the shadow banking sector has been state-regulated since 2008, so it is assumed that the Chinese government could intervene if necessary. Due to regulations, for example, the cap on deposit interest rates and reserve requirements for banks, the shadow banking sector market declined in 2016–2017, while social finance, that is, financing via private individuals, is growing (Schillak 2016, p. 61; Tobin and Volz 2019, p. 24).

The real estate sector is seen as another problem of public debt in the PRC. Many apartments are overpriced. Property prices have risen 15-fold in some cases over the past 15 years. However, 70–80 percent of most Chinese citizens' wealth is in real estate. Many loans in the PRC have real estate as collateral; if real estate prices fall, many loans will have no collateral (Dieter 2019, pp. 20–21). Xiang Songzuo, an economist at the People's University in Beijing, commented on the debt situation in Shanghai in January 2019:

> [The Chinese people] played around with leverage, debts, and finance, and eventually created a mirage in a desert that will soon entirely collapse. (Quoted from Kawase 2019)

11.1.3 Institutions and Banks

The banks are state-owned. The central bank is the People's Bank of China (hereafter PBC), which sets monetary policy and lends to commercial banks (Tobin and Volz 2019, p. 17). With WTO accession in 2001, foreign banks were gradually allowed. The commercial banks are called "the Big Four": Agricultural Bank of China (ABC), China Construction Bank (CCB), Bank of China (BOC), and Industrial and Commercial Bank of China (ICBC), as shown in Fig. 11.1.

The PBC was established in 1948 and is under the direction of the Ministry of Finance (Tobin and Volz 2019, p. 17). The Banking Law passed in 1995 confirms its function and powers as a central bank (Bell and Feng 2013, pp. 98–99). The Chinese central bank is responsible for monetary and exchange rate policy in the PRC. It is responsible for maintaining the price level and monetary stability. It also holds foreign exchange reserves and refinances commercial banks. Banknotes are issued by it. The State Administration of Foreign Exchange (SAFE) is subordinate to it as the guardian of foreign exchange. In the early 1980s, the Big Four were established as commercial banks: The Industrial and Commercial Bank of China (ICBC) is the largest bank in China and the second largest in the world. It is responsible for loans and deposits in cities. On October 28, 2005, it was officially converted into a private company with equity of RMB 248 billion. Half of the capital is held by the Ministry of Finance of China and half by the state-owned Central SAFE Investments Limited. Agricultural Bank of China (ABC) is a partially state-owned bank responsible for rural loans and deposits. It went public in 2010. China Construction Bank (CCB) handles project finance and investment planning. The Bank of China (BOC) is responsible for foreign trade and exchange business (Tobin and Volz 2019, p. 20).

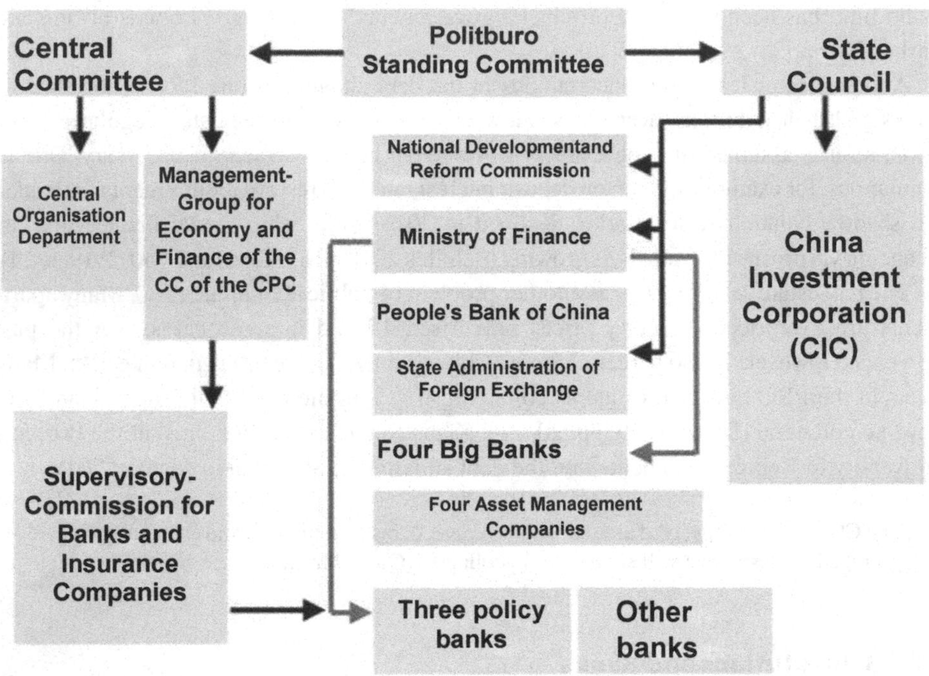

Fig. 11.1 Overview of banks. (Source: Own representation)

The Big Four provided two-thirds of Chinese bank loans until the early 2000s. During this period, they acted more as state financial agents than as normal banks. In the 1990s, a crisis emerged – also relevant for the banks – because state-owned enterprises under the control of local governments began to invest heavily in production, even though there were no markets for the products at all; debt was created between state-owned enterprises and banks that lent to the state-owned enterprises. Moreover, the Asian crisis led to further indebtedness as infrastructure projects were initiated for which banks had to lend. In 1998, the Big Four had debts equal to one-third of their total GDP. This meant that they were effectively insolvent. A key event in 1999 was the collapse of Guangdong International Trust and Investment (GITIC); this led to such a high level of distrust both in China and internationally that politicians decided not to allow any more banks or financial institutions to become insolvent. Prime Minister Zhu Rongji, who was in office at the time, asked who should take over debts of this kind. Many credit institutions were closed at that time (Hess 2014, p. 783).

In 1998, the commercial banks were recapitalized by the Huijin and thus the POB. An "asset management company" (AMC) was established for each of the four banks in the year. The plan was to let these banks exist for 10 years. They were financed by the Ministry of Finance, which issued government bonds for a 10-year term. The AMCs issued 10-year bonds, which were purchased by their respective banks. After 10 years, the bonds issued by the Asset Management Companies were rolled over for another 10 years because the write-offs on the bonds would have led to bankruptcies of the AMCs. The Central Bank or the Treasury would have had to step in to cover the heavy losses. This is a kind of recycling of debt. The banks are in a safe position as the finance ministry will not sell them. The cost of financial stability is borne by the People's Bank of China. Currently, the system can be maintained by citizens due to high savings deposits, but with a shift from an export-oriented nation to more domestic consumption and thus higher private consumption, this may become problematic (Walter and Howie 2011, pp. 123–141). This problem will increase in the context of the trade conflict with the US. In the end, the losers are the citizens who pay government bonds through taxes.

The Big Four were also given RMB 270 billion in funding to meet the Basel conditions of an 8 percent capital adequacy ratio. For this purpose, special government bonds were issued by the Ministry of Finance, which the banks had to buy, and the proceeds were then granted to them as a loan for the equity ratio. In this way, the recapitalization of the banks was generated by bonds issued by the Ministry of Finance. The asset management companies then bought the non-performing loans (Bell and Feng 2013, pp. 270–271). By 2014, the Big Four had reduced their debt by 40 percent, in part because they adopted modern risk management (Kroeber 2016, pp. 128–131). All are among the world's largest companies in terms of revenue (Forbes 2019).

For special policies, three banks were established in 1994 to provide loans according to policy guidelines, not only to state-owned enterprises. These are the China Development Bank, Agricultural Development Bank, and Export-Import Bank of China (Bell and Feng 2013, p. 268). The China Development Bank is well known because it implements the PRC's development policies and is particularly involved in Africa. It is described as the largest development bank in the world. Domestically, it finances large infrastructure projects. The government of the PRC is liable for loans. China grants loans to countries such as Venezuela, Brazil, India, Ghana, and Argentina through the China Development Bank (Hecking 2014).

In 2003, Central Huijin Investment Ltd. was established to lead the shareholder reform and listing of commercial banks. Central Huijin Investment Ltd. is a Chinese investment company of the central government of the PRC. It is a wholly-owned subsidiary of the China Investment Corporation, which has its board of directors and supervisory board. China Investment Cooperation (CIC) is a PRC sovereign wealth fund established in 2007. The principal shareholder rights of Central Huijin are exercised by the State Council. Central Huijin is an organization with which the Chinese government acts as a shareholder for the Big Four (Bell and Feng 2013, p. 251).

In 1998, the Central Finance Working Commission was established. It was dissolved in 2002 and its successor was established in 2003 as the China Banking Regulatory Commission (CBRC). The three administrative commissions China Banking Regulatory Commission, China Security Regulatory Commission, and China Insurance Regulatory Commission have supervisory and command authority. There are overlaps of authority and disputes between the supervisors and regulators, so in 2005 the National Reform and

Development Commission and the Ministry of Finance stopped the reforms (Bell and Feng 2013, p. 251). With one bank and three supervisory commissions, the PRC had a structure like many Western countries. With the state organization reform in March 2018, the China Banking Regulatory Commission and the China Insurance Regulatory Commission were merged into one commission, the Banking and Insurance Regulatory Commission (Banking and Insurance Regulatory Commission 2018). The strategic leadership was taken over by the Steering Group for Economy and Finance, which had already been established in 1980 under the Standing Committee of the CCP Central Committee (Tobin and Volz 2019, p. 19).

11.1.4 Shadow Banks

Lending by state-owned banks is almost exclusively to state-owned enterprises. Therefore, a market for microcredit has formed, also known as the shadow banking sector, where small and medium-sized enterprises can borrow at high-interest rates. Since 2008, microcredit has been legalized by the central bank, but strict conditions have been set. Therefore, the term shadow banking sector is not appropriate, as the sector is tolerated and regulated by the state. In the years from 2010 to 2013, the tighter monetary policy of the central bank led to a decrease in bank deposits and bank loans granted. Due to this situation, many private companies borrowed from shadow banks, which led to a huge growth in this sector. Due to the risk, it was decided by the State Council in 2013 that no further loans from banks could be lent in risky investments. This provision is called a "safe loan" (Schillak 2016, pp. 53–54). Most institutions that operate in the shadow banking sector are state-owned banks because they often use a sub-entity as a vehicle for the shadow banking sector. This means that, in the end, all loans are made by the state-owned banks again after all (Kroeber 2016, pp. 139–140). The Chinese shadow banking sector is closely intertwined with commercial banks, and there is no sharp distinction between shadow banks and regulated banks (Dieter 2019, pp. 16–17).

Overall, the Chinese government is trying to regulate risky financial transactions as quickly as possible. The shadow banking sector is comparatively small compared to other countries and has growth potential. However, it is becoming apparent that the regulations are steering investors into the completely private uncontrolled sector (Schillak 2016, p. 63).

11.1.5 Characteristics of the Financial System

The Chinese financial system relies primarily on formal banks. Capital is not distributed through the market, but through the administration and thus through politics. Individual and corporate savings are largely held by banks, with low-interest rates. Interest rates are kept slightly above the rate of inflation. China uses highly regulated interest rates, as is the practice of other Asian banks, and there are also restrictive controls on banks and other financial institutions, as well as capital controls. In some cases, the interest rate is negative. The government during the tenure of Hu Jintao and Wen Jiabao kept interest rates low to fund ambitious infrastructure programs. Moreover, very low-interest rates help banks to regenerate and they encouraged, for example, the construction boom in the 2000s (Kroeber 2016, pp. 131–133). From 2005 onwards, no major reforms were tackled as a more conservative government took over with the new leadership under Hu and Wen, after Jiang and Zhu. There was political concern that the reform program would cause income inequality to rise exorbitantly. However, support from the leadership government is a prerequisite for successful reforms. Subsequently, the 2008 global economic crisis was accompanied by large investment measures that negatively affected the reforms (Hess 2014, pp. 784–785).

China's capital markets are still comparatively underdeveloped at present; the bond markets consist mainly of government bonds, and state banks are being asked to buy these government bonds. Trading turnover is low. Stock markets are referred to as policy markets because they are highly controlled by the CCP (Hess 2014, pp. 787–790). Insurance has been slow to be introduced. In 2015, a long-awaited insurance system for savings deposits was introduced. The last restrictions on interest rates were lifted in July 2015 (Kroeber 2016, p. 141). In July 2013, regulations on loan interest rates were lifted, and with them the minimum interest rate (Hess 2014, pp. 787–790).

Financial reforms were initiated in 2014 and completed by 2016. This reform program includes the development of a multi-tier stock market, unification of the bond market, promotion of the private market, promotion of the futures market, improvement of the competitiveness of the securities and futures services industry, expansion of capital market opening, prevention and resolution of financial risks, and creation of a good development environment for the capital market. It also aims to liberalize the exchange rate and interest rate policies (State Council of the PRC 2014).

In July 2014, a relaxation of the components used to calculate the loan-to-deposit ratio was implemented to include loans and deposits to non-deposit-taking institutions. In addition, deposit insurance was implemented in the event of a bank failure. Then, in March 2015, the cap on loans was lifted. A deposit insurance scheme was also introduced and the deposit insurance cap was removed. In October 2015, it was announced that the cap on deposit rates was lifted to allow credit institutions to compete (Tobin and Volz 2019, pp. 21–24). The determination of lending rates was reformed again in 2019 (Reuters 2019c). China changed its system for deciding interest rates on bank loans to allow the central bank to effectively control interest rates. Under the new system, 18 lenders have been selected by the People's Bank of China to report their interest rates, from which the central bank generates a reference rate for the entire banking sector each month. This includes foreign banks (Xin and Liu 2019).

The minimum reserve requirements for banks have been gradually reduced by the People's Bank of China. This deposit insurance applies to all deposits managed by Chinese financial institutions. The amount covered per investor and institution is a maximum of

RMB 500,000, while the People's Bank manages the associated deposit insurance fund (Schillak 2016, p. 61).

11.1.5.1 Inflation

Inflation rose by 10 percent at the beginning of 1988 and was 28 percent by the beginning of 1989 (Lu 2014, p. 203). This high inflation – especially on basic foodstuffs – was one of the reasons for the protests in 1989. Since then, the Chinese government has taken strict care to keep inflation under control. In the years from 1993 to 1997, the PRC had inflation, then deflation until 2002, and inflation again in 2003 (Bell and Feng 2013, p. 179). The inflation rate averaged 2.3 percent in 2019 (Ankenbrand 2018). However, these are official figures that even Chinese politicians and economists believe only to a limited extent. Li Keqiang, the current prime minister, for example, consults the price of pork to arrive at a realistic estimate (Bell and Feng 2013, p. 193). Problematically, at the moment, the price of pork is rising exorbitantly due to African fever, which is rampant in China at the moment, and in some cases has grown by over 80 percent compared to the previous year (Wang 2019). In 2018, the price of vegetables increased by 4 percent year-on-year, and rental prices in metropolitan areas such as Shanghai and Beijing are also increasing disproportionately. If the trade conflict with the US continues, this could lead to a further increase in food prices (Ankenbrand 2018).

11.1.5.2 Exchange Rate

Until the late 1970s, the exchange rate of the RMB was a significantly overvalued fixed rate under the planned economy. In 1994, the exchange rate was unified, as from 1979 to 1994 there was a currency called Foreign Exchange Certificates (FEC, also called Waibi – foreign money) for foreigners and the RMB for Chinese citizens. The relative undervaluation led to the expansion of the export sector. In the Asian crisis of 1997, the RMB was adjusted relative to the USD and the exchange rate remained almost unchanged until 2005. Under the exchange rate regime introduced in 2005, the exchange rate was allowed to fluctuate within a certain predetermined range. In the wake of the global economic crisis in 2008, China pegged the RMB to the USD again and did not allow a tolerated range of fluctuation until June 2010. The appreciation of the RMB by almost 40 percent in the period from 2005 to 2014 reduced the exchange rate and price-based competitiveness of China's export sector (Hess 2014, pp. 794–796).

To make the RMB an international reserve currency, the RMB was decoupled from the USD by the central bank in 2015 and has since been determined by a currency basket in which the USD, euro, and yen are the main components. In 2015, the RMB was declared a global reserve currency by the International Monetary Fund and in 2016 became the fifth currency to be included in the International Monetary Fund's global currency basket. While the RMB is a global currency, it cannot currently replace the USD. Its international use is limited at the moment (Kroeber 2016, p. 147). In the context of the trade conflict with the US, the central bank lowered the RMB in August 2019, which is interpreted as currency manipulation by the US side (Giesen 2019). To what extent the currency will be used as a further area of conflict in this dispute remains to be seen.

11.1.6 Stock Market

Until 1992, the central bank was responsible for stock markets and thus for the development of stock exchanges. It was replaced by the China Securities Regulatory Commission. The China Securities Regulatory Commission is under the State Council, which is why conflicts of competence arise with the central bank. After the Asian crisis in 1997, the responsibilities of the Securities Regulatory Commission were expanded and the Commission was upgraded and placed under the direct authority of the State Council. In 1998, the Securities Law was passed, which regulates the supervision of the capital market and the issuance of approvals for the listing of companies, as well as sanctioning powers (China Securities Regulatory Commission 2008).

In 1990, the stock markets in Shanghai and Shenzhen were opened, which are still the only stock markets in the PRC today. The introduction of the stock market is attributed to the former Premier Zhu Rongji, who wanted to use shares to restructure state-owned enterprises (Heilmann 2016, pp. 205–206). Chinese shares are divided into three different classes, A-shares, B-shares, and H-shares. In addition, Hong Kong has its stock market. A-shares are traded on the mainland stock exchanges in Shanghai and Shenzhen and are primarily reserved for Chinese investors. Restrictions on foreign investors were partially lifted in 2006, so trading in A-shares is generally open to foreigners. Unlike A-shares, when investing in B-shares, the trading currency is dollars. In Shanghai, stocks are traded in USD, while in Shenzhen they are traded in Hong Kong dollars. Trading in B-shares is open to foreign private investors. Chinese stocks traded on the Hong Kong Stock Exchange are called H-shares. Trading in these stocks is open to all investors. Both Shanghai, and Shenzhen have free trade zone projects where controlled offshore currency trading is possible. As a result, the stock exchanges in both cities have become world-class international financial centers (Tobin and Volz 2019, p. 26).

Due to a lack of investment assets, as interest rates on banks are very low and do not cover the inflation rate, many Chinese citizens have invested in real estate or shares (Cao 2015b). In 2015, there were strong stock market movements with some losing 8 percent in one day, which had an impact on international stock markets. The Chinese government felt compelled to act, so a regulation was passed in which trading is suspended if the share price falls by more than 7 percent in one day. In addition, major shareholders are only allowed to sell a maximum of 1 percent of the shares they hold in a company every 3 months. Moreover, these plans must be announced 15 days in advance (Freiberger et al. 2016). Furthermore, capital controls and only conditional convertibility of the RMB prevail, meaning that exchanges do not realize their full potential (Tobin and Volz 2019, p. 27).

In summer 2019, the "Star Market" was opened in Shanghai, which is to become a pedant to the Nasdaq. Young companies are to be allowed to access funding via shares so that they can test new technology in their businesses. China's political leadership hopes that this new market will promote the growth of China's players in technology areas (Ren 2019).

11.2 FinTech

In terms of FinTech, China has emerged as the world's leading nation, with the potential to determine the global development of the FinTech market. A 2017 survey by the accounting firm KPMG, which identified the world's top 100 FinTech companies, listed five firms from China in the top 10, with all three top positions held by Chinese FinTechs (Heap and Pollari 2017, p. 3). While in the West FinTech has often been perceived only in connection with start-ups or as a niche market, the FinTech market in China has already revolutionized the entire financial sector (Hunter and Percy 2019).

Alibaba's FinTech offshoot, Ant Financial Group (formerly Alipay), raised approximately USD 14 billion from Chinese and foreign investors in a funding round in the second quarter of 2018, according to its information, bringing the company's valuation to approximately USD 150 billion (Shane and Pham 2018). This valuation exceeds the market capitalization and thus the stock market value of many traditional financial institutions, such as Deutsche Bank (USD 23 billion) (McGeever 2018). Since 2013, FinTech companies have been opening up the Chinese financial services market at a pace never seen before. They have been able to achieve financial inclusion of a large part of the Chinese population, who until then were unable to participate in the financial market. This is putting competitive pressure on traditional financial institutions with their products and services. FinTech has developed much faster and established itself more successfully in China than in other countries.

Due to the topicality of the subject, some challenges arise in the source analysis. There are only a few book sources and scientific papers available that accurately reflect the most current state of the constantly and rapidly developing FinTech market. Due to the lack of empirical research, the primary sources relied upon were internet sources and grey literature in the form of reports from reputable accounting firms, management consultancies, investment houses, etc. Although these companies prepare their reports thoroughly and provide up-to-date figures, the reports are not free of their business interests.

11.2.1 Definition of FinTech

At present, there is still no legally valid definition for the term FinTech, which makes a uniform definition difficult (Deutsche Bundesbank 2016). In the following, the term is explained by describing the characteristics of FinTech companies and the FinTech market.

The term "FinTech" is composed of the initial syllables of the words financial services and technology and is often interpreted as an abbreviation for the term financial technology. It stands as a collective term for products, companies, and markets that combine modern, innovative technologies with financial services (Mackenzie 2015, pp. 50–53). In this way, they exploit potential that traditional financial institutions have so far paid little attention to, and thus digitally and dynamically build up shares in the market for financial products and services. Other characteristics of FinTech companies and applications are that products and services are designed to be application-oriented and internet-based and combine the advantages of efficient functioning, automation as well as high user-friendliness to revolutionize areas from the value chain of the financial industry (Dorfleitner and Hornuf 2016, p. 5). Thus, FinTech can be distinguished from other offerings, such as online banking, by the fact that there is no latest technology that changes a process in the financial industry. The question arises as to whether today's FinTech products still fall into the FinTech category in the medium or long term, provided they do not evolve or are overtaken by other financial technology.

In the following, FinTech companies are understood to be those companies that use new, innovative technologies to offer financial services and products via the Internet, mostly through mobile devices. In this context, the offerings and business models of FinTech companies target the classic value-added segments of traditional financial institutions (payment transactions, financing, asset management, insurance) and change these areas by combining them with the latest technology.

11.2.2 Market Situation

In the following, we first take a look at the development of the FinTech industry in China, then identify the largest and most important segments of the FinTech market in China, and then highlight the significant market players.

11.2.2.1 Development

The development of the Chinese FinTech industry is closely linked to that of the e-commerce sector in China. Before the turn of the millennium, the Chinese financial industry was considered to be backward, underdeveloped, and had to deal with several problems compared to many other countries. Financial infrastructures (e.g., credit markets, financial institutions, payment systems) were inadequate and payment-related fraud was a non-negligible risk (Shim and Shin 2016, p. 171). The inconsistent credit card and the banking system at the time made it difficult to approve transactions between financial houses, as each bank used different credit card networks, which was a barrier to the development of domestic trade and the Chinese economy (Lovelock and Ure 2002, p. 183). In addition, the state of the art in China at that time in terms of information technology (IT) was low. Key technologies, such as hardware components and software used in banking and online trading, were provided by foreign firms (Yu et al. 2003, p. 4). In addition, the China government-controlled led many activities related to the financial system, such as banking and online trading. Therefore, government interference in the business activities of companies was characteristic of the situation in China. Even today, many approvals and government support are required before a company can engage in online trading (Ou et al. 2007, p. 21).

To improve domestic payment systems, the Golden Card Project was launched by the Chinese central government in 1995. This was intended to create a national, unified credit card system to support the spread of credit and debit cards, thereby promoting the development of online commerce (Lovelock and Ure 2002, p. 183). The late 1990s saw the establishment of several, now very significant, Chinese internet companies that established themselves in online commerce. Among them is Alibaba, founded in 1999, whose success is closely linked to the rise of e-commerce in China and the concomitant development of the FinTech market. The company started as a business-to-business (B2B) online platform for small and medium-sized enterprises (SMEs), helping them to catalogue and offer their products online. As a result, Alibaba became the main export hub for Chinese SMEs abroad in China and the link between manufacturers and wholesalers online (Clark 2016, pp. 117–118). The biggest hurdle to Internet commerce remained the payment system, as money transfers, online as well as offline, were then only possible through the post office or the two banks, China Merchants Bank and Industrial and Merchants Bank of China (Shim and Shin 2016, p. 171). Thus, the cash continued to play a major role in Chinese society and payment on delivery initially remained the most common form of payment for orders placed through online platforms (Wang 2012, p. 53). Another problem was that most national banks and credit cards only charged in the Chinese national currency, RMB, and were not suitable for payments in international markets. If payments in international currencies were to be accepted or made to China, special merchant accounts had to be opened, which in turn required further approvals, posing problems for SMEs' import and export activities (Shim and Shin 2016, p. 171). On March 26, 2002, the China UnionPay Network (UnionPay) was established under the auspices of the Chinese central government. UnionPay was the only banking and credit card network in China that processed payments between different banks and held a virtual monopoly (Wright 2015).

One of Alibaba's most important tools in competition with eBay at the beginning of this millennium was the online payment platform Alipay, which the company launched in 2005 (Sun 2018, p. 40). This was expected to revolutionize the way payments were made in China and herald far-reaching changes in the financial sector. As the credit card industry in China was still in its early stages at the turn of the millennium, only about 1 percent of the population had a credit card at the time (Tse 2015, p. 34). Transactions were often conducted with cash or via bank transfers, provided the business partners used the same banking network. The risk of defaulting on payments or services was still a barrier to growth in the Chinese e-commerce sector, and eBay was unable to use PayPal's online payment system in China to date due to strict regulations on foreign banks (Clark 2016, p. 174). Alipay, on the other hand, took advantage of the fact that Chinese customers were used to making transfers through their bank accounts and allowed customers to link their bank accounts to their accounts on Alipay and transfer their money to them. Alipay's partnerships with major Chinese banking networks allowed it to circumvent the problem of transferring money between different banking networks (Tse 2015, p. 35). In addition, Alipay acted as a trustee for online retail payments. If a transaction was made on the Alibaba or Taobao platforms and the payment was processed through Alipay, the payment platform held the money in escrow until the goods reached the customer and the customer confirmed that the goods were correct. Thus, in one fell swoop, Alipay removed the two biggest obstacles to the development of e-commerce in China, namely the risk of fraud and the obstacles to the settlement of payment between merchants and customers. Thus, Alibaba had gained a great advantage over eBay by increasing the liquidity of its online trading platforms.

In 2006, eBay officially withdrew from the China business (Shim and Shin 2016, p. 172). The battle for B2B and B2C e-commerce markets between eBay and Alibaba permanently changed the nature of payment in China. From 2005 to 2010, online retail sales increased by an average of 250 percent, while the amount of money transferred through so-called Third Party Online Payment (TPOP) providers such as Alipay grew sixty-fold over the same period (Shim and Shin 2016, p. 172). Chinese society transformed from a country where cash traditionally dominated to a modern digital economy where people, especially in modern Tier-1 cities such as Beijing, Shanghai and Guangzhou, primarily pay digitally through TPOP applications using their smartphones (Mittal and Lloyd 2016, p. 17). The success of Alibaba and its TPOP application Alipay created a working payment system across China and inspired more internet companies to venture into the fledgling FinTech market. Since 2013, which is considered the starting year of the FinTech boom in China, the market volumes of the most important segments have doubled or tripled every year (Sheng et al. 2017, p. 5).

11.2.2.2 Market Segments

The FinTech industry in China can be divided into four overarching market segments, analogous to the traditional value-added areas of universal banks and financial institutions: Third-Party Online Payment (TPOP), Wealth Management, Financing Services, and Other FinTech Services. The following section provides a more detailed insight into the most important segments of the Chinese FinTech market and how they have developed since 2013 (Ngai et al. 2016, p. 3).

Third-Party Online Payment (TPOP)

The term "third party payment" describes the process of settling payments via a third-party provider. The three parties, buyer, seller, and a bank as intermediary, are participants in the original process, which has been expanded by the "online" component due to the Internet factor. Since banks could not come up with a functioning TPOP system that would inspire confidence and security among Chinese users in time, private companies such as Alibaba and Tencent entered the market with their TPOP platforms as intermediaries to act as an interface between buyers, sellers, and banks. In effect, four parties are involved in the common processes of TPOP payment processing in China.

Transaction volume in the TPOP sector increased from RMB 7.3 trillion in 2013 to RMB 54.5 trillion in 2016 (Sheng et al. 2017, p. 5). According to 2016 figures, 75 percent of these transactions were conducted via mobile devices such as smartphones and tablets, and only a quarter via desktop devices. Moreover, there were already over 3.4 billion accounts with TPOP providers in China in 2016, with Alipay (51 percent) and Tencent's Tenpay (33 percent) accounting for the majority and the remaining 16 percent split between the rest (Sun et al. 2017, pp. 23, 47). Originally, TPOP was closely linked and grew with the growth and spread of online shopping and e-commerce in China, so TPOP quickly made its way into users' offline lives. While debit and credit cards struggled to make inroads in retail, restaurants, or physical stores in China, TPOP quickly gained traction there. Electricity, water, and gas bills as well as public transportation can be paid easily as well as in a few steps via smartphone and even the roadside vegetable vendor has a scannable QR code through which micropayments can be made with a TPOP app (Abkowitz 2018). Sending money among users of the same TPOP platform works quickly, and easily, and is independent of which bank the linked account is located. Because of these advantages, TPOP began to increasingly replace cash in China in recent years. Forty percent of retail payments were already made through providers like Alipay and WeChat Pay in 2016. A full 80 percent of all retail transactions under RMB 5000 were made through TPOP providers (Sun et al. 2017, p. 24).

The success of TPOP in China paved the way for other FinTech-based offerings. Companies that were not active in the TPOP market were still able to benefit from the higher liquidity in the market and the new form of payment in China. FinTech companies that offered wealth management, financing, or insurance services, for example, were able to access capital and the trust of customers more easily by working with TPOP platforms, as they were already easily using financial services via the internet or mobile devices (Mittal and Lloyd 2016, p. 20).

Asset Management

The FinTech market segment asset management describes companies that offer advice, investment, or management of assets and combine these with modern, innovative technologies on the one hand and whose products are application-oriented and internet-based on the other (Mackenzie 2015, p. 15). It is the second largest FinTech market segment in China with a market volume of RMB 10.825 trillion in 2016 (Sheng et al. 2017, p. 5). Established TPOP providers offer wealth management services such as money market funds, trust funds, index funds, or the possibility to invest directly in shares via their internet platforms and on mobile devices. The largest provider in this regard is the Alibaba subsidiary Ant Financial, which has also included Alipay since 2013 (Ding et al. 2018, p. 28). Ant Financial's own wealth management platform, Yu'E Bao, could already boast over 1.43 trillion RMB in assets under management and 300 million users in July 2017, four years after its founding. The next largest FinTech wealth management platform, Licaitong, is owned by Tencent, manages RMB 140 billion in assets, and has 100 million users in China. By connecting to Tenpay or QQ Wallet, Tencent's TPOP platforms, users can access Licaitong's offerings (Sun et al. 2017, pp. 52, 59). Unlike traditional wealth management providers, which require clients to have a minimum investment of RMB 50,000 on average, FinTech providers allow users to invest in products as low as one RMB (Mittal and Lloyd 2016, p. 32). Moreover, no additional authentication is required, as this is done automatically by connecting to the TPOP platform where users are registered. In addition, Yu'E Bao and Licaitong offer T+0 liquidity provided that the money is transferred between the in-house TPOP platform and the wealth management platform. This allows clients to invest or withdraw their assets in products within the same day (Ngai et al. 2016, p. 3). Yu'E Bao boasted over 6 percent returns in the first few years after its founding in 2013 and quickly attracted many customers with its low barriers to entry for users. However, the company has been critically reviewed by Western rating agencies (Lee 2017).

Financing Services

Financing services are services provided by FinTech companies that enable financing for companies or private individuals. This segment can be further divided into the subsegments of crowdfunding, peer-to-peer lending, consumer loans, and business loans (Dorfleitner and Hornuf 2016, p. 5).

Crowdfunding is a form of financing in which the financial means to achieve a goal are raised through the support of a large number of people. A managing FinTech company, the crowdfunding portal, acts as an intermediary for the collected capital instead of a classic bank. The supporters expect a return for the invested capital, which is not always a mere return (Dorfleitner and Hornuf 2016, p. 6).

Peer-to-peer lending, or often referred to as peer-to-peer loans in German, are loans that are granted directly from private individuals to private individuals as personal loans. Lenders and borrowers come together via a kind of marketplace, which in this case is the respective FinTech platform, and can thereby choose for themselves to whom or from whom they want to lend money. Lufax, CreditEase, Renrendai, and Yirendai are the largest and most competitive players in the Chinese market. Most peer-to-peer platforms in China specifically cater to a clientele that falls outside the customer base of traditional banks, mostly lending small amounts of money. As a result, these providers were not initially seen as competition for the traditional lending business of the large Chinese banks (Patwardhan 2018b, pp. 391, 401–402). However, the mass of customers in China here, as with Yu'E Bao and the wealth management business, ensures that considerable amounts of capital are transferred. Compared to the outstanding loan balance of RMB 31 billion in 2014, the value increased to RMB 856 billion in 2017 (Sheng et al. 2017, p. 6).

Consumer loans are loans granted by banks to enable a consumer to purchase goods or services. Entrepreneurial loans refer to loans made by banks to finance businesses (Petermann 1963, p. 22). In China, FinTech companies are increasingly playing an important role in these two lending segments by taking over the role of the traditional bank in the lending process and offering loans via innovative software solutions for desktop computers or mobile devices. The most important providers here are Tencent's WeBank and MYBank, which is run by Alibaba's FinTech division Ant Financial. With WeBank, Tencent launched China's first private bank to conduct business exclusively online in January 2015 (Mittal and Lloyd 2016, p. 18; Ding et al. 2018, p. 32). Through WeBank's Weilidai loan service, users can borrow up to 200,000 RMB without depositing collateral or guarantees, without going through traditional credit checks, and without waiting periods. This is made possible because Tencent's WeChat and its TPOP feature TenPay evaluate user data in advance (e.g. restaurant, e-commerce, taxi spending). Based on this data, creditworthiness checks are carried out and decisions are made as to whether loans can be granted and at what conditions (Ding et al. 2018, p. 33).

MYBank, which was established by Ant Financial in 2015, also acts as a private bank that mainly provides loans to consumers as well as SMEs. Here, the company uses the user data of Ant Financial's system to process it fully automatically and in a few minutes through Big Data technologies to determine creditworthiness (Ding et al. 2018, p. 32). Here, the advantage over traditional banks and loans is that customers can apply for loans as low as two RMB, while traditional bank loans start at 2000 RMB (Zhang and Woo 2018). Since its establishment, MYBank and its predecessor company Ali Small Credit have already granted over four million loans with a total volume of over 700 billion RMB (Ding et al. 2018, p. 32).

Other FinTech Services

Other FinTech services describe FinTech companies that cannot be assigned to the other three FinTech segments TPOP, asset management, and financing services, and do not fall under the traditional banking functionalities. Worth mentioning for the market in China is the insurance sub-segment, where offering FinTech companies are often referred to as InsurTechs (Dorfleitner and Hornuf 2016, p. 10). It is worth highlighting that in 2016, 35 percent of people in China who owned insurance managed or purchased it through a FinTech provider (Mittal and Lloyd 2016, p. 10). The main InsurTechs in China is the top insurer Ping An's InsurTech of the same name, and ZhongAn, which was formed in February 2013 from a joint venture between Alibaba and Tencent in a partnership with Ping An and operates exclusively online (Dula and Lee 2018, p. 12). The majority of insurance sold online by these companies is sold directly through and in conjunction with the offerings of e-commerce and wealth management platforms (Mittal and Lloyd 2016, p. 11). On November 11, which is considered the biggest online shopping day in China, companies such as ZhongAn 2014, were able to turn over RMB 100 million in just one day by the InsurTechs offering 50-cent insurance for shipping online orders (Dula and Lee 2018, p. 12). In 2016, the total revenue of InsurTechs in China was already RMB 235 billion (Sheng et al. 2017, p. 5).

11.2.2.3 Participants in the FinTech Market

The continued growth of the FinTech market in China since 2013 and the seemingly myriad opportunities it offers have led to different market participants from different industries attempting to enter this market. In the literature, these entrants have been classified into three categories: Internet companies revolutionizing the financial sector, financial institutions undergoing digital transformation, and companies not originally from the financial sector but nevertheless seeking to participate in the success of the fledgling FinTech market (Ngai et al. 2016, pp. 9–12).

BAT

Internet companies represent the largest and most media-savvy group of participants in the FinTech market. The Chinese internet sector is dominated by a few large companies, led by Baidu, Alibaba and Tencent, commonly referred to as BAT. In recent years, these companies have entered the financial sector with innovative software solutions and computer systems, creating their own interconnected online financial ecosystem that builds on each other, shaping the FinTech market in China. These companies were able to leverage their large customer base and their core business data, combined with low acquisition costs, to distribute new financial products via the Internet (Ngai et al. 2016, p. 9). The difference between these companies lies in the focus of their actual core business through which they build their financial empire and the target customer group to which they distribute their financial products.

Alibaba, as a globally successful e-commerce company, uses the leverage of its online trading platforms to pave the way for its financial division Ant Financial. The company expanded its business first into a payment system (Alipay) and then from there later into wealth management (Yu'E Bao) and financing services (MYBank) (Ding et al. 2018, pp. 28–29).

Tencent, in turn, used the social networking component of its successful WeChat and QQ Messenger and the myriad of integrated applications covering every aspect of daily life in China to establish its online financial products. By introducing a "red packet" feature, Tencent enabled its users to send digital money envelopes to each other via WeChat in 2014 (Sun et al. 2017, p. 43). In Chinese society, it is customary to present monetary gifts in red envelopes (Hong Bao) at New Year, weddings, or birthdays. The feature became a huge success and helped Tencent to get a large proportion of WeChat users to sign up for the built-in payment feature. Just 2 days after the feature was introduced, it was already used by five million users to send over 20 million envelopes (Zhou et al. 2018, p. 50). By Chinese New Year 2015, the number of envelopes sent was one billion and increased to 8.9 and 14.2 billion in each of the following 2 years (Sun et al. 2017, p. 43). Based on the social network, Tencent first expanded its payment system (Tenpay) and also its wealth management (Licaitong) and financing services (WeBank) (Zhou et al. 2018, p. 50).

China's leading internet search engine, the third major technology giant in the Chinese market Baidu, unlike Alibaba and Tencent, only offers a smaller, own product portfolio of financial products, but invests a lot of capital in other FinTech companies and focuses on entering into partnerships with traditional financial institutions (Patwardhan 2018a, p. 75). The most prominent partnership among them is the joint venture called Baixin Bank, which was jointly entered into with Citic Bank. The strengths of both partners, Citic Group's expertise and assets, and Baidu's hundreds of millions of users are to be combined to be successfully positioned in the market (Zhou et al. 2018, p. 51).

Apart from the three digital giants BAT, there are other small internet companies that are looking for their niches in China's FinTech market. Primarily, these are trying to expand in the area of peer-to-peer lending and asset management (Ngai et al. 2016, p. 10).

Traditional Financial Institutions

Traditional financial institutions, such as banks and insurance companies, have recognized the potential of the Chinese FinTech market and the need to digitize their business models and are also trying to participate in its success. However, compared to internet companies, they are subject to strict regulations and hindered by their conservative mindset as well as low innovation capacity (Dula and Lee 2018, p. 10). Traditional financial institutions try to compensate for these disadvantages by forming strategic partnerships with internet companies, long-standing financial market expertise, and a comprehensive product portfolio (Ngai et al. 2016, p. 12). One example is the partnership of China's largest insurance group PingAn Group with Alibaba and Tencent to form FinTech ZhongAn (Dula and Lee 2018, p. 12). Under a new internet finance strategy called "e-ICBC", the Industrial and Commercial Bank of China (ICBC) offers its FinTech products (Ngai et al. 2016, p. 13). With physical branch offices, these financial institutions have the advantage of being able to provide more personalized services for complicated and personalized financial needs.

Companies Outside the Financial Sector

The third major category of participants in the FinTech market are companies that are based in traditional industries such as retail (e.g. Suning Commerce Group) or construction (e.g. Wanda Group) and originally had few points of contact with the financial services or internet industries (Mittal and Lloyd 2016, p. 30). With a 3-in-1 business model, these companies are entering the Chinese FinTech market. Here, offline resources are supplemented with online platforms and financial products and services to take advantage of the companies' particular strengths in the FinTech market. Advantages include low customer acquisition costs for their FinTech offerings through massive offline customer traffic, strong industry expertise, and data collection along the entire value chain. Unlike internet companies, traditional industrial companies have a large customer base of offline retailers and customers, which enables them to quickly acquire customers for new products and services at a low cost. Wanda Group, for example, had offline customer traffic of 4.6 billion visits in 2015 across all of its businesses, which were spread across the company's shopping, dining, cinema, and other entertainment offerings (Ngai et al. 2016, p. 15).

While internet companies and traditional financial institutions struggle to attract customers, companies such as Wanda Group have already infiltrated many areas of customers' daily lives and can digitally connect them with their financial products and services. The industry expertise of these companies is advantageous in this regard. When lending to SMEs, industrial companies can better cater to the needs of their industrial and retail borrowers because they can understand and evaluate their business better than internet companies and traditional financial institutions. For example, retail giants such as Suning Commerce Group or GOME Electrical Appliances are more familiar with the financing requirements of the components of a company's value chain, which enables them to offer specially adapted financial products for retail customers (Mittal and Lloyd 2016, p. 30).

11.2.3 Success Factors of Chinese FinTech Companies

In recent years, the FinTech market in China has developed rapidly, overtaking its Western competitors in terms of the volume of capital transferred, assets under management or loans granted (Sheng et al. 2017, p. 5). In 2016, the PR already outpaced the rest of the world in terms of volumes of venture capital invested in FinTech companies. While USD 0.1 billion was invested in Chinese FinTech companies in 2013, the value increased to USD 6.4 billion by 2016 compared to USD 4.6 billion in the US and USD 2.6 billion in the rest of the world (Fortnum et al. 2017, pp. 48, 86).

11.2.3.1 Market Regulation in China

At the beginning of the FinTech boom in China in 2013, the Chinese government supported the market development and initially refrained from strict regulations and distracting interventions. The Banking Regulatory Commission promoted digital financial solutions (Zhou et al. 2018, p. 52). Similarly, the People's Bank of China, as the central bank of China, expressed its support for technology companies that developed innovative technology solutions for financial services. These two institutions are the main regulators of financial institutions in China. Premier Li Keqiang also made several calls on behalf of the government in 2014 and 2015 to support young FinTech companies and promoted entrepreneurship and innovation in the financial sector at the World Economic Forum in Davos (Patwardhan 2018b, p. 401). Loose regulations until 2015 ensured that internet companies offering financial solutions were given an advantage, allowing them to create their FinTech market or capture share in existing markets. In the TPOP space, the government gave market participants a free hand to test and commercialize their products in the market to generate a critical mass of users before regulations were enacted. For example, after the introduction of online money transfers by Alipay, it took a full 11 years for the government to set a maximum volume for a transaction, and 5 years after the introduction of transactions via QR codes for the government to publish an official standard with specifications on this (Woetzel et al. 2017, p. 14).

Traditional financial institutions, which have historically been subject to stricter requirements, tighter controls, and regulation by authorities, therefore initially faced a disadvantage. Traditional banks could mostly only lend to Chinese state-owned enterprises or large corporations, as they had to avoid investment risks. Thus, FinTech companies that lent to the "little guy" as well as SMEs did not face competition from traditional financial institutions and were able to expand in this market segment (Sun et al. 2017, p. 14). However, free regulation brought risks and subsequently caused problems. In the peer-to-peer lending market, one-third of online financing platforms ran into financial difficulties by the end of 2015, and about 40 percent exited the market by April 2016 (Sun et al. 2017, p. 14). Ezubao, one of the best-known and largest online lending platforms at the time, turned out to be a classic Ponzi scheme. Investors were lured with annual double-digit expected returns, growing the company into the largest online finance portal in China in just 18 months (Mittal and Lloyd 2016, p. 26). In reality, however, 95 percent of the loans issued were bogus, and when the company went bankrupt in early 2016, approximately one million customers lost a total amount of USD 7.6 billion (Patwardhan 2018b, pp. 403–404).

The government felt compelled to intervene and published comprehensive frameworks and regulations to cover all business areas of the FinTech market in China, defining how and who regulates these areas. This is to better enforce business practices, financing models, consumer safety, and legal compliance. On May 15, 2017, the central bank unveiled a new FinTech committee responsible for reviewing and revising FinTech legislation to, among other things, minimize potential risks (Sun et al. 2017, p. 15). The National Internet Finance Association is led by the central bank and consists of 400 members from the finance and FinTech sectors (Mittal and Lloyd 2016, p. 27).

Tighter regulation has so far led to a strengthening of the position of traditional financial institutions. A recently published legal regulation prohibited internet loan companies from entering the student loan market and instead allowed traditional banks to offer this service, which is an area that banks were not allowed to serve under the previous legal situation (Sun et al. 2017, p. 15).

Regulatory measures in the FinTech market can be divided into the years before 2015 and the years after. Before 2015, the market for internet companies and their online financial products was unregulated. This led to the emergence of many FinTech innovations and rapid growth of the FinTech market, which became a success factor for FinTech in China (Ngai et al. 2016, pp. 4–5). After 2015, the Chinese government intervened more strictly to avoid situations such as the Ezubao case and to be able to steer the growth of China's FinTech sector in a sustainable and healthy development direction in the long term, where digital solutions are integrated into the financial sector (Patwardhan 2018b, p. 412).

11.2.3.2 High Financial Demand from the Population

Another factor in the success of FinTech companies in China is that a high proportion of the population has large financial needs that traditional financial institutions have been unable to adequately meet. In 2018, about 40 percent of the population lived in rural, less developed areas of China (World Bank 2019). However, the financial sector is significantly underdeveloped in these regions compared to the situation in cities in the modern eastern coastal regions. For example, the total number of bank branches at the township level in 2010 was 127,000, of which 75,935 branches were in rural areas. In 2010, 8723 small towns had only one and 2868 villages in China's rural provinces did not have a single bank branch. While cities such as Shanghai and Beijing have a staff density of 26.01 and 20.55 financial services staff per branch, this figure is 9.43 and 7.8 financial services staff per branch in Gansu and Guanxi provinces, respectively (Zhang 2013, pp. 36–37). In general, in 2014, China had 8.1 bank branches and 55 ATMs per 100,000 population, while this ratio was 28.2 to 222 in the United States (Mittal and Lloyd 2016, p. 12).

In addition, due to the historically developed and poorly developed credit card system, payments for small amounts in everyday life were only made with cash until the early 2000s. In 2015, the number of credit cards per person in the country was 0.3, compared to

3.6 bank cards per person, but paying with cards was difficult because most point-of-sale systems were not designed for it or the different banking networks of buyers and sellers led to high fees when paying with a bank card or credit card (Sun et al. 2017, pp. 36, 39). These difficulties in accessing the simplest financial services, such as opening accounts, processing transfers at a bank branch, or withdrawing money from an ATM, led customers to switch to online or mobile offerings. Today, in percentage terms, more people in western regions use their smartphones to make payments than in eastern regions (Ding et al. 2018, p. 28). However, there is also an undersupply of Chinese customers' needs in the area of lending and wealth management.

At the beginning of the FinTech boom, the regulations for granting loans were much stricter for traditional financial institutions. As a result, state-owned banks were unable to meet the high credit needs of SMEs, among others, because they had to minimize risk. Only 20–25 percent of corporate loans extended by banks in China in 2016 went to SMEs, despite the fact that they were responsible for 60 percent of GDP in the same period, employed 80 percent of the urban population, and provided 50 percent of tax revenues in the country. Retail customers are also underserved in terms of credit in China. The 20 percent credit penetration rate among retail customers is one of the lowest by international standards (Mittal and Lloyd 2016, p. 13). In China's rural areas, only 16.4 percent of loans extended were from formal financial institutions, suggesting a pronounced shadow banking sector (Zhang 2013, p. 36). FinTech providers thus encountered an undersupplied credit market in the period before the regulations described above, whose customers had a high demand for financing services and which was also free of any competition from traditional financial institutions.

The situation is similar for wealth management services. Nearly half of households in China fall into the "low-income" category, meaning that they have an annual income of less than RMB 100,000. However, wealth management offerings at traditional financial institutions have an investment threshold of at least RMB 50,000, which prevented a large portion of the population from accessing such offerings (Ngai et al. 2016, pp. 6–7). Combined with a traditionally high savings rate of 40 percent among consumers in China and potentially investable assets of seven trillion USD among the population, this led to high unmet demand for wealth management services (Patwardhan 2018a, p. 76). FinTech offerings such as Yu'e Bao or Licaitong were able to quickly achieve great success as the investment thresholds for users were extremely low (Ngai et al. 2016, p. 3).

11.2.3.3 TPOP Replaces Cash

Chinese society is using less and less cash and prefers to make payments with mobile devices or via the Internet. A cash alternative must be cheap, convenient, and available anywhere and anytime. The digital infrastructure in China makes it possible to have internet or mobile phone reception almost everywhere in the country, allowing customers to access FinTech applications at any time using their mobile devices. When the costs of transactions with bank cards, credit cards, or TPOP in China are considered, it becomes apparent that paying with TPOP is significantly cheaper than the other two alternatives.

Additionally, C2C transactions are free with most providers. When the cost of transactions at Alipay and Tenpay are compared with the cost at the leading Western TPOP provider PayPal, it is found that the transaction cost is about 5 times higher than the Chinese providers (Sun et al. 2017, pp. 37–38).

When it comes to the user-friendliness of FinTech applications, Chinese providers are at the top of the world. Before the introduction of TPOP, payments over the internet were cumbersome to make. Multiple verification steps had to be gone through either with SMS messages, TAN generators, or security tokens for each transaction. In the TPOP process, on the other hand, once a customer has linked their TPOP account to a bank account, they only need to have their QR code, which can be accessed in the smartphone application, scanned at checkout, or have the QR code of another person to whom money is to be transferred scanned with their smartphone camera (Sun et al. 2017, p. 44). Following a 2016 survey that asked 2000 Chinese people about the reasons why they prefer FinTech offerings over traditional banks, it was found that the more attractive fee structure, better internet functionality combined with high user-friendliness, more innovative offerings, and quality of service were considered the most important reasons (Mittal and Lloyd 2016, p. 13). TPOP-FinTech companies in China fulfilled the requirements to act as cash substitutes. These encouraged the emergence, spread, and integration of other FinTech services.

11.2.3.4 Influence of BAT and Banks on the FinTech Market

The development of the FinTech market in Germany and Europe was mainly driven by start-up companies in the early years, while traditional financial service providers have hardly implemented FinTech solutions to date because they rarely cooperate with FinTech companies or show little interest in investing financial resources (Dorfleitner and Hornuf 2016, pp. 51–56). In China, however, the large internet companies BAT were the starting point of the most successful FinTech solutions and support the development of these with financial and human resources (Mittal and Lloyd 2016, pp. 18–19). In general, the digital economy in China is largely supported by BAT. Forty-two percent of venture capital investments in China's digital economy in 2016 were made by BAT, whereas only 5 percent were made by the dominant internet giants there, Facebook, Amazon, and Google, compared to the US market. Furthermore, according to 2016 figures, 50 percent of the top 50 startups in China are spin-offs from BAT, founded by BAT alumni, or supported by BAT with investments (Woetzel et al. 2017, p. 12). Examples of FinTech companies spun out of BAT include the TPOP service providers described earlier, Alipay (later Ant Financial), from Alibaba's e-commerce ecosystem, and Tenpay and WeBank from Tencent's social networking empire. The large internet companies offered their FinTech offshoots the advantages of an existing customer base that was easy and quick to acquire, leading to the rapid growth of these platforms (Ngai et al. 2016, pp. 10–11).

In addition, BAT offers their FinTech offshoots large amounts of customer data from the respective core business of the internet companies, which the FinTech providers can use, for example, to determine the creditworthiness of their customers (Woetzel et al. 2017, p. 11). The social credit system is controversial because, among other things, the creditworthiness of each Chinese citizen is determined based on financial and social behaviour. However, Ant Financial has already established its credit sy called "Sesame Credit", based on Alibaba Group's customer data, for several years. The system analyzes data from over 300 million customers and 37 million businesses generated on Alibaba's online commerce platforms and assigns a credit rating based on that data. Customers with better ratings don't have to put down a deposit or security when renting cars or apartments, for example, or can expect faster processing when booking travel. Tencent has already implemented a similar system to use WeChat and the partner applications integrated into it as well as to be able to better assess the creditworthiness of customers (Mittal and Lloyd 2016, pp. 35–36).

In China, after the regulations in 2015, banks have intensively invested in or sought cooperation with FinTech companies, which can be seen as a supporting factor to success (Chuen and Teo 2018, pp. 27–28). The reason for these investments is the high returns on equity of Chinese banks. In contrast to EU or US banks, which had a return on equity of three to nine percent from 2007 to 2014, Chinese banks were able to show 15–20 percent during this period (Ngai et al. 2016, p. 7). These profits enabled them to invest more long-term in innovative and digital business ideas, which require start-up time for their business models and initially generate more costs than revenue.

11.2.3.5 Market Size and Customer Characteristics

The number of Chinese internet users is approximately 800 million users, which is three times larger than the US with twelve times the number of mobile payment users (Deng 2018). Further, as of 2016, there were 22 cities in China with populations over five million, compared to only one in the US and four in the EU. This economy of scale is a factor in the success of FinTech providers in China, as the massive base of potential users allows for rapid commercialization and scalability of new business ideas and models, which additionally attracts new investors and entrepreneurs. This leads to FinTech companies being able to achieve profitability in their business operations more quickly and easily (Woetzel et al. 2017, pp. 4–5).

A good example of this is Yu'E Bao, which receives only small investment contributions from its customers on average but makes up for this with a mass of investors. However, the large number of potential customers is not the only advantage of the Chinese market. The special feature of these customers is above all their willingness to adopt innovations and use digital technologies in their daily lives.

FinTech Generation

One reason for the high propensity to adopt innovations is that China has a large number of digital natives among internet users in society. Over 280 million users fall under this category, comprising almost 40 percent of all internet users in the country, while the same figures in the US and EU are 26 and 31 percent in percentage terms and 75 million and 135 million users in absolute terms, respectively (Woetzel et al. 2017, p. 5). This young group of tech-savvy consumers accounts for 45 percent of online consumption and is characterized by a higher willingness to take financial risks than older generations (Mittal and Lloyd 2016, p. 20). Having grown up in a digital world, they are accustomed to receiving offers tailored to the individual, with a high level of user-friendliness and accessibility at any time. They are enthusiastic to use digital tools and willing to try new things (Hirn 2018, pp. 172–173).

Furthermore, traditional banks have relatively little relevance for this user group in terms of the brand value of the banks or in financial offers and advice that go beyond the safekeeping of money (Schlich et al. 2016, pp. 4–8). China's digital natives are not afraid to switch from traditional banks to the new digital offerings of FinTech companies, provided that they are more responsive to their financial needs, offer more convenient services, and promise higher profits (Gomber et al. 2018, p. 230). Therefore, it is little wonder that more and more young consumers in China prefer financial services and products from FinTech platforms over traditional financial institutions (Mittal and Lloyd 2016, p. 20).

Open-Mindedness Towards Innovations

A 2014 study on the influence of Chinese consumers' cultural characteristics showed that several factors are involved (Song 2014, pp. 339–440). Willingness to adopt the innovation was defined as the customer's intention to use the product in the next 6 months (Song 2014, p. 344). The most important factors influencing consumers in China in this regard are utilitarian perception, hedonistic perception, cost, and quality concerns, social pressure, and along with it, the concept of "saving face" or "getting face" (Song 2014, p. 346). These factors can be analyzed in terms of their influence on the adoption of FinTech products in Chinese society.

Utilitarian perception is described as the degree of benefit that the use of innovation provides to its user and is one of the strongest factors in predicting whether an innovation will be adopted in society (Venkatesh and Brown 2001, p. 83). The use of FinTech applications offers many benefits to Chinese consumers, whether it is to be better connected to financial services, to be able to take advantage of more offers or to enjoy cheaper transaction fees in online commerce.

Hedonic perception describes the degree of pleasure and enjoyment that using an innovation gives the consumer (Song 2014, p. 341). First WeChat and then Alipay have taken advantage of this factor and, by implementing the "red envelope" function, combined FinTech applications with a play and fun factor that ties in with Chinese traditions (Sun et al. 2017, p. 43).

Cost and quality concerns represent perceived barriers to adopting an innovation. The cost dimension describes concerns about the high costs of switching to an innovation, benefiting from it, or whether the costs of benefiting from it are even worth it (De Marez et al. 2007, p. 84). Chinese FinTech companies offer financial services that are significantly cheaper than those offered by traditional financial institutions, so cost concerns are not barriers to Chinese consumers' willingness to adopt FinTech innovations. The same can be said about the quality dimension, where the lack of quality of an innovative product may prevent its adoption. With regard to FinTech offerings in China, it can be seen that due to the highly developed digital infrastructure in China, the highest quality standards are met in terms of the availability of the offerings. In addition, better conditions can be offered in the area of lending and asset management than with traditional financial institutions. Quality concerns, therefore, do not represent barriers to the willingness to accept FinTech offerings.

Social pressure is an important factor in regard to the willingness of Chinese consumers to adopt innovations (Song 2014, p. 344). It describes the pressure that an individual feels from his social environment to show a certain behavior. The higher the pressure, the higher the willingness to adopt and use innovations. Furthermore, almost 70 percent of all Chinese internet users use TPOP applications when making payments (Woetzel et al. 2017, p. 4). This makes it difficult for individuals to avoid FinTech innovations. In addition, there is another cultural feature, the concept of face.

The face represents a central concept in Chinese culture and describes an individual's perception of his or her position in society and the need to present or elevate that position. Maintaining face and thus elevating one's social position is as important as keeping one's face and maintaining one's social status (Chu 2008, p. 34). A 2014 study describes that consuming or using innovative products can have a positive effect on an individual's face in China, highlighting one's status and obtaining recognition from others (Song 2014, p. 341). Thus, it can be concluded that the concept of face in Chinese culture also has a positive effect on the willingness to adopt innovations and thus influence the success of FinTech applications in China.

11.2.3.6 Dealing with Data Protection

In the digital financial ecosystems created by BAT and in the core business of the companies, masses of user data are collected, stored, and shared within their system (Ngai et al. 2016, p. 11). It is possible to track down to the smallest detail when, where, and what the participants of this system have processed. On the one hand, this enables FinTech companies to address their customers with offers in an even more personalized way and to better assess risks; on the other hand, the customer loses large parts of his privacy and is relatively unprotected against possible misuse or unauthorized disclosure of his data. While Western customers are less willing to accept such FinTech offers due to their concerns in this regard, Chinese users are far more generous with their data and place less value on data protection in comparison (Mittal and Lloyd 2016, p. 20). According to a study from 2015, Chinese users worldwide place the least value on the protection of their personal data, while consumers in Germany, the UK, and the USA place the most value on its protection (Morey et al. 2015). This permissiveness with one's own data with regard to activities on the internet, especially payment transactions and online trading, is a positive factor influencing the success of e-commerce and digital financial services of all kinds in China.

11.2.3.7 Risks and Security of FinTech Applications

Another factor that influences the adoption of innovations is the perceived risk involved in using them as well as the security of the products. In a 2009 study on online banking adoption, it was found that perceived risks had a higher impact on willingness to use than the perceived benefits of using it. These perceived risks were further categorized into security risks, financial risks, fulfillment risks, and social risks. This shows why due to the highly developed digital infrastructure in the country, fulfillment risks (poor server performance, long waiting time, etc.) can be ruled out as a cause for rejection. Furthermore, the aspect of social risks has already been addressed and identified as a reason for increased willingness to adapt to FinTech. Financial risks are less considered by Chinese users of the FinTech market. The greatest potential barrier to the adoption of online banking, and thus applied analogously to the adoption of FinTech, is security risks and concerns (Lee 2009, pp. 138–139).

Chinese users are the most afraid of identity theft and fraud in the world (Morey et al. 2015). Looking at the impact of security risks on the situation of FinTech in China, it can be seen that it is very low. To open a user account with FinTech providers, a strict verification process must be followed. Each customer has to identify himself in person at a bank branch and present a valid mobile phone number to link his bank account to a TPOP account, for example (Sun et al. 2017, p. 40). If a customer wishes to change their mobile number, they must apply in person with valid identification documents at their bank branch. In China, the strict verification process of banks cannot be circumvented and each user account of a FinTech application can be accurately traced back to a citizen. Risk control and hedging are high priorities for FinTech companies, and Alibaba and its FinTech offshoots have been extremely successful in this area for years and have the trust of their customers. 1500 of the 7000 employees at Ant Financial alone, deal solely with monitoring, analyzing, and managing risk (Ding et al. 2018, p. 31). The strict identification process to participate in the FinTech ecosystem and the way FinTech companies deal with security risks act as a positive influencing factor on the success of FinTech in China.

11.3 Crowdfunding

In the wake of the 2008 financial crisis, a new form of capital formation emerged, largely due to entrepreneurs' difficulties in raising capital. With traditional financial markets reluctant to lend, entrepreneurs began to look elsewhere for capital. As an internet-enabled way for businesses or other organizations to raise money in the form of donations or investments from multiple people, crowdfunding is gaining an increasing position as a financial tool. It plays an important role not only in financing but also in risk sharing (The World Bank 2013, p. 8). While before the financial crisis crowdfunding was mainly used by charities and actors in the creative industries, it has become an alternative financing tool for start-ups, entrepreneurs, and small and medium-sized enterprises (SMEs) in all sectors. According to a 2013 World Bank report, the total market potential is estimated to be between 621.35 billion and 662.77 billion renminbi (RMB) per year by 2025, surpassing traditional venture capital financing. The greatest potential for crowdfunding is in China, where this amounts to RMB 317.58 to 345.20 billion of the total, followed by the rest of

East Asia, Central Europe and Latin America, and the Caribbean (The World Bank 2013, p. 10).

Although crowdfunding did not exist in China before 2011, its exponential growth rates reflect its underlying potential. China is the world's largest market for online alternative finance, with a transaction volume of RMB 638.79 billion in 2015. This represents almost 99 percent of the total volume in the Asia-Pacific region (Zhang et al. 2016, p. 19). In comparison, the total volume of the German market for alternative online financing amounted to just under two billion RMB in 2015 (Statista 2019a). The Chinese market for alternative online financing grew from RMB 38 billion in 2013 to RMB 168 billion in 2014 and increased to RMB 702 billion in 2015. Thus, between 2013 and 2015, the average growth rate was 328 percent. P2P consumer loans form the largest market segment in China with RMB 362 billion, followed by P2P business loans with RMB 274 billion and real estate loans with RMB 38 billion. Equity-based crowdfunding recorded nearly seven billion RMB and rewards-based crowdfunding increased to nearly six billion RMB in 2015 (Zhang et al. 2016, p. 19). Not only the increasing growth rates could make China the largest crowdfunding market in the world in the near future, but moreover, three other fundamental criteria that are specifically applicable to China: the underdeveloped informal financial sector, the agile internet sector, and the support from the Chinese government. Because China's formal financial sector is underdeveloped, state-directed, and unable to serve the entire market, Chinese people are accustomed to using alternative sources of financing. As a result, the Chinese population is open to new forms of financing. Crowdfunding is further encouraged by the agile internet sector, which is driven by e-commerce giants and social media providers. In addition to their core business, these companies offer online payment systems and financial services such as crowdfunding. As a result, banks and other traditional financial institutions face additional competition. As a third factor, the Chinese government is promoting the development of internet finance, as this is seen as a driving force for reforms in the traditional financial sector (Funk 2018, pp. 2–3).

11.3.1 Definition of Crowdfunding

The concept of crowdfunding is derived from the broader concept of crowdsourcing (Belleflamme et al. 2014, p. 585). The term "crowd" is used as a resource to obtain ideas, feedback, and solutions to develop business activities. Crowdfunding refers to any type of capital formation where funding needs and purposes are generally communicated via an open call in a forum. As it stands today, crowdfunding is often carried out via internet-based applications and platforms (Belleflamme et al. 2014, p. 604).

Historically, the term crowdfunding is new, but not a modern-day phenomenon. There have been collective fundraising campaigns for centuries, such as membership models for letterpress printing in the seventeenth century, public subscriptions to fund British parks in the nineteenth century, and the collective funding of Bollywood films in the 1970s. A famous example of an "offline" crowdfunding campaign is the granite base of the Statue of Liberty of the United States of America. Publisher Joseph Pulitzer made a public appeal for donations in the newspaper "The New York World". What was impressive was the speed with which the money was raised, the number of small donations, and the fact that the whole process was managed by one agent, namely the newspaper (BBC News 2013).

However, the Chinese view is that despite crowdfunding being around for almost a decade, the term is not known to the masses or there are disagreements about its exact meaning and scope. Some take the term crowdfunding literally, counting as crowdfunding all ventures realized through the collective investment of a specific group of people. Others do not distinguish between group buys, crowdfunding, and rebates, or conflate them with pre-sales or lotteries. But crowdfunding has become a trendy word in China, as evidenced by the rapid presence of the term in Chinese daily life. Given the confusion over the meaning of crowdfunding, it is challenging to clarify its exact application and operation (Funk 2018, p. 151).

11.3.1.1 Crowdfunding Models

Crowdfunding models refer to both the non-financial return forms of donation- and reward-based crowdfunding, as well as the crowdfunding models with financial returns. Here, loan-based, debt-based, equity-based, and royalty-based crowdfunding can be mentioned. Non-financial crowdfunding and financial crowdfunding differ primarily in the allocation of contributions from individuals to expectations of a financial return at a later date. Each type of crowdfunding has different implications for campaign solicitation and investor or donor engagement, as well as the institutions, infrastructure, and regulations required to support the different crowdfunding models. Five different variants of crowdfunding are distinguished (Esposti 2015, pp. 34, 40–45):

1. Donation-based crowdfunding

This usually involves requesting funds for independent projects or initiatives that appeal to others with similar interests. As long as there is no expectation or legal obligation for the money provided to be returned, the transfer is considered a donation or gift and is therefore referred to as donation-based crowdfunding. Donation-based crowdfunding in China includes both grassroots platforms and a smaller number of new platforms from e-commerce companies. In 2015, RMB 978.21 million was raised for donation-based crowdfunding campaigns in China (Zhang et al. 2016, p. 38).

2. Reward-based crowdfunding

In a reward-based campaign, members of the crowd, known as crowdfunder, provide capital to support a campaign in exchange for some kind of benefit or reward. This can be an offered product in pre-sale or at a discounted price. According to Zhang et al. (2016, p. 40), reward-based crowdfunding is much practiced and established in mainland China, with a volume of RMB 5.73 billion in 2015. Since 2013, a number of Chinese e-commerce companies have launched reward-based crowdfunding platforms. This has allowed the new reward-based platforms to tap into large, pre-existing supplier and customer bases to optimize their offerings and funding campaigns. The average growth rate of performance-based financing between 2013 and 2015 was 2.349 percent. Overseas platforms comprise less than 1 percent of the total amount of rewards-based crowdfunding in China.

3. Crowdfunding on a loan basis

In credit-based crowdfunding, investors receive a debt instrument that specifies the terms of future repayment. In essence, the campaign owner agrees to repay the funds provided by the investor, which typically consist of the face value of the investment amount and a fixed interest rate. Interest rates are not always fixed in loan-based crowdfunding contracts and allow investors freedom. Loan-based crowdfunding can include peer-to-peer (P2P) and peer-to-business (P2B) loans. P2P consumer loans are the largest category of online alternative finance in mainland China. In 2015, the market volume was RMB 362 billion. The year before, the market volume was RMB 98.73 billion, and in 2013, it was RMB 26.58 billion. Consumer loans have been able to grow rapidly as increasingly diverse financial needs of individuals and households in China – especially in urban areas – have had to be met. They include all forms of consumer credit, from personal loans to auto and home loans to student loans. Consumer credit platforms have been able to grow rapidly by offering a variety of unsecured microloans and consumer loans on more flexible terms than traditional banks, which have been slow to respond to rapidly diversifying consumer finance needs. While the dominant share of these loans is consumer finance, individual and family businesses are using P2P lending platforms for their short-term working capital needs. The average annual growth rate was nearly 270 percent between 2013 and 2015 (Zhang et al. 2016, p. 38).

4. Equity-based crowdfunding

With equity-based crowdfunding, the investor is entitled to a share in the ownership or future profits and thus in the success of the business. This model, therefore, offers higher returns as the value of the equity increases with the success of the business. However, it poses a greater risk to the shareholders. In the event of a venture's failure, investors are usually subordinate to creditors in the repayment line. This segment was unregulated until 2015 and operated in part as angel investing, "club investing" or was subject to a private placement mechanism for equity investments using both online and offline channels. In total, equity-based crowdfunding recorded RMB 6.55 billion in 2015 (Zhang et al. 2016, p. 39).

5. Royalty-based crowdfunding

This type of crowdfunding offers a crowdfunder, for example, a contract in which they are guaranteed a percentage or fixed amount of the revenue stream. This is usually associated with a royalty interest in a company's intellectual property and the resulting revenue. However, in doing so, companies have the option of no longer using the originally funded intellectual property or selling it to a competing company. This causes complexity in the royalty-based model, leading to lower usage. Nevertheless, the model offers an interesting example of innovation in the crowdfunding market. Crowdfunding platforms received a total of RMB 260.48 million from investment backers in mainland China in 2015 (Zhang et al. 2016, p. 40).

Donation and reward-based crowdfunding carry fewer risks than financial crowdfunding. As with all types of crowdfunding, crowdfunders have the risk of falling for fraudulent campaigns and must factor in cybersecurity risks. Reward-based campaigns expose crowdfunder to fulfillment risk, as the campaign owner may not fulfill the promise to deliver a reward.

11.3.1.2 Types of Crowdfunding Donors

During the development and early stages of a company, several types of capital providers may approach the company. With the help of capital and expertise, they support the existence and growth of the company. For this purpose, different capital providers can be classified, which differ from each other in their characteristics and pursued goals. In the beginning, the founder(s) bear the costs of developing the business idea. Since a high capital requirement has to be covered at high risk in the start-up phase, the founders turn to family members and friends in this phase. Under the premise of a higher basis of trust and personal relationship, these support the founder's idea and help to bridge the first financing gap. The first category also includes fans and fools who identify with the business idea and willingly provide their capital. These five donors are referred to as the 5-Fs. Similar to fans, followers and idealists are enthusiastic about the idea of the project but do not have a personal relationship with the initiator. Most often, supporters are found in communities, political parties, and social networks, and idealists share a similar worldview. Successful entrepreneurs, artists, politicians, banks or non-profit organizations support projects for social or political reasons. Since they are in the public eye in this way, they thereby advertise a supposedly positive image. People and institutions acting under such aspects can be called philanthropists and altruists. Less self-interested are the beneficiaries and promoters who spend money to support an enterprise or project that they consider rationally useful, but benefit themselves. This type of investor includes lobbyists, politicians, and entrepreneurs who tend to stay in the background once they are involved. The final type of capital provider is business angels. These can be any of the types of people listed above, excluding family members, as long as they invest as an individual. Their interest may serve a social purpose, but need not necessarily be anthropological in nature. Business angels focus on financial returns, which they seek to achieve by providing capital, expertise, and their network (Hemer et al. 2011, p. 34).

11.3.1.3 Financing Phases

The life cycle of a company can be divided into three superordinate phases: the foundation phase, the expansion phase, and the maturity phase. These phases are further divided into several sub-phases. These range from five to seven phases depending on the literature (Barthelmess 2010, p. 49; Weitnauer 2018, p. 320; Brehm 2012, p. 11). The start-up phase begins with the "seed stage", in which legal preparations are made for the establishment of the company. In the next stage, the market analysis is completed and results in the development of a prototype. By creating a business plan, the "proof of principle or concept" (Brehm 2012, p. 11) and the business model are precisely formulated in this phase. The aim is to evaluate the potential of the company or product idea and to identify future sales markets, customers and competitors. The "seed stage" is highly risky due to a lack of sales and cash flow, which is why informal capital from the 5-F's is usually used. One approach to capital acquisition is to launch a crowdfunding campaign. This attempts to build a crowd early on and find potential buyers. Following this, the "start-up stage" finalizes the formal entrepreneurship formation (Brehm 2012, p. 11). Since the activities in the start-up phase cause high costs and a growing need for liquidity must be covered, the development of new sources of financing is essential for the existence of the company. This is where government funding, incubators, venture capital companies, and business angels intervene. The latter not only supports young companies with capital but also advises management and provides the founders with an infrastructure. The start-up phase is completed with a successful market entry and the company moves into the expansion phase (Kollmann and Kuckertz 2003, p. 37).

During the expansion phase, the company grows. The goal is to develop a product that is ready for mass production to eventually reach the break-even point and stabilize cash flow (Barthelmess 2010, p. 49). Monetarily, start-ups cannot fully self-finance and rely on further capitalization funding. As the investment risk decreases with the maturity of the company and collateral has to be presented, informal capital alternates with formal capital in this area. Private equity and venture capital firms provide equity capital from this stage. Traditional, institutional debt capital, such as bank loans, is also used to ensure the company's liquidity (Hemer et al. 2011, p. 28).

The next phase is called the "Bridge Phase" because it serves as a "bridge" between the Expansion Phase and the "Later Stage", also called the "Exit Phase". Here, follow-up innovations, product or business area diversifications, and the development of new sales markets are targeted. This phase prepares the company for its exit, where venture capitalists and investors seek a profitable exit. The following options can be used as an exit strategy:

1. Initial Public Offering (IPO): Venture capital companies are separated from the investment structure with the help of investment banks, and their shares are sold to institutional and private shareholders.
2. Sale to a strategic investor.
3. Sale to another financial investor.
4. Initiator buys back the shares in the company.
5. Liquidation: liquidation of all assets with the termination of the business (Brehm 2012, p. 12).

11.3.2 Crowdfunding Platforms

Although the basic principle of crowdfunding has been adapted from America to the Chinese market, a kind of hybrid model has developed over the years from the American guidelines and Chinese culture. China's long history of informal financing, the advanced internet sector, and online payments can be cited as particularly strong influencing factors (Funk 2018, pp. 149–150).

China's first crowdfunding platform is Demohour, which went online in 2011 in response to the 2008 financial crisis (The World Bank 2013, p. 10). After 2 years of operation, Demohour received more than 7000 project applications. They had raised funds for more than 2000 projects from small online contributors in the absence of access to bank loans or wealthy private investors. The majority of the successful Demohour projects are film animations, including: Big Fish and Begonia (2013) with a raised amount of nearly 1.6 million RMB from over 3596 backers and revenue of 430 million RMB. A Hundred Thousand Bad Jokes (2014): the campaign raised nearly 1.4 million RMB from 5300 backers. The film grossed 120 million RMB at the Chinese box office (Zhang et al. 2014, p. 13).

In 2013, Alibaba Group launched Yu Le Bao as a crowdfunding platform for the entertainment industry. With Yu Le Bao, it is possible to finance TV, film, and online game projects by investing an amount of 55.23 RMB and earning an annual return of 7 . The platform aims to bring Chinese people closer to the country's cultural activities and initiatives. Yu Le Bao investments can be made online through the mobile app of Taobao, which is also part of the Alibaba Group. Following the success of Yu Le Bao, Alibaba launched another alternative financing platform called Zhao Cai Bao in April 2014. As a P2P platform, it is designed for investment and financial products offering universal insurance, structured funds, and loans to SMEs and individuals (Uhm et al. 2018, p. 304). Jing Dong (JD), another leading e-commerce platform in China, established Coufenzi in July 2014, which allows users to provide funds for individual projects and products. It also established JD Equity Crowdfunding in 2015. It targets entrepreneurs who need to find early-stage investors (Bischoff 2014). Baidu launched Baifa Youxi in December 2013 (Xiang 2015), followed by Tencent's Licaitong in January 2014 (Reuters 2014). Within the first 5 years, the number of platform providers has grown rapidly. In January 2016, there were already 3383 platforms registered (Liu 2018). Tencent Lejuan (722 projects), JD Crowdfunding (351 projects), and Taobao's Crowdfunding (261 projects) are currently the three largest crowdfunding platforms in China in terms of several projects. Duocaitou, Kashiba, and JD Crowdfunding rank in the top three in China in terms of the amount of revenue generated, as can be seen in Table 11.1. Zhongchou, founded in 2013, and Suning Crowdfunding, founded in 2015, are also among the top platforms in China, having achieved a high market share (Zhongchoujia 2018).

11.3.3 Regulatory Intervention

The tremendous growth of crowdfunding in general, and P2P lending in particular, has been made possible by the trial-and-error approach of the relevant authorities in regulating and supervising the nascent industry. From 2007 to 2014, the government allowed P2P platforms and other online finance companies to operate without establishing a legal and

Table 11.1 Number of projects and amount of funds raised by April 2018 (in RMB)

Duocaitou	92	543 million
Kaishiba	62	197 million
JD Crowdfunding	351	127 million
Zhongtou8	189	127 million
Taobao crowdfunding	261	90 million
MI Crowdfunding	13	84 million
Idianchou	152	24 million
Fenfentou	5	24 million
Jumi Crowdfunding	14	20 million
Renrenhehuo	1	20 million
Weiming1898	3	18 million
Ezc360	196	16 million
Tencent Lejuan	722	13 million
Hehuo8	1	13 million
Suning Crowdfunding	71	12 million
MINIPO	1	12 million

Source: Adapted from Zhongchoujia (2018)

regulatory framework. On November 19, 2014, Premier Li Keqiang stated for the first time during a State Council meeting that China was initiating a pilot crowdfunding program. The resulting measures for the administration of equity-based crowdfunding were published for test implementation by the Securities Association of China (SAC) a month later (Cao 2015a). However, these regulations do not apply to donation-based and premium-based crowdfunding. Before this, equity-backed investments were either trading in an informal area or a grey area, if not illegal (Funk 2018, p. 151). The lack of regulations, on the one hand, contributes to the exponential development of platform providers, and on the other hand, practices are used that should be critically assessed. Examples include:

- Pooling, slicing, and packaging of underlying loans (a process similar to securitization).
- Guarantee of repayment and financial returns without proven ability to deliver.
- Shadow banking-like maturity transformation (Aveni and Jenik 2017, p. 2).

The initial combination of self-regulation and ad hoc supervision by local authorities led to inconsistencies. Many platforms are operated under different regulations issued by the city or provincial authorities. Due to failures and scandals, including the E'zubao Ponzi scheme, regulations were tightened in 2015. Due to the regulatory changes, such as the mandatory opening of an account with a bank, the number of P2P platforms decreased from more than 3000 platforms in 2015 to 2448 platforms in 2016 (Aveni and Jenik 2017, pp. 2–3). A further reduction took place in 2018 (1021 providers). The number is expected to decrease even further in 2019, leaving only 50–200 providers (Holmes 2019). In China, more than half of the platforms surveyed, except equity-based crowdfunding, considered existing regulations on funding models to be insufficient and too loose (Zhang et al. 2016, p. 20).

11.3.4 China's Peer-to-Peer Fraud Cases

China's "purge" of corruption and denunciations and imprisonments of political opponents sparked the fight against the Chinese mafia and triads. In the course of this campaign, according to Guo Shengku, secretary of the Central Commission for Politics and Law, pornography, gambling, illegal drugs, pyramid-like organized ventures, kidnapping, and human trafficking are at the top of the list of criminal activities (Shi 2018). In the face of financial instability and social unrest, combating so-called pyramid schemes is a top priority for the police and government. This highlights the urgency to take action and stop such activities. While China's economy has grown at a rapid pace, large segments of the population have remained cut off from the country's advancement and unfair distribution of wealth has emerged (Leng 2017).

The naivety of the population and the desire for high returns have contributed to the proliferation of fraudulent investment schemes in China. Although sales in pyramid-style companies have been illegal since 2005, fraud cases continue to occur. Such as the case of E'zubao, where police arrested the 30-year-old owner of the P2P lending company (Tang 2017). The growth of P2P lending almost quadrupled in 2015, reaching RMB 982 billion compared to RMB 253 billion in the previous year. During this period of growth, E'zubao was brought to the market, with subsequent investigations suggesting that the amount of fraud was approximately 8 percent of the total capital invested (Symonds 2016). In China, the People's Bank of China (PBoC) and the Banking and Insurance Regulatory Commission (CBNEditor 2018) are responsible for financial supervision and therefore also for the supervision of fraud cases.

The Ponzi scheme is a fraudulent investment option that exploits trust between two parties, an investment fund company, and an investor. The achievement of a high-profit margin for the investor is to be achieved by means of his contributed capital at a later date by the intermediary. The deposits of new investors are used to pay the interest of the other investors. This creates a cycle that can be maintained as long as more new investors can be attracted than investors who receive interest. Due to its pyramid-like structure, the Ponzi scheme is referred to as a Ponzi scheme in an economic-scientific context, but in legal terms, there is as yet no demarcation (Kilian 2009, p. 286).

11.3.4.1 The Case of E'zubao

From late 2015 through 2016, the E'zubao case led to a great deal of attention in the Chinese and Western media (Gough 2016). The term E'zubao consists of three parts: "E" stands for electronic as in e-commerce, "Zu" means (to) rent, and "Bao" is translated as precious treasure or jewel. This term refers to an internet platform that operates like a Ponzi scheme, also known as a pyramid in China, by promising investors high returns of nine to 15 percent, only to disappear with their money afterward. According to the Legal Daily, a party-affiliated legal daily, several firms were involved in this fraudulent platform. One of them was Shenzhen Yu Cheng Wealth Management Ltd Zhaoqing, which split off a corporate branch and accepted illegal donations to set up a branch of more than RMB 16 million. Investigative authorities have identified Mega Fubon Information Consultancy, which operated under Kwak Management Co, Ltd Zhaoqing, and Yu Cheng Group, whose wholly owned subsidiary Jin Yirong Network Technology Co, Ltd was involved, as the main actors in the E'zubao platform. Ding Ning, E'zubao's founder, launched the platform on February 24, 2014. The latter suddenly ceased its operations in December 2015 (Zhang 2016). The abrupt closure of the website caused concern among investors. To prevent social unrest and quell demonstrations, authorities announced in early February 2016 that E'zubao customers could register their complaints on the Ministry of Public Security's website to capture the full scope of the offense (Symonds 2016).

According to a report, suspects had stashed some 1200 documents and other evidence related to the scheme in 80 bags and buried them six meters underground at a site on the outskirts of Hefei, the capital of Anhui province. This enabled 21 people involved to be convicted and arrested (China.com 2016). By December 8, 2015, the total transaction volume was RMB 74.6 billion and the total number of deceived investors was 909,500 people. With this volume, Jin Yirong was ranked fourth in the highest investment volume in the internet finance industry.

E'zubao gained popularity, attention, and perceived creditability through publicly aired advertisements on television, the Internet, in newspapers, and in brochures (Zhang 2016). Sohu, one of the most visited Internet portals engaged in online advertising on the internet, search engines, and multiplayer online games, published an article on September 18, 2015, actively promoting E'zubao. First, the article addressed that Chinese ministries are initiating the reform of internet banking and online payment methods, and these are considered key industries. In this context, E'zubao was described as competitive because it uses the innovative A2P model to distinguish itself in the market. It then explained what was behind this so-called A2P approach: the acronym stands for asset-to-person, meaning assets (A) in the form of leases or loans to people (P). According to the ad, after a supposedly rigorous selection and high-quality corporate bonds, information was passed on to investors through the platform so that they can invest and gain a profitable return on investment. The potential of this approach was backed up by the Secretary General of the Beijing Online Loan Industry Association, Guo Dagang, who advocated low risk and called for the promotion of such innovative models. To make this model appear more trustworthy, E'zubao was portrayed as transparent about the use of the investment and the origin of the returns. Second, the ad reported that the risk of default is reduced because a clear line is drawn between the investor's ownership and entitlement and the company's ownership and use. Professional regulators, the article said, would review the relevant qualifications before licensing. However, more specific details of which authority and what criteria the A2P was concerned with were not given. Finally, the stable returns were praised and the advantages of financial openness, access, convenience, simplicity, and speed of investment with low risk in a safe investment environment were highlighted (Newshoo 2015).

The misappropriated money was used by the founder Ding Ning to supplement the income of family members on the one hand, and for his benefit on the other. By doing so, he had raised his brother's monthly income from 18,000 RMB to one million RMB. Beyond this, a total of another RMB 800 million was shown on E'zubao's payroll 1 month before closing. Ding Ning had invested in real estate alongside designer clothes, cars, and other luxury goods (Gough 2016). In the process, a sum of 27 million Singapore dollars ($17.87 million) was illegally transferred from China to Singapore. This alerted Chinese authorities, who then collaborated with Singaporean police to investigate this transfer. The 2-year collaboration concluded that Mr. Ning had bought a villa in Singapore, and the money seized as a result was returned to China to be refunded to the investors in accordance with Chinese laws. Furthermore, 26 suspects were sentenced to prison terms ranging from 3 years to life (Ng 2018). Refunds to Chinese citizens were not mentioned in any article. However, from the reactions and demonstrations of the citizens, it can be concluded that they had suffered heavy losses and no compensation had been made.

Despite government intervention and tightening laws, 9183 cases came to court in 2018, a 57 percent increase from 2015. Although only about 1000 platforms remain online, the Chinese are investing in fundraising for online loans, wealth management, private equity, cryptocurrency, and elder care services, apart from traditional outlets such as product marketing, real estate investment, and education (Leng 2019).

11.3.4.2 Social Implications of the Ponzi Scheme

Why can such frauds happen on the scale they do in China? For one thing, the number of university graduates is steadily increasing. In 2017, 7.36 million students graduated, compared to 7.04 million students the previous year and 4.48 million students 10 years ago. Over the last 10 years, the annual number of Chinese graduates increased by an average of 5.14 percent (Statista 2019b). Assuming an unemployment rate of 3.8 percent (Statista 2019c), without taking age or sector into account, this means that, in percentage terms, nearly 280,000 graduates who enter the labor market each year are unemployed. This situation can contribute to vulnerability to opaque deals. Graduates, for example, are lured to the city by means of a job offer. The advertised company subsequently turns out to be an illegal, pyramid-like enterprise, also called chuan xiao (multi-level marketing). They, in turn, manipulate and brainwash the students into forcing them to work for them and recruit new recruiters. Bodies turning up in lakes and rivers, set investigations in motion where it is determined that the respective victims were involved in the machinations of such organizations and may have wanted to leave or not participate. But it is not only students who are moving to cities for better job prospects; many people from rural areas are also moving. China's economic growth widened the inequality between rural and urban populations. The income gap is almost impossible to close for people in rural areas, which is why many move to the cities in the hope of "getting a piece" of China's rise. The poorest 25 percent of the population owns only 1 percent of the country's total wealth, while the richest own 1 percent and thus one-third of the wealth. In July 2017, several investors gathered in Tiananmen Square to protest the government's ruling on the Shanxinhui company. This protest is considered the largest since the Falun Gong movement branded a cult in 1999 (He 2017). The company was deemed dangerous and a pyramid scheme. Since such businesses are illegal in China, the founder Mr. Zhang Tianming and other employees were then arrested and business activities were stopped. This meant all investors had been deceived and robbed of their savings. Therefore, the protesters, demanded the release of the founder and the resumption of business activities as they hoped to recover their money (Buckley and Ramzy 2017). More than five million people were reportedly registered with Shanxinhui. However, this case is not unique and some are reported to have lost tens of millions of USD. Between the years 2015 and 2017, fraud cases increased by 19 percent, reaching several 2800 cases per year (Hernández and Zhao 2017). In the summer of 2018, controls on P2P platforms were tightened again and a quarter of these platforms disappeared (Dieter 2019, p. 17).

11.4 Conclusion

The Chinese financial market has developed rapidly. The Big Four banks are among the largest in the world and the bond market is the third largest. In the FinTech market, China has the largest market (Tobin and Volz 2019, p. 35). However, this official data and ranking should not hide the fact that the Chinese financial market is state-controlled; relinquishing this control would be a loss of power for the CCP.

Growing local government and corporate debt, a real estate bubble, and a growing shadow banking sector could endanger the Chinese financial system – and thus economic growth. However, reliable forecasts are hardly possible. When economic crises occur, the central government continues to demand that local governments invest in infrastructure. The extent to which this increases their debt burden and could lead to a collapse is unclear.

The shadow banks are closely intertwined with the formal banks, and a crisis could spill over to them. However, this is a rather unlikely scenario, as the central bank would cushion the crisis. China has so far been able to withstand the financial strains of high debt by maintaining high economic growth. However, it is becoming increasingly difficult to prevent credit from growing if economic growth stagnates (Kroeber 2016, pp. 137–138).

FinTech has developed very rapidly in China and has become an important part of the country's financial market. While the development of FinTech is improving the financial inclusion of the Chinese population and enabling them to participate in financial activities previously reserved for them, traditional financial institutions are losing market share in the financial market due to the FinTech activities of Internet companies and industry companies from outside the sector. As the world moves from a workbench to capital-intensive and high-quality products and services, the FinTech sector is a testament to the potential for innovation and technology, and thus an important factor in the PRC's economic growth. Within the last few years, it has become the center of global FinTech innovation and adaptation, producing companies that rank top internationally in terms of stock market values, customer traffic, and capital volumes. While in the West, FinTech innovations have struggled to find their way into their respective financial markets and have been slow to be adopted by customers, there seems to be little limit to the growth and speed of development of the FinTech sector in China. New combinations of financial services and technologies are creating a variety of business models and products in China.

It remains to be seen whether the success in China will transfer abroad, as previous efforts by Alibaba and Tencent to expand abroad, for example, have been slow. Many factors that have influenced the development in China only have an effect within China or only occur in connection with the market situation in the PR. Examples include the almost non-existent regulation in the early days of the FinTech boom in China and a large number of potential customers.

Crowdfunding has become a mainstay in China in a relatively short period. The vast majority of companies accessing crowdfunding channels in China are SMEs, as crowdfunding platforms are more individually focused and represent a direct investment model. There are many variations of crowdfunding in China, depending on the goals of the players providing the funding and those seeking funding. However, recent scandals such as E'zubao show that the high risk involved in online financing requires regulatory protection. The legal and regulatory system in China has responded both regionally and nationally to promote economic development and prevent questionable business practices.

In the international market, Chinese banks are increasingly interconnected, so the question arises whether, if China were to fall into a financial crisis, this would mean a crisis for everyone. On the other hand, the trade conflict with the USA in particular is causing the financial sector to open up more quickly, with limits on foreign holdings in insurers and brokers set to fall as early as 2020 (Reuters 2019a).

References

Abkowitz, Alyssa. 2018. The cashless society has arrived – Only it's in China. *The Wall Street Journal.* https://www.wsj.com/articles/chinas-mobile-payment-boom-changes-how-people-shop-borrow-even-panhandle-1515000570. Accessed 25 July 2019.

Ankenbrand, Hendrik. 2018. In China wird die Inflation zum Thema. *Frankfurter Allgemeine Zeitung*. https://www.faz.net/aktuell/finanzen/in-china-wird-die-inflation-zum-thema-15781746.html. Accessed 20 Aug 2019.

Aveni, Tyler, and Ivo Jenik. 2017. *Crowdfunding in China: The financial inclusion dimension*. https://www.cgap.org/sites/default/files/Brief-Crowdfunding-in-China-Jul-2017_0.pdf. Accessed 28 Oct 2019.

Banking and Insurance Supervisory Commission. 2018. *Tasks and functions*. Available online at http://www.cbirc.gov.cn/cn/list/9101/910101/1.html, Last verified 19 August 2019.

Barthelmess, Philipp. 2010. *Einfluss einer Private Equity Beteiligung auf den Erfolg von Unternehmen.* Diss. der Wirtschafts- und Sozialwissenschaftlichen Fakultät der Universität Rostock. https://d-nb.info/1001604172/34. Accessed 28 Sep 2019.

BBC News. 2013. *The Statue of Liberty and America's crowdfunding pioneer*. https://www.bbc.com/news/magazine-21932675. Accessed 28 Sep 2019.

Bell, Stephen, and Hui Feng. 2013. *The rise of the People's Bank of China. The politics of institutional change*. Cambridge, MA: Harvard University Press.

Belleflamme, Paul, Thomas Lambert, and Armin Schwienbacher. 2014. Crowdfunding: Tapping the right crowd. *Journal of Business Venturing* 29 (5): 585–609.

Bischoff, Paul. 2014. *China's second-biggest ecommerce firm JD launches Kickstarter-like crowdfunding site.* https://www.techinasia.com/chinas-secondbiggest-ecommerce-firm-jd-launches-kickstarterlike-crowdfunding-site. Accessed 28 Sep 2019.

Brehm, Christian. 2012. *Das Venture-Capital-Vertragswerk: Die Bedeutung für Management und Strategie des Zielunternehmens.* Diss. der Universität Passau. Wiesbaden: Springer Gabler.

Buckley, Chris, and Austin Ramzy. 2017. Pyramid investigation has investors protesting near heart of Beijing. *The New York Times.* https://www.nytimes.com/2017/07/24/world/asia/china-beijing-protest-shanxinhui.html. Accessed 10 June 2019.

Bundesministerium für Finanzen. 2015. Abkommen zwischen der Bundesrepublik Deutschland und der Volksrepublik China zur Vermeidung der Doppelbesteuerung auf dem Gebiet der Steuern vom Einkommen und vom Vermögen. https://www.bundesfinanzministerium.de/Content/DE/Standardartikel/Themen/Steuern/Internationales_Steuerrecht/Staatenbezogene_Informationen/Laender_A_Z/China/2015-12-29-China-Abkommen-DBA.html. Accessed 18 Aug 2019.

Cao, Jacob. 2015a. Sie müssen mich retten. *Die Zeit*, Nr. 31/2015 vom 30.07.2015. https://www.zeit.de/2015/31/china-boerse-kapitalmarkt-aktien/seite-2. Accessed 10 Aug 2019.

Cao, Olivia. 2015b. China crowdfunding market research report 2015 is released; encouragement from the State Council presents huge opportunities for crowdfunding. http://en.pedaily.cn/Item.aspx?id=220256. Accessed 26 May 2019.

CBNEditor. 2018. New regulatory framework set to accelerate Chinese financial reforms. http://www.chinabankingnews.com/2018/04/09/new-regulatory-framework-set-accelerate-chinas-financial-reforms/. Accessed 10 June 2019.

China Securities Regulatory Commission. 2008. About CSRC (China Securities Regulatory Commission). http://www.csrc.gov.cn/pub/csrc_en/about/. Accessed 20 Aug 2019.

China.com. 2016. "E zū bǎo" 1 nián bàn xī zī 500 duō yì qiān yú cè zhèngjù máicáng jiāowài. https://news.china.com/domestic/945/20160201/21379203.html. Accessed 10 June 2019.

Chu Wei-Chi Rodney. 2008. The dynamics of cyber China: The characteristics of Chinese ICT use. *Knowledge & Policy* 21 (1): 29–35.

Chuen, David Lee Kuo, and Ernie G.S. Teo. 2018. The game of Dian Fu: The Rise of Chinese Finance. In *Handbook of Blockchain, digital finance, and inclusion*, ed. D.K.C. Lee and R.H. Deng, 1–36. London: Academic.

Clark, Duncan. 2016. *Alibaba – The house that Jack Ma built.* New York: HarperCollins Publishers.

De Marez, Lieven, Patrick Vyncke, Katrien Berte, Dimitri Schuurman, and Katrien De Moor. 2007. Adopter segments, adoption determinants and mobile marketing. *Journal of Targeting, Measurement and Analysis for Marketing* 16 (1): 78–95.

Deng, Iris. 2018. Chinese internet users surge to 802 million in test of government's ability to manage world's biggest online community. *South China Morning Post.* https://www.scmp.com/tech/china-tech/article/2160609/chinese-internet-users-surge-802-million-test-governments-ability. Accessed 26 Sep 2019.

Deutsche Bundesbank. 2016. FinTechs – Finanztechnologie Unternehmen. https://www.bundesbank.de/Redaktion/DE/Standardartikel/Aufgaben/Bankenaufsicht/FinTechs.html. Accessed 15 Aug 2018.

Dieter, Heribert. 2019. *Chinas Verschuldung und seine Außenwirtschaftsbeziehungen. Peking exportiert ein gefährliches Modell.* Deutsches Institut für Internationale Politik und Sicherheit. https://www.swp-berlin.org/fileadmin/contents/products/studien/2019S18_dtr_Website.pdf. Accessed 29 Sep 2019.

Ding, Ding, Guan Chong, David Kuo Chuen Lee, and Tan Lee Cheng. 2018. From ant financial to Alibaba's rural taobao strategy – How FinTech is transforming social inclusion. In *Handbook of blockchain, digital finance, and inclusion, Bd. 1*, 19–35. London: Academic.

Dorfleitner, Gregor, and Lars Hornuf. 2016. *FinTech – Markt in Deutschland*. Bundesministerium für Finanzen. http://www.bundesfinanzministerium.de/Content/DE/Standardartikel/Themen/Internationales_Finanzmarkt/2016-11-21-Gutachten-Langfassung.pdf%3F__blob%3DpublicationFile. Accessed 28 Sep 2019.

Dula, Christopher, and David Kuo Chuen Lee. 2018. Reshaping the financial order. In *Handbook of blockchain, digital finance, and inclusion, Bd. 1*, 1–18. London: Academic.

Esposti, Carl. 2015. 2015CF The crowdfunding industry report. http://www.smv.gob.pe/Biblioteca/temp/catalogacion/C8789.pdf. Accessed 19 June 2019.

Eurostat. 2019. Öffentlicher Bruttoschuldenstand. https://ec.europa.eu/eurostat/de/web/government-finance-statistics/statistics-illustrated. Accessed 18 Aug 2019.

Fischer, Doris, and Christoph Müller-Hofstede, eds. 2014. *Länderbericht China*. Bonn: Bundeszentrale für politische Bildung.

Forbes. 2019. The world's top 25 companies. https://www.forbes.com/pictures/mgl45fkfj/the-worlds-top-25-companies/#66eaaa075295. Accessed 19 Aug 2019.

Fortnum, Dennis, Warren Mead, Ian Pollari, Brian Hughes, and Arik Speier. 2017. *The pulse of FinTech Q4 2016 – Global analysis of investment in FinTech*. KPMG International Cooperative. https://assets.kpmg/content/dam/kpmg/xx/pdf/2017/02/pulse-of-fintech-q4-2016.pdf. Accessed 28 Sep 2019.

Freiberger, Harald, Christoph Giesen, and Stephan Radomshy. 2016. Der Crash zeigt: Chinas Börse ist kaputt. *Süddeutsche Zeitung*. https://www.sueddeutsche.de/wirtschaft/weltkonjunktur-der-crash-zeigt-chinas-boerse-ist-kaputt-1.2808026. Accessed 20 Aug 2019.

Funk, Andrea S. 2018. *Crowdfunding in China: A new institutional economics approach*. Diss. der Universität Würzburg. Cham: Springer Nature Switzerland AG.

Giesen, Christoph. 2019. China schlägt zurück. Peking wertet die Währung stark ab und verkündet den Stopp von Agrareinfuhren aus den USA – ein Affront gegen US-Präsident Donald Trump. Amerikas Börsen reagieren prompt, die Kurse stürzen ab. *Süddeutsche Zeitung* 15, 06. August 2019.

Gomber, Peter, Rober J. Kauffman, Chris Parker, and Bruce W. Weber. 2018. On the Fintech revolution: Interpreting the forces of innovation, disruption, and transformation in financial services. *Journal of Management Information Systems* 35 (1): 220–265.

Gordon, Roger, and Wei Li. 2018. Chinese tax structure. In *The Oxford companion to the economics of China*, ed. Shenggen Fan, Ravi Kanbur, Shang-Jin Wei, and Xiaobo Zhang, 178–182. Oxford: Oxford University Press.

Gough, Neil. 2016. Online lender Ezubao took $7.6 billion in Ponzi scheme. *The New York Times*. https://www.nytimes.com/2016/02/02/business/dealbook/ezubao-china-fraud.html?_r=1. Accessed 05 June 2019.

He, Huifeng. 2017. The perils of pyramid schemes: A dark corner of China's economic miracle. https://www.scmp.com/news/china/policies-politics/article/2108031/perils-pyramid-schemes-dark-corner-chinas-economic. Accessed 17 June 2019.

Heap, Ben, and Ian Pollari. 2017. *2017 FinTech100 – Leading global Fintech innovators*. H2Ventures & KPMG. https://home.kpmg/content/dam/kpmg/qm/pdf/H2-Fintech-Innovators-2017.pdf. Accessed 28 Sep 2019.

Hecking, Claus. 2014. Mächtig wie kaum eine andere. Die China Development Bank ist Pekings Türöffner zur Welt – und könnte bald noch viel einflussreicher werden. *Zeitonline*. https://www.zeit.de/2014/09/cdb-china-development-bank. Accessed 19 Aug 2019.

Heilmann, Sebastian. 2016. *Das politische System der Volksrepublik China*, 3. Aufl. Wiesbaden: Springer VS.

Hemer, Joachim, Uta Schneider, Friedrich Dornbusch, and Silvio Frey. 2011. *Crowdfunding und andere Formen informeller Mikrofinanzierung in der Projekt- und Innovationsfinanzierung*. Karlsruhe: Fraunhofer.

Hernández, Javier C., and Iris Zhao. 2017. As China's economy slows, 'business cults' prey on young job seekers. *New York Times*. https://www.nytimes.com/2017/09/27/world/asia/china-pyramid-schemes.html. Accessed 10 June 2019.

Hess, Patrick. 2014. Reformen, Status, Perspektiven des chinesischen Finanzsystems. In *Länderbericht China*, ed. Doris Fischer and Christoph Müller-Hofstede, 775–802. Bonn: Bundeszentrale für politische Bildung.

Hirn, Wolfgang. 2018. *Chinas Bosse – Unsere unbekannten Konkurrenten*. Frankfurt/New York: Campus.

Holmes, Chris. 2019. The rise and fall of P2P lending in China. https://www.finextra.com/blogposting/17107/the-rise-and-fall-of-p2p-lending-in-china. Accessed 03 June 2019.

Hua, Sha. 2019. Droht China die Rückkehr in die Schuldenfalle? – Sorgen vor neuer Finanzkrise wachsen. *Handelsblatt*. https://www.handelsblatt.com/politik/konjunktur/nachrichten/konjunktur-droht-china-die-rueckkehr-in-die-schuldenfalle-sorgen-vor-neuer-finanzkrise-wachsen/24055426.html?ticket=ST-3956860-A1bzP9Mp5APE5PJNbrZy-ap6. Accessed 19 Aug 2019.

Huang, Yiping. 2018. Local government debts. In *The Oxford companion to the economics of China*, ed. Shenggen Fan, Ravi Kanbur, Shang-Jin Wei, and Xiaobo Zhang, 207–211. Oxford: Oxford University Press.

Hunter, John Stanley, and Rosie Percy. 2019. A $2.7 billion German fintech startup backed by Facebook VC Peter Thiel is about to launch in the US. *Business Insider Deutschland*. https://www.businessinsider.com/n26-german-digital-bank-backed-by-peter-thiel-us-launch-2019-7?r=US&IR=T. Accessed 26 July 2019.

Kawase, Kenji. 2019. China's housing glut casts pall over economy. *Financial Times*. https://www.ft.com/content/51891b6a-30ca-11e9-8744-e7016697f225. Accessed 29 Sep 2019.

Kilian, Robert. 2009. Zur Strafbarkeit von Ponzi-schemes – Der Fall Madoff nach deutschem Wettbewerbs- und Kapitalmarktstrafrecht. *Onlinezeitschrift für Höchstrichterliche Rechtsprechung zum Strafrecht Juli* 10: 26–310.

Kollmann, Tobias, and Andreas Kuckertz. 2003. *E-Venture – Unternehmensfinanzierung in der Net Economy – Grundlagen und Fallstudien*. Wiesbaden: Springer Gabler.

Kroeber, Arthur. 2016. *China's economy. What everyone needs to know*. Oxford: Oxford University Press.

Lee, Georgina. 2017. China's giant Yu'e Bao money market fund riskier than US rival, Fitch says. *South China Morning Post*. https://www.scmp.com/business/money/markets-investing/article/2124465/chinas-giant-yue-bao-money-market-fund-riskier-us. Accessed 28 Sep 2019.

Lee, Ming-Chi. 2009. Factors influencing the adoption of internet banking: An integration of TAM and TPB with perceived risk and perceived benefit. *Electronic Commerce Research and Applications* 8 (3): 130–141.

Leng, Sidney. 2017. China's dirty little secret: Its growing wealth gap. *South China Morning Post*. https://www.scmp.com/news/china/economy/article/2101775/chinas-rich-grabbing-bigger-slice-pie-ever. Accessed 10 June 2019.

———. 2019. Cryptocurrency and pyramid schemes add to US$44.5 billion surge in illegal fundraising in China. *South China Morning Post*. https://www.scmp.com/news/article/2184372/china-grapples-surge-illegal-fundraising-online-lenders-bend-rules. Accessed 10 June 2019.

Liu, He, and Wei Li. 2013. *Xīn yī lún gǎigé de zhànlüè hé lùjìng*. Beijing: CITIC.

Liu, Jiefei. 2018. The rise and fall of China's online P2P lending. https://technode.com/2018/08/02/the-rise-and-fall-of-chinas-online-p2p-lending/. Accessed 28 May 2019.

Lovelock, Peter, and John Ure. 2002. E-Government in China. In *China's digital dream – The impact of the Internet on Chinese society*, ed. Junhua Zhang and Martin Woesler, 177–200. Bochum: Bochum University Press.

Lu, Feng. 2014. China's inflation in the post-reform period. In *The Oxford companion to the economics of China*, ed. Shenggen Fan, Ravi Kanbur, Shang-Jin Wei, and Xiaobo Zhang, 200–206. Oxford University Press.

Mackenzie, Annette. 2015. The FinTech revolution. *London Business School Review* 26 (3): 50–53.

McGeever, Jamie. 2018. Commentary: Deutsche's dwindling market cap belies still-huge global footprint. *Reuters*. https://www.reuters.com/article/us-europe-markets-deutsche/commentary-deutsches-dwindling-market-cap-belies-still-huge-global-footprint-idUKKCN1J21YP. Accessed 28 Sep 2019.

Ministry of Finance of the PRC. 2019. Annual report 2018 on the budget of the departments of the Ministry of Finance. http://bgt.mof.gov.cn/zhengwuxinxi/gongzuodongtai/201907/P020190719290491574621.pdf. Accessed 18 Aug 2019.

Mittal, Sachin, and James Lloyd. 2016. *The rise of FinTech in China – Redefining financial services*. EY Asian Insights Office, DBS Group Research. https://www.coursehero.com/file/32659164/ey-the-rise-of-fintech-in-chinapdf/. Accessed 28 Sep 2019.

Morey, Timothy, Theodore Forbath, and Allison Schoop. 2015. Customer data: Designing for transparency and trust. *Harvard Business Review*. https://hbr.org/2015/05/customer-data-designing-for-transparency-and-trust. Accessed 28 Sep 2019.

Münzel, Frank. 2007. Chinas Recht 5/16.3.07/2 – Unternehmenseinkommenssteuergesetz der VR China. http://www.chinas-recht.de/inhalt.htm. Accessed 19 Aug 2019.

National People's Congress of the PRC. 2018. Zhōnghuá Rénmín Gònghéguó Gèrén Suǒdéshuì Fǎ. http://www.gov.cn/xinwen/2018-10/20/content_5332913.htm. Accessed 18 Aug 2019.

Newshoo. 2015. E zū bǎo A2P móshì de jiānshǒu yǔ chuàngxīn. http://business.sohu.com/20150918/n421425224.shtml. vam 05 June 2019.

Ng, Huiwen. 2018. Over $27 million linked to Ezubao Ponzi scheme recovered in Singapore: Police. https://www.straitstimes.com/singapore/courts-crime/over-27-million-linked-to-ezubao-ponzi-scheme-recovered-in-singapore-police. Accessed 28 Sep 2019.

Ngai, J.L., J. Qu, N. Zhou, X. Liu, J. Lan, X. Fang, F. Han, and V. Chen. 2016. *Disruption and connection: Cracking the myths of China Internet finance innovation*. McKinsey Greater China FIG Practice.

Ou, C.X., C.L. Sia, and P.K. Banerjee. 2007. What is hampering online shopping in China? *Journal of Information Technology Management* 18 (1): 16–32.

Patwardhan, A. 2018a. Financial inclusion in the digital age. In *Handbook of blockchain, digital finance, and inclusion*, ed. D.K.C. Lee and R.H. Deng, 57–89. London: Academic.

———. 2018b. Peer-to-peer lending. In *Handbook of blockchain, digital finance, and inclusion*, ed. D.K.C. Lee and R.H. Deng, 389–418. London: Academic.

Petermann, G. 1963. *Marktstellung und Marktverhalten der Verbraucher*. Wiesbaden: Springer Fachmedien.

Ren, Daniel. 2019. China officially launches technology innovation board, with trading expected to begin within two months. *South China Morning Post*. https://www.scmp.com/business/companies/article/3014290/china-officially-launches-technology-innovation-board-does-not. Accessed 29 Sep 2019.

Reuters. 2014. China's Tencent latest online platform to launch fund product. https://www.reuters.com/article/tencent-finance/chinas-tencent-latest-online-platform-to-launch-fund-product-idUSL3N0KP1BY20140116. Accessed 16 June 2019.

———. 2019a. China will seinen Finanzsektor schneller öffnen. *Frankfurter Allgemeine Zeitung* vom 20.07.2019. https://www.faz.net/aktuell/finanzen/china-will-seinen-finanzsektor-schneller-oeffnen-16294551.html. Accessed 18 Aug 2019.

———. 2019b. Studie – Chinas Schuldenberg über 300 Prozent der Wirtschaftsleistung. https://de.reuters.com/article/china-schuldenberg-idDEKCN1UD139. Accessed 19 Aug 2019.

———. 2019c. Zinssatz in China fällt nach Reform nicht so stark wie erwartet. https://de.reuters.com/article/china-zinsen-idDEKCN1VA0JF. Accessed 20 Aug 2019.

Schillak, Jan. 2016. Schattenbanken in China – Bedrohung oder Chance? *ASIEN* 141:52–67.

Schlich, B., D. Ebstein, T. Schrezenmaier, and A. S. Turner. 2016. The relevance challenge – What retail banks must do to remain in the game. http://www.fmvoe.at/uploads/eytherelevancechallenge2016_725_DE.pdf. Accessed 13 Nov 2019.

Shane, D., and S. Pham. 2018. Jack Ma's online payments firm is now worth more than Goldman Sachs. https://money.cnn.com/2018/06/08/technology/ant-financial-ipo-valuation/index.html. Accessed 18 Aug 2018.

Sheng, C., J. Yip, and J. Cheng. 2017. FinTech in China – Hitting the moving target. https://www.oliverwyman.com/content/dam/oliver-wyman/v2/publications/2017/aug/FintechInChina_Hitting-theMoving-Target.PDF. Accessed 13 Nov 2019.

Shi, J. 2018. Xi Jinping puts China's mafia in cross hairs, but fears of judicial abuse remain. https://www.scmp.com/news/china/policies-politics/article/2130629/xi-puts-chinas-mafia-cross-hairs-fears-judicial-abuse. Accessed 10 June 2019.

Shim, Y., and D. Shin. 2016. Analyzing China's FinTech industry from the perspective of actor–network theory. *Telecommunications Policy* 40 (2–3): 168–181.

Song, J. 2014. Understanding the adoption of mobile innovation in China. *Computers in Human Behavior* 38 (1): 339–348.

State Council of the PRC. 2014. Guówùyuàn guānyú jìnyībù cùjìn zīběn shìchǎng jiànkāng fāzhǎn de ruògān yìjiàn. http://www.gov.cn/zhengce/content/2014-05/09/content_8798.htm. Accessed 20 Aug 2019.

Statista. 2019a. Marktvolumen alternativer Online-Finanzdienste in Deutschland von 2013 bis 2016 nach der Finanzierungsform (in Millionen Euro). https://de.statista.com/statistik/daten/studie/408548/umfrage/marktvolumen-online-finanzdienste-in-deutschland-nach-finanzierungsform/. Accessed 15 May 2019.

———. 2019b. Number of university graduates in China between 2007 and 2017. https://www.statista.com/statistics/227272/number-of-university-graduates-in-china/. Accessed 18 June 2019.

———. 2019c. China: Arbeitslosenquote von 2008 bis 2018. https://de.statista.com/statistik/daten/studie/167111/umfrage/arbeitslosenquote-in-china/. Accessed 18 June 2019.

Sun, Mancy, Piyushi Mubayi, Tian Lu, and Stanley Tian. 2017. The rise of China FinTech – Payment: The ecosystem gateway. Goldman Sachs' Global Investment Research division. https://hybg.cebnet.com.cn/upload/gaoshengfintech.pdf. Accessed 28 Sep 2019.

Sun, Tao. 2018. Balancing innovation and risks in digital financial inclusion – Experiences of Ant Financial Services Group. In *Handbook of blockchain, digital finance, and inclusion*, 37–43. London: Academic.

Symonds, Peter. 2016. The Ezubao scam: A sign of deeper problems in China's financial system. https://www.wsws.org/en/articles/2016/02/05/chin-f05.html. Accessed 10 June 2019.

Tang, Frank. 2017. Why Ponzi schemes are thriving in China despite crackdowns. *South China Morning Post*. https://www.scmp.com/news/china/money-wealth/article/2104062/chinese-ponzi-schemes-feed-publics-lack-financial-knowledge. Accessed 10 June 2019.

The World Bank. 2013. Crowdfunding's potential for the developing world. https://www.infodev.org/infodev-files/wb_crowdfundingreport-v12.pdf. Accessed 28 Sep 2019.

The World Bank Group. 2019. Rural Population (% of total population) World Bank staff estimates based on the United Nations Population Division's World Urbanization Prospects: 2018 Revision. https://data.worldbank.org/indicator/SP.RUR.TOTL.ZS?locations=CN. Accessed 26 July 2019.

Tobin, Damian, and Ulrich Volz. 2019. The development and transformation for the financial system in the People's Republic of China. In *Routledge handbook of banking*, ed. Ulrich Volz, Peter Morgan, et al., 15–38.

Tse, Edward. 2015. *China's disruptors – How Alibaba, Xiaomi, Tencent, and other companies are changing the rules of business*. New York: Portfolio/Penguin.

Uhm, Chul Hyhn, Chang Soo Sung, and Joo Yeon Park. 2018. Understanding the accelerator from resources-based perspective. *Asia Pacific Journal of Innovation and Entrepreneurship* 12 (3): 258–278.

Venkatesh, Viswanath, and Susan Brown. 2001. A longitudinal investigation of personal computers in homes: Adoption determinants and emerging challenges. *MIS Quarterly* 25 (1): 71–102.

Volz, Ulrich, Peter Morgan, and Yoshino Naoyuki, eds. 2019. *Routledge handbook of banking in finance in Asia*. London/New York: Routledge.

Walter, Carl E., and Fraser J.T. Howie. 2011. *Red capitalism. The fragile financial foundation of China's extraordinary rise*. Singapore: Wiley.

Wang, Helen H. 2012. *The Chinese dream – The rise of the world's largest middle class and what it means to you*. Charleston: Bestseller Press.

Wang, Orange. 2019. China's pork imports surged almost 80 per cent in August to cover gap left by African swine fever. *South China Morning Post*. https://www.scmp.com/economy/china-economy/article/3029961/chinas-pork-imports-surged-almost-80-cent-august-cover-gap. Accessed 26 Sep 2019.

Wang, Zhijie, and Gisela Färber. 2016. Lokale Verschuldung in China und Deutschland im Vergleich. *Speyerer Arbeitsheft*, Nr. 224. https://www.uni-speyer.de/files/de/Forschung/Publikationen/Arbeitshefte/224-Faerber-Wang.pdf. Accessed 02 Aug 2019.

Weitnauer, Wolfgang. 2018. *Handbuch Venture Capital: Von der Innovation zum Börsengang*. München: C.H.Beck oHG.

Woetzel, Jonathan, Jeonming Soeng, Kevin Wang, James Manyika, Michael Chui, and Wendy Wong. 2017. *China's digital economy – A leading global force*. McKinsey Global Institute. https://www.mckinsey.com/featured-insights/china/chinas-digital-economy-a-leading-global-force. Accessed 28 Sep 2019.

World Bank Group. 2019. Rural Population (% of total population) World Bank staff estimates based on the United Nations Population Division's World Urbanization Prospects: 2018 Revision. URL: https://data.worldbank.org/indicator/SP.RUR.TOTL.ZS?locations=CN. Accessed 26 July 2019.

Wright, Davis. 2015. Chinese Government's announcement to open bank cards network market. https://www.paymentlawadvisor.com/2015/03/13/chinese-governments-announcement-to-open-bank-cards-network-market/. Accessed 05 Aug 2018.

Xiang, Tracey. 2015. China's crowdfunding market as of 2014. https://technode.com/2015/01/23/chinas-crowdfunding-market-2014/. Accessed 16 June 2019.

Xin, Zhou, and Pearl Liu. 2019. China changes the way bank loan rates are set, with HSBC left out of rate-setting club. *South China Morning Post*. https://www.scmp.com/economy/china-economy/article/3023256/china-changes-way-bank-loan-rates-are-set-hsbc-left-out-rate. Accessed 18 Aug 2019.

Yu, Jianlong, Meiqing Han, and Yu Liu. 2003. *China – E-Commerce Development*. International Trade Center. https://www.intracen.org/WorkArea/DownloadAsset.aspx?id=52019. Accessed 28 Sep 2019.

Zhang, Bryan, et al. 2016. Harnessing potential – The Asia-Pacific alternative finance benchmarking. https://assets.kpmg/content/dam/kpmg/pdf/2016/03/harnessing-potential-asia-pacific-alternative-finance-benchmarking-report-march-2016.pdf. Accessed 18 June 2019.

Zhang, Joe. 2013. *Inside China's shadow banking. The next subprime crisis?* Hongkong: Enrich Professional Publishing.

Zhang, Junhua, and Martin Woesler, eds. 2002. *China's digital dream – The impact of the Internet on Chinese society*. Bochum: University Press.

Zhang, Ningdan. 2016. Lìyòng "e zū bǎo" fēifǎ xī cún jìn 1800 wàn. https://web.archive.org/web/20160216094719/http://www.legaldaily.com.cn/index/content/2016-01/13/content_6442679.htm. Accessed 28 Sep 2019.

Zhang, Shu, and Ryan Woo. 2018. Alibaba-backed online lender MYbank owes cost-savings to home-made tech. *Reuters*. https://www.reuters.com/article/us-china-banking-mybank/alibaba-backed-online-lender-mybank-owes-cost-savings-to-home-made-tech-idUSKBN1FL3S6. Accessed 28 Sep 2019.

Zhang, Tao, Christine Yip, G. Wang, and Q. Zhang. 2014. China crowdfunding report. http://www.ied.cn/sites/default/files/CIF%20China%20Crowdfunding%20Report_Final.pdf. Accessed 16 June 2019.

Zhongchoujia. 2018. Zhōngguó zhòng chóu hángyè fāzhǎn bàogào 2018. http://www.zhongchoujia.com/data/31205.html. Accessed 16 June 2019.

Zhou, Weihuan, Douglas W. Arner, and Ross P. Buckley. 2018. Regulating FinTech in China: From permissive to balanced. In *Handbook of blockchain, digital finance, and inclusion, Bd. 2*, 45–64. London: Academic.

Agricultural Policy and Food Supply 12

Sarah Kliem

After the CCP came to power in 1949, it swiftly implemented its agricultural policy at the national level (Aubert 2003, p. 424). The CCP's goal was and still is, on the one hand, to secure China's independent food supply and, on the other, to mobilize rural resources to create the conditions for rapid industrialization (Zhong 2009, p. 5). In terms of food security, the main focus was on grain self-sufficiency (Ghose 2014, p. 87). To achieve these goals, the government introduced a restructuring of all Chinese agriculture in five phases from 1950 to 1976 (Aubert 2003, p. 424). As a result, there were major famines and food supply problems. As a result, today a good food supply has a very high value in Chinese society and at the same time is an integral part of Chinese culture (Gandhi and Zhou 2014, p. 117).

Since 2010, the policy has focused on the sustainable development of the agricultural sector. The promotion of food quality and food safety has also been set as another priority, with an overall high level of food self-sufficiency remaining a priority. The National Medium and Long Term Framework sets the achievement of 100 percent self-sufficiency level in cereals and 95 percent self-sufficiency level in the remaining food items by the year 2020. To enable the implementation of these targets, the agricultural sector will be supported by other industry sectors. In addition, further reforms are being initiated on the part of policymakers. These include, on the one hand, the introduction of the minimum purchase price and target price, direct subsidies of agricultural products and agricultural materials, the abolition of agricultural taxes, the approval of superior crop varieties, and agricultural insurance premiums. In 2014, a campaign on the new food security strategy began in China. Until this year, the focus was on the independent supply of grain from its rural resources. In the new strategy of 2014, a change is supposed to take place at this

S. Kliem (✉)
University Mannheim, Mannheim, Germany

B. Darimont (ed.), *Economic Policy of the People's Republic of China*,
https://doi.org/10.1007/978-3-658-38467-8_12

point. Since then, food security is to be achieved through a combination of China's own and international agricultural resources. In this way, China is simultaneously pursuing more sustainable development of agriculture, as its resources are relieved. Nevertheless, the main part of the food for food security is to be obtained from China's rural resources. The remaining share can be covered by moderate imports and the application of new technologies and science (OECD 2018, pp. 132–134).

12.1 Validity of the Data

Data regarding current food demand as well as projections of future demand vary widely across Chinese and international sources. This is partly due to the difficulty of recording and forecasting food demand. Due to this, individual data may differ from each other. With regard to the forecast of future food demand, it should be noted that it depends on many different variables and therefore cannot be determined exactly. The following forecast should therefore only be regarded as a trend in the further development of food demand in China.

In addition, the data on production volumes in different sources are very different and also vary considerably within individual sources. The production volumes presented below are therefore derived from several sources. In each case, the volume with the greatest agreement within the various sources is listed. Due to the inconsistencies in production volumes and demand forecasts, it is difficult to determine the degree of self-sufficiency of individual foodstuffs precisely.

12.2 Food Requirements of the Chinese Population

China boasts the largest population in the world with a total of 1.4 billion inhabitants (OECD 2018, p. 40). Although China's population is expected to peak in 2030, the number of able-bodied people has been decreasing since 2015. One reason for this demographic change is the introduction of the one-child policy in 1980, as after this reform each family was only allowed to have one child (OECD 2018, p. 42). Despite the low birth rate, the Chinese population will continue to grow steadily until 2030, as the current relatively young population structure can compensate for the low birth rate. However, as the proportion of the young population class continues to decline, population growth will again decrease in the long term (Banister et al. 2012, p. 121).

In addition, average life expectancy is rising, which is further driving the aging of society. In 2015, 10.5 percent of the total population was already over 65 years old. Due to the continued decline in the birth rate and the increase in average life expectancy, the proportion of the population over 65 years of age is expected to increase by 20 percent by 2035 and 30 percent by 2050. Furthermore, the migration of the rural population to the cities continues to increase. In 1980, the urban population grew at a rate of 20 percent, whereas by 2010, the growth rate had already increased to 50 percent. The further growth rate of the urban population is projected to reach 75 percent by 2050. The rural population is therefore projected to halve by 2050. The expected increase in demographic change, as well as the further increase in migration of the rural population to the cities, are important factors in determining China's current and future food needs (OECD 2018, p. 42).

12.2.1 Development of Eating Habits

In the course of the last decades, the consumption of animal products as well as fruits and vegetables has increased due to the widespread increase in prosperity among the Chinese population. However, due to the inequality between the urban and rural population, the income difference between the population strata (von Braun 2007, pp. 1–2) and regionally determined different food preferences, the demand for different foods is very different in the individual regions of China.

The disparity between China's urban and rural populations is evident, among other things, in their respective dietary habits and overall food requirements. The urban population has a much higher consumption of high-quality and animal-based foods than the rural population. This is partly due to the income difference between urban and rural areas, and partly due to the poorer accessibility of these foods in rural areas (Zhou et al. 2014, pp. 9–10). Figure 12.1 shows the comparison between rural and urban household food consumption and average consumption per capita in 2017.

The food consumption of fish, meat, eggs, and dairy products of the urban population is overall significantly higher than the consumption of the rural population. The urban demand for fish and dairy products is almost twice as high as in rural areas. Only in the case of cereal consumption does the rural population have a higher consumption level than

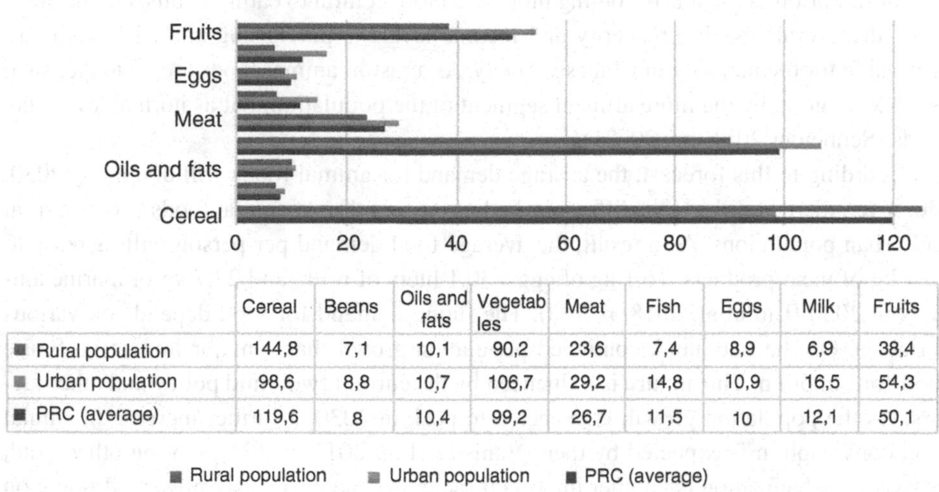

	Cereal	Beans	Oils and fats	Vegetables	Meat	Fish	Eggs	Milk	Fruits
Rural population	144,8	7,1	10,1	90,2	23,6	7,4	8,9	6,9	38,4
Urban population	98,6	8,8	10,7	106,7	29,2	14,8	10,9	16,5	54,3
PRC (average)	119,6	8	10,4	99,2	26,7	11,5	10	12,1	50,1

Fig. 12.1 Food consumption per capita (kg) in 2017. (Source: National Bureau of Statistics of China 2018)

the urban population (Jun 2018, p. 21). The difference in food consumption is due to urban-rural inequality on the one hand, and income divergence among different strata of the population on the other. As income increases, there is not only a change in the quantity and quality of food demand but also a restructuring of the demand for each food. Therefore, increasingly greater consumption of animal foods can be seen among the high-income class of the population (Zhou et al. 2014, p. 10). Although the consumption of animal foods varies within the different population strata and regions of China, an overall sharp increase in the consumption of animal products can be observed over the past few years regardless of this. In 1930, an average of 97 percent of calories were consumed through the consumption of grains and vegetables. In 2003, this proportion is only 63 percent. The proportion of calories consumed through meat consumption continued to increase over the years (Schneider 2011, pp. 20–21). Accordingly, the diet of the Chinese population is shifting from a semi-vegetarian to an animal product-dominant diet (Lam et al. 2013, p. 2046). The Chinese population consumes a total of 27 percent of the meat produced in the world. Thus, China records a higher consumption of meat than the United States. Pork demand accounts for the majority of total meat consumption, totaling 75 percent. The consumption of poultry is about 15 percent and the consumption of beef and mutton is about 10 percent of the total meat demand. Furthermore, the Chinese population consumes a quarter of the fish produced worldwide each year (Ghose 2014, pp. 90–91).

12.2.2 Forecast of Future Food Demand

The trend is for the demand for animal foods to continue to rise in the future. On the one hand, this is due to the trend towards an increasingly westernized diet, which means that the consumption of meat is becoming more and more central to eating habits. On the other hand, the overall rise in prosperity and income levels is purchasing animal foods more affordable for the majority of Chinese society. As a result, animal foods are no longer seen as a luxury good by the more affluent segment of the population, but as normal, everyday foods (Schneider 2011, pp. 20–21).

According to this forecast, the average demand for animal foods will double by 2030. Moreover, there is still a large difference in the consumption of animal foods between rural and urban populations. As a result, the average food demand per person will increase to 65.7 kg of meat products, 16.1 kg of eggs, 30.1 liters of milk, and 23.7 kg of marine animals in 2030 (Liu et al. 2018, p. 125). The future demand for food depends on various variables. On the one hand, continued population growth has a major impact on future food consumption. This in turn is influenced by the current two-child policy and other factors. As the population growth is expected to peak in 2030, a further increase in animal food consumption is expected by then (Banister et al. 2012, p. 121). On the other hand, increasing urbanization is another important factor. As the previously presented figure on food consumption in 2017 shows, there is a large difference between urban and rural populations in terms of animal food consumption (National Bureau of Statistics of China 2018).

If migration to urban areas continues to increase as expected and the rural population decreases as a result, it is likely that there will be an overall increase in the income of a further proportion of the population. In addition, an increasingly larger proportion of the population will have easier access to animal-based foods (Zhou et al. 2014, pp. 9–10). Therefore, it is very likely that in the long term, a large proportion of the Chinese population will consume food according to the consumption pattern of the previous urban population. Consequently, there may be a further increase in the consumption of animal foods (Gandhi and Zhou 2014, p. 122). At the same time, lower demand for cereals for consumption will be caused as the overall demand shifts towards higher quality food products such as animal products, vegetables, and fruits (OECD 2018, p. 42). Consequently, in the long run, the demand for livestock feed will increase even further (Gandhi and Zhou 2014, p. 122). However, due to demographic change, the population structure will shift towards an older Chinese society in the long run (OECD 2018, p. 42). The older segment of the population in China is consuming fewer animal foods overall and more plant foods. This could slow the trend towards an increase in animal food consumption. Overall, the demand for animal foods in China will continue to increase in the future, but the exact extent depends on various variables (Zhou et al. 2014, p. 11).

12.3 Classification of the Land Use Plan

In 2016, per capita, available arable land in China was 0.086 ha, whereas the global average is 0.192 ha (World Bank Group 2019). This illustrates the unbalanced ratio of available arable land to the PRC's large population. The total area of land suitable for agricultural use in China is 135 million ha, which is roughly 7–9 percent of the world's arable land. Of this, only 40 percent is high-quality arable land suitable for growing crops (Lam et al. 2013, p. 2045). Of the total arable land, 70 percent is used for growing cereal crops. Of this 70 percent of the total land area, 60 percent is devoted to cereals, 6 percent to soybeans, and 4 percent to tuber crops, of which potatoes account for half. In the category of cereal crops, the arable land of maize comprises the largest share with 25 percent. This is followed by acreage of rice (18 percent) and wheat (14 percent). The remaining 30 percent of the total arable land is divided between vegetables and melons (13 percent), oil crops (7 percent), and cotton, sugar crops, and other agricultural products (National Bureau of Statistics of China 2018) in small proportions.

Overall, the PRC can be geographically divided into nine different agricultural regions, which are distinguished by different temperature and precipitation levels and other specific characteristics (National Bureau of Statistics of China 2018). The 40 percent of high-value farmland is located in the Northeast China Plain, the North China Plain, and the Yangtze Plain (Lam et al. 2013, p. 2045). The Northeast China Plain includes the provinces of Jilin, Liaoning, as well as Heilongjiang, and has the second largest arable land in China, with a total of 27.8 million ha. Heilongjiang Province has the largest arable land in all of China, with approximately 15.9 million ha (National Bureau of Statistics of China

2018). In the Northeast China Plain, potatoes, rapeseed (China Agriculture Press 2014, pp. 102–103), and maize (Ministry of Agriculture and Rural Affairs of the PRC 2016) are the main crops grown. In addition, donkey, beef, and buffalo cattle breeding (China Agriculture Press 2014, pp. 124–125) is practiced, and egg-laying batteries are run (China Agriculture Press 2014, p. 129). In addition, Liaoning province is home to much fishing (China Agriculture Press 2014, p. 90) and poultry farming (China Agriculture Press 2014, p. 127), as well as vegetable farming (China Agriculture Press 2014, p. 111). Heilongjiang produces much of China's milk and dairy products (China Agriculture Press 2014, p. 128) and rapeseed (China Agriculture Press 2014, p. 103).

The second agricultural region includes Inner Mongolia. Here, cattle, buffalo, sheep, and goats are the predominant livestock (China Agriculture Press 2014, pp. 124–125). In addition, a lot of milk and dairy products are produced (China Agriculture Press 2014, p. 128) and oilseeds and sugar crops are cultivated (Ministry of Agriculture and Rural Affairs of the PRC 2016).

The provinces of Hebei, Shandong, and the urban areas of Beijing and Tianjin belong to the North China Plain. In this region, arable land is mainly used for growing wheat (China Agriculture Press 2014, p. 98), vegetables (China Agriculture Press 2014, p. 111), and fruits. Shandong Province also has a lot of fishing (China Agriculture Press 2014, p. 90) and livestock raising of poultry, rabbits (China Agriculture Press 2014, p. 127), sheep, and goats (China Agriculture Press 2014, p. 125). Furthermore, a lot of milk and eggs are produced in this entire agricultural region (China Agriculture Press 2014, pp. 128–129).

The Shanxi and Shaanxi regions, which encompass the Loess Plateau, mainly grow fruits (China Agriculture Press 2014, p. 116) and produce milk (Ministry of Agriculture and Rural Affairs of the PRC 2016).

The provinces of Jiangsu, Zhejiang, Anhui, Fujian, Jiangxi, Hubei, Hunan, and Henan, as well as the Shanghai metropolitan area, form the Yangtze Plain with 34.5 million ha of arable land, making it the largest agricultural region in China (National Bureau of Statistics of China 2018). Potatoes, soybeans (China Agriculture Press 2014, pp. 101–102), and fruits are grown in this region (China Agriculture Press 2014, p. 116), and pork (China Agriculture Press 2014, p. 125) and eggs are produced (China Agriculture Press 2014, p. 129). Forestry is most prevalent in Anhui, Fujian, and Jiangxi provinces. In addition, much of China's fisheries are maintained in Jiangsu, Fujian, Zhejiang, and Hubei provinces (China Agriculture Press 2014, p. 90). Jiangsu and Henan provinces are also home to poultry livestock, with Henan producing much milk (China Agriculture Press 2014, pp. 127–128), as well as beef and buffalo meat (China Agriculture Press 2014, p. 124). The arable land of Jiangsu, Zhejiang, Fujian, and Henan provinces is also the main cultivation area for sugar crops (China Agriculture Press 2014, p. 108). A large part of the arable land in Jiangsu, Henan, and Anhui provinces is also used for wheat cultivation (China Agriculture Press 2014, p. 98).

The south-western region of Sichuan, Guizhou, and Chongqing provinces is mainly responsible for livestock production. The livestock production of cattle, buffaloes, pigs

(China Agriculture Press 2014, pp. 124–125), rabbits, and poultry is mostly located in Sichuan Province (China Agriculture Press 2014, p. 127). In addition, Sichuan Province is the largest producer of cereals and oilseeds (Ministry of Agriculture and Rural Affairs of the PRC 2016).

The agricultural region of South China consists of the provinces of Guangdong, Guangxi, Hainan, and Yunnan. This region generates its main revenue from forestry (China Agriculture Press 2014, p. 90) and sugar crops (China Agriculture Press 2014, p. 108). Soybeans, potatoes (China Agriculture Press 2014, pp. 101–102), and fruits are also grown (China Agriculture Press 2014, p. 116). Fishing is primarily practiced in Guangdong province (China Agriculture Press 2014, p. 90), whereas poultry farming is located in both Guangdong and Guangxi (China Agriculture Press 2014, p. 127). The province of Yunnan has one of the largest cattle and buffalo livestock farms in the whole of China (China Agriculture Press 2014, p. 124).

The Gansu-Xinjiang region includes the provinces of Gansu, Xinjiang, as well as Ningxia and cultivates mainly vegetables (China Agriculture Press 2014, p. 111) and fruits (China Agriculture Press 2014, p. 116) on its agricultural land. Sugar crops (China Agriculture Press 2014, p. 108), potatoes, and soybeans are mostly grown in Xinjiang (China Agriculture Press 2014, pp. 101–102). Similarly, livestock production of cattle, buffalo, donkeys, sheep, and goats is mostly located in Xinjiang Province (China Agriculture Press 2014, pp. 124–125).

Qinghai Province and Tibet Autonomous Region form the Qinghai-Xizang Region. After Guangxi Province, Tibet is the second largest sugar-growing region (Ministry of Agriculture and Rural Affairs of the PRC 2016), which also increasingly grows potatoes and soybeans (China Agriculture Press 2014, pp. 101–102). Furthermore, there is a lot of cattle and buffalo cattle farming in this agricultural region (China Agriculture Press 2014, p. 124). Wheat is also grown in both regions (China Agriculture Press 2014, p. 98).

Overall, in addition to the production priorities described above, a large proportion of arable land in all agricultural regions is used for the cultivation of rice (China Agriculture Press 2014, p. 96) and maize (China Agriculture Press 2014, p. 98). The main areas of maize cultivation in this regard are in the northeastern and northern agricultural regions, as well as Henan Province and Inner Mongolia (Ministry of Agriculture and Rural Affairs of the PRC 2016). The North China Plain, along with the provinces of Henan, Anhui, Jiangsu, and the Tibet Autonomous Region, produces the bulk of wheat produced (China Agriculture Press 2014, p. 98). The remaining agricultural regions of China are tending to see a decline in the cultivation of wheat. The grain produced is increasingly used as animal feed, due to the increasing demand for livestock products. Livestock farms have therefore adapted geographically to the location of feed farms over the years. For example, soybeans are now mainly imported, with only a minimal amount produced in mainland China. As a result, the majority of pig and poultry livestock production is increasingly located in coastal regions of China (OECD 2018, pp. 48–49). The main meat production takes place in Shandong province, followed by Henan and Sichuan (China Agriculture Press 2014,

p. 127). Other large livestock farms are located in Guangxi, Guangdong, Hubei, Anhui, Liaoning, and Hebei provinces (China Agriculture Press 2014, pp. 124–125).

12.4 National Agricultural Production in Terms of Imports and Exports

Like the food demand of the Chinese population, the production structure of the agricultural sector in the PRC has changed over the years. From 1978 to 2002, the share of crop production in total production decreased by 22 percent, whereas the shares of fisheries and livestock increased from 2 to 11 percent each, and 15 to 31 percent, respectively (Schüller 2004, p. 517). From 1975 to 2013, the production of rice and cereals remained mostly constant, rising slightly in between and falling again towards the end. The production of meat, fruits, and vegetables increased constantly over the years, with the production volume of fruits and vegetables still exceeding that of meat. The largest overall increase was in milk production (OECD 2018, p. 49). The restructuring of China's agricultural sector can be understood through the theory of comparative cost advantage. To produce different agricultural products, each requires a different amount of land and labour input. Due to the relatively small area of arable land and the high availability of labor, China has a comparative cost advantage in the production of labor-intensive agricultural goods. These goods include, for example, horticulture and the production of processed agricultural products. This infers the disadvantage in the production of land-intensive goods such as the cultivation of cereals. China thus aligns the production of the agricultural sector according to its comparative cost advantage (OECD 2018, p. 61). This explains the decrease in the cultivation of cereals as well as the increase in the production of animal products and the increase in the cultivation of fruits and vegetables.

Figure 12.2 illustrates the agricultural production of crops in 2016.

In 2016, agricultural production totaled 565,381,000 tons of grain. Corn accounts for just under 40 percent of this production volume. Rice accounts for the second largest share, followed by wheat and other cereals. Other major agricultural products are potatoes, soybeans, and fruits and vegetables. Vegetables account for a very small share of the total production volume of fruits and vegetables. Furthermore, China generated 39,294,966 tons of oilseeds (Ministry of Agriculture and Rural Affairs of the PRC 2016). Of the agricultural products produced by arable farming, China mainly exported corn (85,921 tons), mandarins (561,681 tons), apples (1,334,636 tons), rice (1,200,000 tons), vegetables (9,250,000 tons), and soybeans (110,000 tons) in 2017. On the other hand, the main import products of the agricultural sector in 2017 are cereals and cereal flour (25,590,000 t), soybeans (95,530,000 t), wheat (4,420,000 t), rice (4,030,000 t), vegetable oil (5,770,000 t) and sugar (2,990,000 t) (National Bureau of Statistics of China 2018).

The focus of the agricultural sector in the production of animal food products is on the production of pork, aquatic products, eggs, and milk (National Bureau of Statistics of China 2018). In 2016, China produced 83,635,000 tons of meat. Overall, pork accounts for

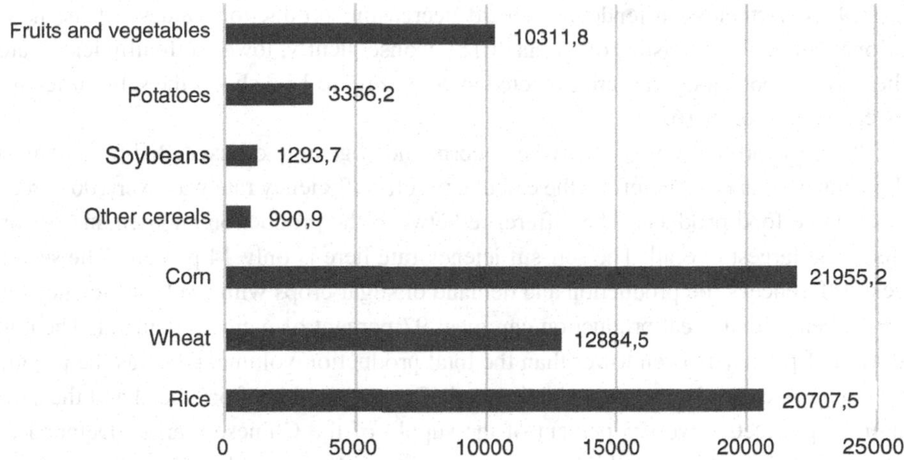

Fig. 12.2 Agricultural production in 2016 in 10,000 t. (Source: Ministry of Agriculture and Rural Affairs of the PRC 2016)

the largest share of the total production volume with 52,991,000 tons. Poultry production accounts for the second largest volume. Compared with the production volume of pork and poultry, mutton and beef account for a relatively small proportion of total meat production (Ministry of Agriculture and Rural Affairs of the PRC 2016). Catches from fisheries are divided almost equally between marine and freshwater animals. The production volume of aquatic animals exceeds the amount of pork produced but is lower than the total amount of meat produced. Furthermore, eggs and milk also register high production volume. In 2017, China exported a total of 1,570,000 live pigs and 2,300,000 live chickens. In addition, frozen pork (51,289 t), beef (922 t), poultry (128,674 t), and aquatic products (4,210,000 t) are exported (National Bureau of Statistics of China 2018). Thus, China is not only one of the world's main consumers of meat and fish, as shown in the previous chapter, but is also the largest producer and exporter of pork (Schneider 2011, p. 5) and aquatic products (Villasante et al. 2013, p. 932).

12.5 Self-Sufficiency Rates for the Various Agricultural Products

To determine the current self-sufficiency rates of the PRC in terms of various food products, the production volume is compared to the total food demand. The food demand data for the Chinese population is from 2015, and since then the population has increased from 1374.62 million in 2015 to 1395.38 million in 2019. Therefore, the food demand is still expected to be higher than the data below, reducing the self-sufficiency rate of various agricultural commodities (Statista 2019a). The production data used is from 2016. The production volume of the respective agricultural commodities decreased slightly from 2015 to 2017. Due to the only minor deviation, similar production volumes are assumed for 2019. Nevertheless, a tendency toward decreasing production values can be noted (National Bureau of Statistics of China 2018). Consequently, lower self-sufficiency rates in the various food categories are expected in 2019. Figure 12.3 shows the self-sufficiency rates calculated for 2016.

In 2016, the production of rice, wheat, corn, and vegetable oil exceeded food demand in the Chinese market. Therefore, the calculated self-sufficiency rate was over 100 percent for the above food products. The difference between the production and demand of soybeans is the largest overall. The self-sufficiency rate here is only 14 percent. The second largest difference is the production and demand of sugar crops with a self-sufficiency rate of 79 percent. Total meat production can meet 97 percent of national demand. The consumption of poultry is even lower than the total production volume, whereas the production of mutton can only just meet the demand. The pork and beef produced and the catch of aquatic products cover 95 percent of the supply of the Chinese market. Demand for dairy products is significantly higher than production volume, resulting in milk registering only an 85 percent self-sufficiency rate (Liu et al. 2018, p. 122; Ministry of Agriculture and Rural Affairs of the PRC 2016; National Bureau of Statistics of China 2018). The imports and exports presented in the previous section provide further insight into the interpretation of the data. In 2017, China's exports of agricultural products were mainly pork, beef, poultry, and aquatic products. These exports therefore also indicate a high rate of self-sufficiency in these food categories. On the other hand, China mainly imports cereals. Here, the import of soybeans shows the largest volume. The high difference between production and demand of soybeans is thus compensated by imports. The import of sugar crops also fills the existing gap between national supply and demand. These imports confirm a low national self-sufficiency rate. Furthermore, cereals and cereal flour, rice, and wheat are imported. This is contradictory to the cited high self-sufficiency rate of the

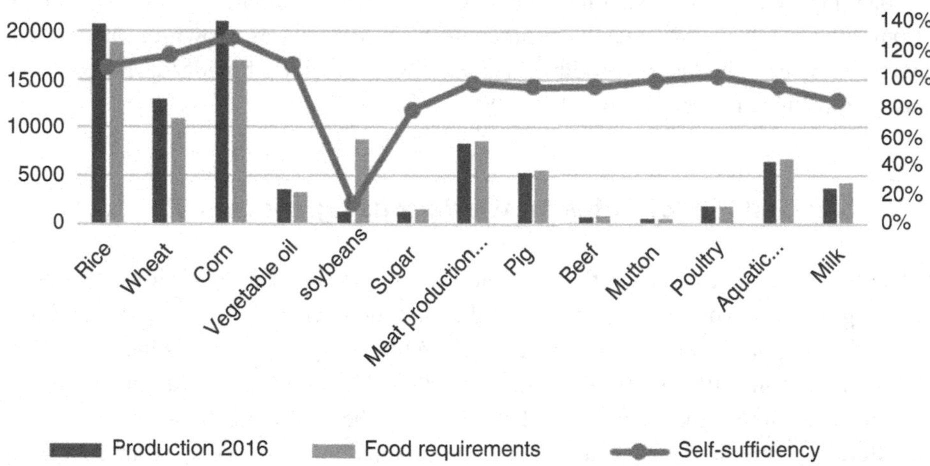

Fig. 12.3 Determination of 2016 self-sufficiency rates. (Source: Ministry of Agriculture and Rural Affairs of the PRC 2016; National Bureau of Statistics of China 2018; Liu et al. 2018, p. 122)

respective agricultural products. Therefore, it can be surmised that the demand for cereals is higher than the national production volume. The large import volume of vegetable oil also indicates a controversy regarding the self-sufficiency rate of this commodity (National Bureau of Statistics of China 2018).

To further verify the self-sufficiency rate of cereals, the following compares the per capita production volume with the per capita demand for cereals. In 2016, 447 kg of cereals were produced per capita (Ministry of Agriculture and Rural Affairs of the PRC 2016). The data presented in the second section show that in 2017, an average of 119.6 kg of grain was consumed per capita for consumption. Accordingly, 327 kg of cereals remained for other purposes. The data also recorded a per capita demand for meat of about 27 kg. This demand was mainly met by the consumption of pork (National Bureau of Statistics of China 2018). A total of 13 kg of feed is required to produce one kilogram of industrially produced pork. Assuming meat demand is met by pork consumption alone, 351 kg of feed would thus be required per capita to produce sufficient pork (Schneider 2011, p. 22). This thus confirms the hypothesis that the actual self-sufficiency rates for cereals, rice, and wheat are lower than the previously determined level. It is unclear whether this difference between production and demand is attributable to the deficit of cereals for direct consumption or for use as animal feed.

While the use of imported grains is unclear, these imports show that self-sufficiency rates for grains, rice, and wheat cannot exceed 100 percent and that there are deficits here as well. Accordingly, it is not possible for the country to be self-sufficient if current livestock production levels are maintained, as China has shifted cultivation from land-intensive to labour-intensive crops (OECD 2018, p. 61) and therefore relies on grain imports (OECD 2018, p. 50). Therefore, while China's overall animal food self-sufficiency rate is high, the national supply of cereals is insufficient (Liu et al. 2018, p. 125; Ministry of Agriculture and Rural Affairs of the PRC 2016).

This also confirms the low self-sufficiency rate of soybeans, which is attributable to the livestock sector's need for feed. As shown in the fifth section, over the past decades the production of animal food has continued to increase, while the national supply of cereals has decreased (Lohmar and Gale 2008, p. 12). Due to this, China changed its grain cultivation structure in the past and became the main importer of soybeans in the world from an exporter in 2003 (Schneider 2011, pp. 10–11). Due to the shift in cropping structure away from land-intensive and towards labour-intensive goods, China has been importing increasing amounts of agricultural products, mainly grains and vegetable oil, since 2000. These goods together account for more than half of agricultural imports (OECD 2018, p. 50). In 1996, three-fifths of total national grain production was already used as livestock feed for livestock (Tian and Chudleigh 1999, p. 396). Therefore, the imports of soybeans listed above are mainly used to meet the demand for livestock feed (Lam et al. 2013, p. 2046).

The focus of the national cereal production is now on the production of maize. The reason for the increasing cultivation of maize, instead of other cereals, is the increased number of livestock farms, as maize is used as one of the main feedstuffs. Government measures severely limit the use of maize in the industry for the production of, for example, sweeteners or ethyl alcohol, so that most of the resources can be devoted to livestock production. Thus, grain production in China still contributes to meeting the demand for livestock feed (Lam et al. 2013, p. 2046). However, coverage by national production alone is no longer possible. Furthermore, there is a presumption that the production of vegetable oil is lower than the national demand and thus also has a lower self-sufficiency rate. Due to the low self-sufficiency rate of sugar crops and milk, China is also likely to depend on imports for these food products (Liu et al. 2018, p. 125; Ministry of Agriculture and Rural Affairs of the PRC 2016; National Bureau of Statistics of China 2018).

12.5.1 Forecast of the Further Development of Self-Sufficiency Rates

The increase in per capita demand for animal foods and the reasons for this increase have already been described. Assuming a constant production volume of meat, the self-sufficiency rate in 2030 is around 90 percent. A change in meat consumption can be observed, as the demand for poultry and beef increases significantly. Pork still has the highest self-sufficiency rate at 96 percent. The future demand for beef and poultry significantly exceeds the current production volume, so self-sufficiency is only 56 percent and 69 percent respectively (Liu et al. 2018, p. 125; Ministry of Agriculture and Rural Affairs of the PRC 2016; National Bureau of Statistics of China 2018). To determine the cereal self-sufficiency rate, data on per capita production volume and per capita demand for cereals and meat are used below, as in the previous section. In 2030, per capita demand for meat in China is projected to be 62 kg (Liu et al. 2018, p. 125), higher than meat consumption in Germany, which fell to 59 kg in 2016 (Heinrich-Böll-Stiftung et al. 2018, p. 13). If the per capita demand for meat increases to circa 65 kg (Liu et al. 2018, p. 125), then the consumption of feed for livestock increases to 845 kg of grain per capita (Schneider 2011, p. 22). As noted above, the current per capita production of grain is 447 kg (Ministry of Agriculture and Rural Affairs of the PRC 2016). Accordingly, the demand for grain would double for livestock alone. Together with normal grain consumption, this would increase the per capita demand for grain to 964 kg by 2030. The calculated demand for animal feed must be put into perspective, however, as it is assumed that the demand will be covered by pork alone. The amount of feed required in the future, however, depends above all on the further development of demand for animal products. Overall, the rearing of chickens, followed by cattle, is less feed-intensive than pig rearing. An increase in the demand for poultry and a reduction in the demand for pork, for example, would therefore have a positive impact on the demand for feed, as this would be reduced (Brown and Ewers 1997, pp. 46–47).

The projected total consumption of milk is only minimally higher than the current demand and the demand for aquatic products has decreased by half compared to 2015. Thus, national production can still meet 85 percent of the demand for dairy products. Accordingly, the overall forecast does not foresee an increase in the total demand for fish and dairy products (Liu et al. 2018, p. 125; Ministry of Agriculture and Rural Affairs of the PRC 2016; National Bureau of Statistics of China 2018). In contrast, the ratio between the per capita food consumption forecast in 2030 and the per capita food demand in 2015 indicates a significant increase in the consumption of dairy products and fish (Liu et al. 2018, p. 125). Studies regarding the future sales volume of different food categories also forecast a further increase in the total demand for all foods. This includes the demand for fish and dairy products (Statista 2019b). The forecast of consumption of dairy products in the Hong Kong market also predicts a further increase over the next few years. Therefore, regardless of the available data, the demand for milk and fish is expected to increase (Statista 2019c). Accordingly, the self-sufficiency level of milk, as well as aquatic products is expected to decrease as the demand far exceeds the current national production. The total consumption of eggs in 2030 is significantly lower than the production volume in 2016, where the calculated self-sufficiency rate is 135 percent. The per capita demand for eggs increased from 10 kg in 2017 to 16 kg in 2030. The per capita production volume of eggs is 22 kg and is therefore consistent with this high self-sufficiency rate. In contrast, the demand for sugar crops in 2030 is significantly higher than the consumption in 2015, causing the self-sufficiency rate to drop even more. In 2016, according to official data, national production is slightly higher than the demand for vegetable oil. However, as shown above, the high imports point to a lower self-sufficiency rate. According to the forecast for the year 2030, the demand increases far above the current production level, reducing the self-sufficiency rate to 77 percent.

According to this, in 2030 national production is clearly below the future demand for poultry, beef, milk, sugar, and vegetable oil. Due to the increase in demand for pork, a lower independent supply is also to be expected here. The ratio of the demand for fish to the available fish catch will depend in the future on the actual demand as well as on the production possibilities, which will be discussed in more detail below. The data above show that the calculated future demand is higher than the current production volume, which causes the self-sufficiency rates of the various food products to fall below the 2016 rates. Only egg production continues to exceed demand, allowing for independent supply in the future (Ministry of Agriculture and Rural Affairs of the PRC 2016; National Bureau of Statistics of China 2018; Liu et al. 2018, p. 125).

12.5.2 Future Development of Production Possibilities

The previous section assumes that production volumes will remain constant, as future production possibilities depend on various variables. In the following, we will consider the development of China's further production possibilities under the influence of various factors. It is assumed that China will continue to pursue the strategy of producing labour-intensive agricultural products, in line with its comparative cost advantage (OECD 2018, p. 61).

12.5.2.1 Inequality in the Distribution of Water Resources

The development of the agricultural sector in China depends, among other things, on various environmental factors. First and foremost among these is water supply. The PRC has about 6 percent of the world's water resources (Wang et al. 2018, p. 903), thus only 2100 m^3 on average is available to the population per capita annually. This is roughly a quarter of the average annual global per capita water resources available. In addition, the extent of water supply varies by region due to differences in the distribution of water resources. Precipitation also varies significantly within different regions, which is why river flow varies. The river basins in the southeast and southwest receive relatively high inflows with an average annual rainfall of 500 mm. In contrast, the average annual precipitation for the northern river basins is only 150 mm. In addition, the average precipitation in northwestern China changes greatly due to seasonal changes. Therefore, a rainy year can correspond to eight times the precipitation of a drier period. Overall, 60–80 percent of total precipitation in China falls in monsoon phases, causing the amount of river water to change seasonally (Wang et al. 2018, pp. 900–901). Of the total water resources in China, 68 percent is used for agriculture, of which 90.3 percent is in turn used for soil irrigation (Wang et al. 2018, p. 903). In 2015, the proportion of cropland with irrigation facilities was equivalent to 48.8 percent. Irrigators consume about 55 percent of the total water resources available annually. The main problem is that not even half of this water is used effectively. Overall, although agricultural water use increased from 378.4 to 385.1 billion m^3 from 2000 to 2015, the agricultural sector's share of total water use fell from 69 to 63 percent over the same period (OECD 2018, p. 43). Due to the growing population as well as better water supply, the water demand of China's population has increased sharply. In addition to the increase in domestic water demand, the demand from industries continues to increase. Due to the overall increase in water consumption, the water resources available for agriculture are decreasing in the long run. Another problem is the contamination of water resources by industry and agriculture, which have already polluted a large part of the water resources. In addition, climate change is expected to increase the inequality of the spatial distribution of water availability. Accordingly, precipitation continues to decrease in northeast China and increase in the southwest and southeast (Wang et al. 2018, pp. 903–904). In addition, the rise in temperature, the melting of the Himalayan glacier, and the rise in seawater level will greatly affect the cropping structure of China's agricultural sector. At this point, climate change should be noted as another risk factor and must be taken into account in further planning for the productivity and sustainability of the agricultural sector (OECD 2018, pp. 44–45).

The most severe water shortages in China are currently experienced by regions in the Yellow River, Huaihe River, and Haihe River basins (Wang et al. 2018, pp. 903–904). The water shortage in the Huaihe River region affects Shandong, Jiangsu, Anhui, and Henan provinces. Due to seasonal rainfall, the annual per capita available water resources are only one-fifth of the national average. Yet this region contributes 16 percent of total agricultural production (Xia et al. 2011, pp. 170–171). The Yellow River is the main water supply for the northern and northwestern agricultural regions. In total, these regions have

2 percent of the water resources available to cultivate 15 percent of the total agricultural land (Wu et al. 2004, p. 34). Due to the lack of water, the northwestern agricultural region is experiencing increased desertification, which is increasingly inhibiting agricultural production (Lam et al. 2013, p. 2045). The Haihe region includes Hebei province, Beijing and Tianjin urban areas, and peripheral areas of Inner Mongolia and Liaoning, Shanxi, Shandong, and Henan provinces. Due to the lack of water in this area, the amount of annual water available per capita is only 325 m^3. Furthermore, the northern regions of Hebei and Inner Mongolia hold 60 percent of the total farmland but have only 19 percent of China's water resources (Wang et al. 2018, p. 900).

12.5.2.2 Reduction of the Area Under Cultivation

Another important factor regarding the development of the agricultural sector is the increasing decrease of agricultural land in China. The reduction in arable land is due to various circumstances, such as urbanization, the expansion of livestock farms, and the long-term increase in the number of aquaculture farms. The gain in arable land in other agricultural regions should be contrasted with the above-mentioned loss. In this context, water scarcity in individual agricultural regions also plays an important role.

Due to economic development, population growth, land degradation, and, most importantly, urbanization (Tan et al. 2005, p. 187), the total arable land in China decreased by 19.2 million ha from 1991 to 2004. This is roughly equivalent to 15.3 percent of China's total arable land (OECD 2018, p. 43). It is assumed that China loses about 400,000 ha of arable land each year due to construction (Deutsche Vertretung in China 2019). Especially in the northern agricultural region, Hebei province, and the Beijing and Tianjin urban areas, both large and small cities are surrounded by fertile farmland. Thus, increasing urbanization in this region and throughout China comes at the expense of fertile cropland (Tan et al. 2005, p. 194). By comparing data on arable land in each province from 2015 and 2016, it is possible to determine which regions are currently losing the most arable land. The largest percentage loss in terms of own cropland is in the urban areas of Shanghai and Beijing, each with a decrease of 45,500 ha (−13.4 percent) and 22,400 ha (−12.9 percent), respectively. They are followed by Shanxi province with 46,900 ha (−1.24 percent), Hubei with 108,900 ha (−1.37 percent), Liaoning with 155,800 ha (−3.69 percent), and Hainan with 22,000 ha (−2.6 percent). After offsetting the loss against the new arable land created in the respective provinces of each agricultural region, the following agricultural regions record the largest decrease in arable land. The northeastern agricultural region loses a total of 26,100 ha of arable land, although the loss is greatly reduced by the new arable land gained in Heilongjiang. As a result, the decrease in arable land in the northeastern province amounts to 889,000 ha. The highest reduction in arable land is in the Yangtze plain. Here, a total of 1,944,100 ha will be lost. This loss of area is only offset by the newly gained area in the provinces of Hunan and Henan, which totals just under 124,000 ha (Ministry of Agriculture and Rural Affairs of the PRC 2016).

As illustrated, the northern and northeastern agricultural regions and the Yangtze Plain are among the most fertile agricultural regions in all of China (Lam et al. 2013, p. 2045).

The increasing loss of land in these areas has a great impact on China's agricultural production. Apart from land loss, these agricultural regions have severely limited water resources. Especially the northern, northeastern, and northwestern agricultural regions are affected by this water scarcity. In addition, there are water shortages in Jiangsu, Anhui, and Henan provinces. Due to these problems, production in the above agricultural regions is expected to decline in the long run (Wang et al. 2018, pp. 903–904). From the above, it can be seen which crops are mainly affected in each agricultural region. In the northeastern agricultural region, a decline in cultivation is mainly at the expense of rapeseed (China Agriculture Press 2014, p. 103), maize (China Agriculture Press 2014, p. 98), and potato (China Agriculture Press 2014, p. 102) production. As a result, the northeastern agricultural region is also able to grow fewer grains, especially corn, oilseeds (Ministry of Agriculture and Rural Affairs of the PRC 2016), and wheat (China Agriculture Press 2014, p. 98). In the Yangtze Plain agricultural region, a reduction in the cultivation of cereals and oilseeds is possible, especially in Anhui Province (Ministry of Agriculture and Rural Affairs of the PRC 2016). In the entire agricultural region, there may also be a reduction in the production of wheat (China Agriculture Press 2014, p. 98), soybeans (China Agriculture Press 2014, p. 101), and sugar crops (China Agriculture Press 2014, p. 108).

The largest increase in the cultivated area due to newly acquired arable land is recorded in the South-West Agricultural Region and the Gansu-Xinjiang Agricultural Region, whereby in the latter region the gain in arable land is mainly attributable to Xinjiang Province. In addition, a significant gain in agricultural land is recorded in Inner Mongolia (Ministry of Agriculture and Rural Affairs of the PRC 2016). The increase in land in the southwest agricultural region would allow for the expansion of sugar crops (China Agriculture Press 2014, p. 108), cereals, and oilseeds (Ministry of Agriculture and Rural Affairs of the PRC 2016). In Xinjiang province, it is also possible to grow more sugar crops (China Agriculture Press 2014, p. 108), maize (China Agriculture Press 2014, p. 98), and potatoes (China Agriculture Press 2014, p. 102). On the additional land in Inner Mongolia, it would be possible to expand the cultivation of sugar crops and oilseeds (Ministry of Agriculture and Rural Affairs of the PRC 2016). However, this assessment is based only on the current cultivation structure of the provinces and agricultural regions. It is unclear to what extent the newly acquired cultivation areas are suitable for the cultivation of the above-mentioned crops and can compensate for the loss of the other agricultural regions.

China is pursuing a strategy of a labour-intensive cropping structure with a focus on the production of livestock products (OECD 2018, p. 61). Based on the assumption that China will still maintain this focus in 2030, it is likely that national livestock production will be further increased to meet the increased demand for animal-based food. This will increase the amount of land needed for livestock production as well as the demand for feed (Gandhi and Zhou 2014, p. 122). Consequently, it is possible that cropland will again decrease due to the increase in livestock farms. For example, the northeastern agricultural region in Heilongjiang Province is home to much of China's dairy production (China Agriculture Press 2014, p. 128) on the one hand, and the main cereal production on the other (Ministry of Agriculture and Rural Affairs of the PRC 2016). Therefore, an expansion of dairy production would likely come at the expense of cereal acreage. Similarly, the northwestern agricultural region is home to the main producers of dairy products (China Agriculture Press 2014, p. 128) and much of the oilseed and sugar production (China Agriculture Press 2014, p. 108). Shandong province in the northern agricultural region also produces the most eggs (China Agriculture Press 2014, p. 129), fish (China Agriculture Press 2014, p. 90), and meat in all of China, in addition to grain and oilseed production. Similarly, Hebei province is among the major grain-growing regions while producing much of the meat (Ministry of Agriculture and Rural Affairs of the PRC 2016), eggs, and milk (China Agriculture Press 2014, pp. 128–129). In the Yangtze Plain agricultural region, the same problem arises as in Henan, Hunan, and Hubei provinces. Henan province is home to the largest acreage of oilseeds and grains, as well as the main production of meat (Ministry of Agriculture and Rural Affairs of the PRC 2016), eggs, and milk (China Agriculture Press 2014, pp. 128–129). Similarly, in Hunan province, oilseed crops and meat production are primarily located. In Hubei province, oilseed crops are mainly located, as well as livestock, aquaculture, and egg-laying farms. Furthermore, Sichuan province has a lot of arable land for grain and oilseed cultivation in the southwest agricultural region. At the same time, however, a lot of meat (Ministry of Agriculture and Rural Affairs of the PRC 2016) and eggs (China Agriculture Press 2014, p. 129) are produced in this province.

An increase in the production of animal foodstuffs would therefore very likely lead to a reduction in the area under cultivation in these provinces and regions. The increase in demand for animal feed (Lam et al. 2013, p. 2045) would then be offset by an increasingly reducing area under cereal crops. At this point, it should be emphasized that the main areas of maize cultivation are in the northeastern and northern agricultural regions, as well as Henan province and Inner Mongolia. Apart from Inner Mongolia, these regions are all experiencing a sharp decline in their cultivated areas (Ministry of Agriculture and Rural Affairs of the PRC 2016) and have severely limited water resources (Wang et al. 2018, pp. 903–904). Therefore, a further reduction in cultivated land due to urbanization or increasing numbers of livestock farms, as well as a reduction in water resources, would be accompanied by a reduction in the production possibilities of maize and other crops (Ministry of Agriculture and Rural Affairs of the PRC 2016). The use of maize is almost exclusively limited to livestock production (Lam et al. 2013, p. 2046). Even partial elimination of maize production would therefore make the national livestock sector dependent on imports of maize, similar to soybeans (Schneider 2011, p. 11).

With regard to the further development of demand and supply of aquatic products, a similar problem arises. The future production possibilities of fish are expected to become more limited. At present, China meets half of its demand for aquatic products from freshwater and half from seawater animals (Villasante et al. 2013, p. 932). Due to the increasing overfishing of the oceans, in the long run, a large part of the demand will have to be met by raising fish on aquaculture farms. This in turn increases the demand for grain. Moreover, maintaining aquaculture farms comes at the expense of China's scarce water and land resources. Therefore, either the national supply of fish or the available land and water resources will decrease. If the national supply is reduced, an increase in the import of aquatic products would be likely (Brown and Ewers 1997, p. 49).

12.6 Conclusion

Food supply in China has changed significantly over the past decades. The overall demand for food, especially animal products, has increased over the years and will continue to increase at least until 2030. Currently, meat production can almost fully meet demand in the national market, but there are shortfalls in the production of milk, sugar crops, and soybeans in particular. A comparison of 2016 production figures with 2017 demand data reveal self-sufficiency rates for rice, wheat, maize, and vegetable oil of over 100 percent each. Import data, in turn, reveal large import volumes of these agricultural products. The resulting controversy suggests that the actual demand is higher than the national production volume after all. This is also confirmed by the comparison between per capita production and per capita demand for cereals. The demand for food will continue to increase in all categories until 2030, including the consumption of fish and milk. Accordingly, while production volumes remain constant, a decline in self-sufficiency rates is observed for all food products. Based on the strategy of comparative cost advantage, it can be assumed that China will further increase the production of animal foods. This greatly increases the demand for feed for livestock. The decisive factor for the amount of feed required here is the future consumption structure or the extent to which there is an actual change in demand from pork to poultry and beef. Overall, however, the current production volume will not be able to cover the future demand for animal feed without an increase in the production of grain. The cultivation structure of the Chinese agricultural sector is now increasingly concentrated on the production of maize. This, along with imported soybeans, serves as the main feedstuff in livestock farming. To meet the demand for animal feed, the national production of maize would have to be increased. On the other hand, there is an increasing loss of arable land due to growing urbanization. Furthermore, as a result of the unequal distribution of water resources, there is water scarcity in the most important cultivation regions in the country. In addition, more land is also required due to an increase in livestock farms. This may result in a further loss of cultivable land. These problems, therefore, suggest a decline in production volume in the long run. In this respect, the cultivation of maize is particularly affected, as it is mainly produced in regions affected by a large loss of cultivable land and water scarcity. Therefore, the question arises whether the PRC will have to rely on the import of corn, in addition to soybeans, in the long run. The projection of declining self-sufficiency rates in 2030, together with the potential loss of production opportunities, suggests a long-term increase in the volume of imported agricultural products. The extent to which this can be circumvented by new techniques, land consolidation, international mergers, genetic manipulation, upgrading of land for agriculture, etc. remains to be seen.

References

Aubert, Claude. 2003. Landwirtschaftspolitik. In *Das große China-Lexikon. Geschichte, Geographie, Gesellschaft, Politik, Wirtschaft, Bildung, Wissenschaft, Kultur*, ed. Brunhild Staiger, 424–427. Darmstadt: Primusverlag.

Banister, Judith, David E. Bloom, and Larry Rosenberg. 2012. Population aging and economic growth in China. In *The Chinese economy. A new transition*, ed. Masahiko Aoki, 114–149. Basingstoke: Palgrave Macmillan.

von Braun, Joachim 2007. *The world food situation. new driving forces and required actions*. International Food Policy Research Institute. http://citeseerx.ist.psu.edu/viewdoc/download?doi=10.1.1.559.576&rep=rep1&type=pdf. Accessed 17 June 2019.

Brown, Lester R., and Traute Ewers. 1997. *Wer ernährt China? Alarm für einen kleinen Planeten*. Hamburg: Deukalion.

China Agriculture Press. 2014. *Chinas Agriculture Yearbook 2014*. http://english.agri.gov.cn/service/ayb/201511/P020151104554814851102.pdf. Accessed 28 May 2019.

Deutsche Vertretung in China. 2019. *Basisinformationen zur chinesischen Landwirtschaft*. https://china.diplo.de/cn-de/themen/wirtschaft/landwirtschaft-basisinformationen. Accessed 28 May 2019.

Gandhi, Vasant P., and Zhangyue Zhou. 2014. Food demand and the food security challenge with rapid economic growth in the emerging economies of India and China. *Food Research International* 63: 108–124.

Ghose, Bishwajit. 2014. Food security and food self-sufficiency in China: From past to 2050. *Food and Energy Security* 3 (2): 86–95.

Heinrich-Böll-Stiftung, Bund für Umwelt und Naturschutz Deutschland, und Le Monde Diplomatique. 2018. *Fleischatlas 2018. Daten und Fakten über Tiere als Nahrungsmittel*, 2. Aufl. https://www.bund.net/fileadmin/user_upload_bund/publikationen/massentierhaltung/massentierhaltung_fleischatlas_2018.pdf. Accessed 09 June 2019.

Jun, Chen Yin, ed. 2018. *Wo guo xin xing liang shi an quan guan yan jiu. Di 1 ban*. Beijing: Zhong guo nong ye ke xue ji shu chu ban she.

Lam, Hon-Ming, Justin Remais, Ming-Chiu Fung, Xu Liqing, and Samuel Sai-Ming Sun. 2013. Food supply and food safety issues in China. *The Lancet* 381 (9882): 2044–2053.

Liu, Yang, Qiyou Luo, Zhenya Zhou, Fei You, Mingjie Gao, and Qu Tang. 2018. Analysis and prediction of the supply and demand of China's major agricultural products. *Chinese Journal of Engineering Science* 20 (5): 120.

Lohmar, Bryan, and Fred Gale. 2008. *Who will China feed?* https://www.ers.usda.gov/amber-waves/2008/june/who-will-china-feed/. Accessed 01 June 2019.

Ministry of Agriculture and Rural Affairs of the PRC. 2016. *Zhongguo nongye tongji ziliao*. http://zdscxx.moa.gov.cn:8080/misportal/public/publicationRedStyle.jsp. Accessed 30 May 2019.

National Bureau of Statistics of China. 2018. *China Statistical Yearbook 2018*. http://www.stats.gov.cn/tjsj/ndsj/2018/indexeh.htm. Accessed 27 May 2019.

OECD. 2018. *Innovation, agricultural productivity and sustainability in China*. Paris: OECD Publishing (OECD food and agricultural reviews). https://read.oecd-ilibrary.org/agriculture-and-food/innovation-agricultural-productivity-and-sustainability-in-china_9789264085299-en#page1. Accessed 22 Aug 2019.

Schneider, Mindi. 2011. *Feeding China's pigs. Implications for the environment, China's smallholder farmers and food security*. Institute for Agriculture and Trade Policy. https://repub.eur.nl/pub/51021/. Accessed 25 May 2019.

Schüller, Margot. 2004. *Chinas Landwirtschaft Neue Entwicklungstrends nach dem WTO Beitritt*. China aktuell. https://www.researchgate.net/profile/Margot_Schueller/

publication/5079995_Chinas_Landwirtschaft_Neue_Entwicklungstrends_nach_dem_WTO-Beitritt/links/547497150cf245eb436de824/Chinas-Landwirtschaft-Neue-Entwicklungstrends-nach-dem-WTO-Beitritt.pdf. Accessed 21 May 2019.

Statista. 2019a. *China: Einwohner (Gesamtbevölkerung) von 2008 bis 2018 (in Millionen Einwohner).* https://de.statista.com/statistik/daten/studie/19323/umfrage/gesamtbevoelkerung-in-china/. Accessed 01 June 2019.

———. 2019b. *Lebensmittel.* China. https://de.statista.com/outlook/40000000/117/lebensmittel/china. Accessed 04 June 2019.

———. 2019c. *Milk Products.* Hong Kong. https://www.statista.com/outlook/40010000/118/milk-products/hong-kong. Accessed 04 June 2019.

Tan, Minghong, Xiubin Li, Hui Xie, and Lu. Changhe. 2005. Urban land expansion and arable land loss in China – A case study of Beijing-Tianjin-Hebei region. *Land Use Policy* 22 (3): 187–196.

Tian, Wei-Ming, and John Chudleigh. 1999. *China's feed grain market: Development and prospects.* https://onlinelibrary.wiley.com/doi/epdf/10.1002/%28SICI%291520-6297%28199922%2915%3A3%3C393%3A%3AAID-AGR7%3E3.0.CO%3B2-E. Accessed 01 June 2019.

Villasante, Sebastián, David Rodríguez-González, Manel Antelo, Susana Rivero-Rodríguez, José A. de Santiago, and Gonzalo Macho. 2013. All fish for China? *AMBIO* 42 (8): 923–936.

Wang, Xiaojun, Jianyun Zhang, Juan Gao, Shamsuddin Shahid, Xinghui Xia, Zhi Geng, and Li Tang. 2018. The new concept of water resources management in China: Ensuring water security in changing environment. *Environment, Development and Sustainability* 20 (2): 897–909.

World Bank Group. 2019. *Arable land (hectares per person).* https://data.worldbank.org/indicator/AG.LND.ARBL.HA.PC?locations=CN-DE. Accessed 31 May 2019.

Wu, Baosheng, Zhaoyin Wang, and Changzhi Li. 2004. Yellow River Basin management and current issues. *Journal of Geographical Sciences* 14:29–37. https://link.springer.com/content/pdf/10.1007%2FBF02841104.pdf. Accessed 10 June 2019.

Xia, Jun, Yongyong Zhang, Chesheng Zhan, and Ai Zhong Ye. 2011. Water quality management in China: The case of the Huai River Basin. *International Journal of Water Resources Development* 27 (1): 167–180.

Zhong, Funing. 2009. China's agricultural policy. In *Feeding the dragon. Agriculture – China and the GMS*, ed. Mingsarn Kaosa-ard, 5–28. Chiang Mai: Chiang Mai.

Zhou, Zhangyue, Hongbo Liu, Lijuan Cao, Weiming Tian, and Ji-Min Wang. 2014. *Food consumption in China. The revolution continues.* Cheltenham: Edward Elgar.

13 Environmental Policy

Janny Tieu, Lucas Bréhéret, and Irini Louloudi

The PRC is struggling with numerous environmental problems that lead to serious consequences for the environment and the health of the population. Rapid industrialization, as well as sloppily implemented environmental monitoring, contribute most to these problems. Moreover, the economic expansion accelerates environmental degradation because industrialization involves significant energy consumption, which in itself causes environmental problems. Accounting for about half of the world's coal consumption, China is the largest greenhouse gas emitter since 2007 and was responsible for 27% of global emissions in 2014, more than the United States and the European Union combined. Air pollution has attracted particularly intense attention since the smog incidents between 2012 and 2013. The large-scale and prolonged heavy air pollution at that time led to great concern among the population. The smog was classified as hazardous over a period of about 3 weeks in total, covering a quarter of China's land area and affecting up to 600 million citizens. The 2015 smog in northern China at the time of the Paris climate summit, in which the concentration of particulate matter with an average diameter of no more than 2.5 micrometers (PM2.5) exceeded the established World Health Organization (WHO) guideline value by several times (Huang 2015b), also led to discontent among the Chinese public regarding air pollution, which was felt in both state and social media (Huang 2015a). Immediately after the 2013 incidents, the government issued a series of policies, including laws and other measures, to control air quality (State Council of the PRC 2013).

J. Tieu
Ludwigshafen am Rhein, Germany

L. Bréhéret (✉)
Heddesheim, Germany

I. Louloudi
Kaiserslautern, Germany

B. Darimont (ed.), *Economic Policy of the People's Republic of China*,
https://doi.org/10.1007/978-3-658-38467-8_13

Water is a scarce resource in China. It is questionable how China intends to meet the increasing demand. One consideration was and is to divert rivers from the Himalayan mountains towards Beijing because that is where the water shortage is greatest. These projects could lead to disputes with India. However, there are difficulties in the implementation, so the People's Republic prefers alternatives.

In recent years, waste has become an issue in the PRC, as a belt of illegal landfills had grown up around cities like Beijing. The Chinese government has begun to address this problem, announcing a plastic import ban, for example.

13.1 Environmental Situation

The environmental situation in China must be considered critical. The quality of groundwater, for example, is classified as either "low" or "very low" in about 60% of major cities, and more than a quarter of China's major rivers are "unsuitable for human contact" (Thinktank-Resources 2016). Arable land is turning into deserts due to lack of water, poor farming practices, and overgrazing. According to a study supported by the China Forestry Administration, about 2.62 million square kilometers of China's land area is already desert land. This affects more than 400 million people (Wang et al. 2012, pp. 97–98).

In addition, lack of waste disposal and inadequate waste treatment aggravates the situation. Pollution and desertification make it difficult to supply the population with food and clean water. China's rapid economic development has led to increased water and soil pollution, which not only severely affects the health of the Chinese population, but also damages the natural ecosystem (Khan and Chang 2018, p. 4).

In addition to water and soil pollution, air pollution problems, in particular, are attracting considerable national and international attention. Due to urbanization, industrialization, and motorization, a large number of Chinese cities are affected by severe air pollution. The level of air pollution is partly due to the heavy reliance on coal for energy production. The main causes are fossil fuels, coal mining, coal-fired power plants, other coal-related uses, and vehicle emissions. Finally, in recent years, power generation accounted for about 60% of China's coal consumption and contributed significantly to China's carbon emissions (Zhu 2015, p. 2). In 2018, China had the fourth largest coal reserves in the world and was number one in global coal production (46.7%). Moreover, China has been a coal importer for years to meet its huge energy demand in first place. As a result, China accounts for another 3.8 percentage points, making it responsible for more than half of the world's coal production (50.5%) (BP p.l.c. 2019, pp. 42, 44–45). Burning coal releases air pollutants such as particulate matter (PM) or carbon dioxide, which not only severely affect the health of the Chinese population, but also the climate. Scientific evidence shows that the increase of carbon dioxide concentration in the atmosphere is due to human activities, leading to climate change, rising sea levels, and more frequent occurrence of extreme weather conditions (Intergovernmental Panel on Climate Change (IPCC) 2013, pp. 1–2). Air pollution and climate change not only negatively affect the quality of life, but can also cause illness, disability, or premature death, and eventually bring economic losses (Zhao et al. 2016, p. 92). Long-term exposure to high concentrations of hazardous PM with an average diameter of 2.5 micrometers or less or 10 micrometers or less (PM2.5 and PM10) is associated with high health risk and chronic diseases (Kan et al. 2012, p. 14). These fine dust particles can penetrate deep into the lungs, and even cross the cell membrane directly into the bloodstream, triggering disease in this way (Anger et al. 2015, pp. 262–263).

According to one report, over 1.6 million deaths in China could be attributed to exposure to polluted air, roughly equivalent to an economic loss of 10.9% of GDP in 2013 (World Bank and Institute for Health Metrics and Evaluation 2016, pp. 58, 93). Since 2013, PM has been the fifth leading cause of death in China, and about 900,000 premature deaths are attributed to exposure to PM 2.5 annually (GBD MAPS Working Group 2016, p. 7), with the highest mortality in Beijing (Liu et al. 2017, p. 78). According to a more recent report, approximately 1.2 million people in China died from air pollution in 2017 (Health Effects Institute 2019, p. 11). This significant health impact extends beyond China's national borders. It also affects neighboring Japan and South Korea, which have frequently raised concerns about air pollution from China (Ryall and Yoo 2013).

13.2 Institutional Development

China participated in the first United Nations Conference on the Environment in 1972 by sending a delegation to the Stockholm conference (Feng and Liao 2015, p. 4). Two years later, the "Environmental Protection Management Team" was officially convened by the State Council. This was the first environmental protection institution responsible for developing national environmental protection plans, policies, and regulations, and enforcing and monitoring environmental protection measures. In 1982, the 23rd Session of the Standing Committee of the 5th National People's Congress decided to establish an Environmental Protection Bureau under the Ministry of Rural-Urban Construction and Environmental Protection (MURCEP). In 1984, an Environmental Protection Commission was established under the State Council and was responsible for formulating guidelines for environmental protection and directing, organizing, and coordinating relevant activities, in addition to the Environmental Protection Bureau of the MURCEP. In the same year, the Environmental Protection Bureau of the MURCEP was further developed into the National Environmental Protection Agency, which continued to be under the guidance of the MURCEP and the Environmental Protection Commission of the State Council. Its main role was to plan, coordinate, supervise and direct national environmental protection efforts. These responsibilities regarding environmental protection were officially transferred from MURCEP to the National Environmental Protection Agency, which was a sub-ministerial level agency, in 1988. It was the competent department of the State Council for environmental management, that is, an agency under the State Council and also the office of the State Council's Environmental Protection Commission. In 1998, the National Environmental Protection Agency was reformed into the State Environmental Protection

Agency (SEPA) and elevated to the ministerial level, thereby abolishing the State Council Environmental Protection Commission. SEPA's elevated ministerial status enabled it to better enforce environmental guidelines and relevant regulations, such as those controlling air pollution. To further strengthen this role, SEPA was eventually upgraded to the Ministry of Environmental Protection (MEP) in 2008 and gained great importance as an essential department of the State Council. The current reform of environmental policy institutions was the restructuring of MEP into the current Ministry of Ecology and Environment (MEE) in 2018, increasing staff from the original 300 in MEP to 500 for MEE (2018b).

In the past, although the rank of MEP was strengthened, there were still other ministries or agencies responsible for relevant environmental functions. Ultimately, the overlapping powers of different government agencies led to coordination problems and ineffective regulations (Wang 2018, p. 113). For this reason, the MEE was established. This was one of many innovations in the State Council's institutional restructuring plan, which was promulgated on March 13, 2018, at the fourth plenary session of the first session of the 13th National People's Congress (Xin 2018). The new reform plan combined the responsibilities of environmental protection, which were previously distributed among different ministries, into the new MEE. The MEE took over the related supervision and enforcement (State Council of the PRC 2018c) to facilitate the coordination of policies. In addition to all responsibilities of the old Ministry of Environment MEP, the MEE takes over the following responsibilities:

- Control of Greenhouse Gases and Combating Climate Change of the National Development and Reform Commission (NDRC).
- Monitoring of groundwater pollution of the Ministry of Soil and Resources (MLR).
- Department of Water Resources (MWR) Water Function Zones, Sewer Setting, and Watershed Protection.
- Control of agricultural pollution from diffuse sources of the Ministry of Agriculture.
- Protection of the marine environment/control of marine pollution of the State Oceanic Administration (SOA).
- Environmental Protection during the implementation of the South-North Water Transfer Project of the relevant construction committee, which was canceled when this responsibility was transferred (State Council of the PRC 2018a).

Thus, the MEE is responsible for compiling and implementing China's environmental policies, including laws, plans, and standards, as well as monitoring the environment. In addition, the MEE regulates nuclear safety, radiation protection, and the organization of inspections by the central environmental protection agencies. According to the CPC Central Committee's "Plan for Deepening the Party and State Institutional Reform," the MEE is empowered to conduct prosecutions regarding environmental protection violations (Zentralkomitee 2015). Integrating law enforcement will help the MEE more efficiently and consistently enforce environmental laws and regulations across China. Although this has unified many functions from different ministries, the MEE will still need to cooperate with other ministries, such as the Ministry of Natural Resources (MNR), the Ministry of Disaster Management, and the NDRC. For example, the NDRC formulates plans for energy efficiency and industrial upgrades, which is why cooperation between the MEE and the NDRC is important for climate change mitigation. Nonetheless, the MEE is better equipped than the previous MEP to address the daunting environmental challenges (Wang 2018, pp. 115–117).

13.3 Chinese Government Policy

Environmental issues have been a top priority for the Chinese government in recent years, and several laws and policies have been enacted to counteract the further deterioration of air quality and improve air pollution control. These include laws such as the "Air Pollution Prevention and Control Law" or the new Environmental Protection Law of 2015, and plans such as the 13th Five-Year Plan or the "Air Pollution Prevention and Control Action Plan" (Action Plan). Together with China's commitment to the 2016 Paris Climate Change Agreement (Tambo et al. 2016, p. 152), and the commitment to reduce air pollution, enforcement and control measures regarding air pollution at the national level seems to be in full swing. Among other things, one goal is to bring the annual PM2.5 concentration below the National Ambient Air Quality Standard Level II (National AQI Level II) of 35 micrograms per cubic meter (μg/m^3) in all cities by 2030 (Standardization Administration of the PRC 2012, p. 3).

13.3.1 The 13th Five-Year Plan

It is evident from the latest five-year plans that the central government is making efforts to mitigate the impact of industrial development on the environment. It has recognized the consequences of the large volume of emissions from inefficient power plants and factories and is emphasizing strategies to reduce greenhouse gas emissions. China's 13th Five-Year Plan (2016–2020) includes commitments to improve air quality and control emissions:

The number of days with heavy air pollution is to be reduced by 25% in cities at the district level or above, and the reduction of particulate matter emissions is to be strengthened in core regions. A monitoring system is to be established to ensure compliance with environmental protection standards for vehicles and fuel oil. The proportion of natural gas users is to be increased in cities. In addition, construction site dust is to be controlled and the burning of straw on open roads is to be banned (NDRC 2016, Section 10, Chap. 44, Paragraph 1).

The five-year plan also includes promoting the reduction of emissions, ensuring compliance with emission standards, and the use of clean energy: Industrial polluters to comply with emission standards. Emission standards will be improved, monitoring of industrial pollution sources will be strengthened, and a blacklist of enterprises will be published.

Those that do not meet emission standards will be asked to correct this within a time limit. All heavily polluting enterprises in cities will either be relocated, upgraded, or closed. There are plans to reform the entire emission control system for pollutants to include more pollutants. Small and medium-sized coal-fired power plants will be replaced to achieve a clean production environment (NDRC 2016, 13th Five-Year Plan, Section 10, Chapter 44, Paragraph 2).

The use of coal, one of China's major sources of pollution, is to be strictly controlled in all regions, with a focus on the Beijing-Tianjin-Hebei region and its surrounding areas, as well as other areas with very poor air quality, i.e. the regions along the Pearl River Delta and the Yangtze River Delta, and northeastern China (NDRC 2016, 13th Five-Year Plan, Section 10, Chapter 44, Paragraph 5). The plan notes the goal of reducing carbon intensity (CO_2 emissions per unit of GDP) by 18% from 2015 to 2020 (NDRC, 13th Five-Year Plan, Section 1, Chapter 3). The 12th Five-Year Plan (2011–2015) already set binding reduction targets for air pollutants (NDRC 2016, 13th Five-Year Plan, Section 1, Chapter 1). Regarding climate change mitigation, the government emphasizes its plan to control carbon emissions in relevant industries, and to participate in climate change mitigation. It intends to promote low-carbon development and introduce a national emissions trading scheme (NDRC 2016, 13th Five-Year Plan, Section 10, Chapter 46, Paragraph 1).

13.3.2 Laws and Action Plans

Measures to protect the environment are introduced at national, regional, and lower county levels to address the challenges of a sustainable environment (Wang et al. 2018, p. 11).

13.3.2.1 Revised Environmental Law

The Environmental Protection Act of 1989 was considered outdated. A revision was therefore initiated in 2011 (Liu 2014). The drafting was delayed due to disagreements so that the revised law could only come into force in 2015. Ultimately, the delay contributed to environmental damage that could have been prevented through legal action (Jabeen et al. 2015, p. 957). The amendments to the Environmental Protection Act removed the fine restrictions on polluting units (Kaiman 2014). The current environmental law has made litigation more specific. There are five important measures in the new law that mainly empower environmental officials and they are as follows: (Albert and Xu 2016) Imposing fines on polluters daily, (Alpermann 2010) Power to seize polluting machinery and equipment, (Anger et al. 2015) Power to limit or stop production in case of excessive pollution over a long period, (Arpi 2011) Detention for serious violations, (Axe 2010) Permission for environmental non-governmental organizations (environmental NGOs) to criminally charge companies (Liu 2014).

13.3.2.2 Prevention of Air Pollution

Among the many laws and policies, the 2013 "Action Plan for Air Pollution Prevention and Control" served as a guideline because it included specific targets and measures. The 2013 Action Plan is important for significantly improving air quality in China because it set specific PM2.5 targets for core regions. The plan includes air quality improvement targets and emission controls in the performance and promotion evaluation system for civil servants, so the incentive to implement air pollution control measures increases significantly (Lin and Elder 2014, pp. 79–81). The action plan emphasizes that protecting the atmospheric environment promotes human welfare, sustainable economic development, and prosperous society. The plan set nationwide sub-targets for the period from 2012 to 2017, focusing on the Yangtze River Delta, the Pearl River Delta, and the Beijing-Tianjin-Hebei region. By 2017, urban PM10 concentrations should decrease by 10% compared to 2012, and consequently, the annual number of days with relatively good air quality should increase. PM2.5 concentrations in the three core regions should decrease by about 25% (Beijing-Tianjin-Hebei), 20% (Yangtze River Delta), and 15% (Pearl River Delta), respectively. Among these regions, the Beijing-Tianjin-Hebei region was the most polluted, so the annual PM2.5 concentrations in Beijing should be kept below 60 $\mu g/m^3$. The main tasks of this plan included industrial restructuring, renewable energy generation, coal and oil quality management, small coal boiler control, industrial emissions, municipal dust control, vehicle pollution control, air pollution control investment, energy saving, and heat measurement management for buildings, and atmospheric environment management (Wang et al. 2018, p. 4).

13.3.2.3 Three-Year Action Plan for the Victory of the Blue Sky War

After the expiration of the action plan in 2017, the new "Three-Year Action Plan for Winning the Blue Sky War" was published by the State Council. This set targets for further improving air quality by 2020 (State Council of the PRC 2018b). An important change in the Three-Year Action Plan is the definition of the three core regions where air pollution is severe. The previous core region along the Pearl River Delta is replaced by the new core region at the "Fen Wei Plains" (Section 1, Paragraph 1). The "Fen Wei Plains" describe the lowlands along the Wei He River in Shaanxi Province and the lowlands along the Fen He River in Shanxi and Henan Provinces. In addition, the Beijing-Tianjin-Hebei core region has been expanded to include the surrounding areas (Section 1, Paragraph 3). The new action plan includes specific targets for reducing emissions of sulfur dioxide and nitrogen oxides by 2020, with both to be reduced by at least 15% compared to 2015 levels. For cities where existing PM2.5 standards have not been met, PM2.5 concentrations are to decrease by at least 18% compared to 2015. The annual number of days with relatively good air quality is to increase to at least 80%, and heavily polluted days are to be reduced by at least 25% compared to 2015 (Section 1, Paragraph 2). The plan focuses more on controlling ozone compared to the previous action plan by setting specific targets not just for nitrogen oxide, but for volatile organic compounds. This is because ozone is formed when organic compounds react with nitrogen oxides. By 2020, organic compounds are to decrease by 10% compared to 2015 (section 2, paragraph 7).

13.3.3 Implementation of Industrial Restructuring

A major challenge is the implementation of environmental plans and laws (Beyer 2006, p. 209). Local governments are given more autonomy to set environmental strategies and priorities, but in reality, environmental costs are not sufficiently taken into account as rapid economic growth is preferred (Florence and Defraigne 2012, pp. 2–3).

13.3.3.1 Implementation Process

The implementation of industrial restructuring in the Beijing-Tianjin-Hebei region followed a top-down approach from the central government to the local government. The central government enacted legislation on emission standards and emission reduction plans. Provincial governments set emission reduction targets and delegated implementation to local governments. These in turn have enacted more specific regulations and delegated implementation to local industries. In parallel, local governments conducted routine monitoring and assessment of industrial emissions through three channels: routine inspections, emission monitoring platforms, and public reporting involving third parties such as NGOs. The government monitoring platform was for internal assessment, and for the industrial self-emission monitoring system, the emission data can be selectively released to the public depending on the consent of each industry. The network platform can monitor the emission behavior of major polluting industries in real-time, which is not publicly available. The major industries have their emission monitoring networks which are selectively available to the public. The National Monitoring Centre conducts spot checks on local industries and compares the emission data with industry self-monitoring data. It then reports to the MEE. Industries that failed to meet emission standards had licenses suspended and heavy fines imposed until they adjusted emissions. The "Pollution Prevention and Control Department" of the MEE is responsible for these measures. In addition, the "North China Environmental Protection Supervision Center", one of six "Environmental Protection Supervision Centers", also conducts supervision in the Beijing-Tianjin-Hebei region. If industrial emissions still did not meet standards after deadlines were set, those industries were permanently closed. This particularly affected industries with low technology and resulted in high emissions. Social problems, including unemployment, resulted from the closures of industries due to the lack of immediate alternative employment opportunities for often low-skilled workers (Wang et al. 2018, pp. 12–13).

There are still companies that violate emission standards, which can be partly attributed to the selective measures taken by local governments. In the case of large state-owned enterprises, the powers of local governments were limited, so they only partially implemented laws and policies (China Council for International Cooperation on Environment and Development 2014, p. 39). In addition, some industries evaded routine government inspections. For example, industrial facilities that operated only at night or were located in remote rural areas were often energy-intensive and had very high emissions (Wang et al. 2018, p. 13).

13.3.3.2 Regional Resistance

As stricter environmental emission standards apply in certain regions, industries have been relocated to neighbouring areas. This created regional resistance. As Beijing upgraded its industry, the relocation of less technologically advanced manufacturing to neighboring regions brought employment opportunities and economic growth, but also led to pollution (Liu et al. 2016b, p. 294).

In addition, social problems intensified in mid-2017 when the "Action Plan for Comprehensive Control of Autumn and Winter Air Pollution in Beijing-Tianjin-Hebei and Adjacent Regions 2017–2018" (MEE 2017) was released by the then MEP. A monitoring group of 1400 people was set up to conduct strict accountability monitoring in the Beijing-Tianjin-Hebei region, Shandong province, Shanxi province, and Henan province. It was also decided to close about 30% of the aluminum smelting volume for coal consumption and aluminum production in Beijing, Shandong, Shanxi, and Henan between November and March each year (Lian and Burton 2017). This significantly reduced the number of pollutants emitted in the autumn and winter of 2017 (Wang et al. 2018, p. 15).

13.3.4 International Position on Climate Policy

In 2016, China committed to the Paris Agreement, which is about the long-term reduction of fossil fuels to promote climate protection. Human and natural processes are responsible for air pollution that affect climate change. To mitigate this change and limit the increase in global average temperature to well below 2 °C above pre-industrial levels, the Paris Agreement aims to enforce policies, regulations, action plans, and programs (Zhao et al. 2016, pp. 92, 98, 103). In line with the climate change agreement, investments have already been made on the part of China to switch to renewable energy sources such as nuclear, solar, and wind power. As the world's largest emitter of CO_2 and the world's second-largest economy, China has become a major force influencing the failure or success of climate change cooperation. Therefore, it is focusing on climate change mitigation measures to reduce coal-fired power plant emissions by 50% from 2015 to 2020 (Tambo et al. 2016, p. 153).

Some critics from the US felt that China's efforts were inconsistent with its increased economic strength and corresponding responsibility for climate change (Browne 2015). At the UN Climate Change Conference in Copenhagen in 2009, China was accused of blocking the achievement of a more substantive agreement (Dimitrov 2010, p. 796; Qi and Wu 2015). But this has changed with the 2016 Paris Climate Change Agreement, which followed the 2015 UN Climate Change Conference in Paris. Through the US-China bilateral agreement on climate change, China played a key role in concluding the Paris Agreement in 2014. China did not block the 2015 climate negotiations but even called for a strong legally binding agreement (Dimitrov 2016, pp. 3, 9). China was praised by both domestic and foreign press for its active and constructive role in achieving the Paris climate summit (Flitton 2015). China has signed several bilateral agreements on climate change and clean energy with Germany, the UK, France, India, and, most importantly, the US (Qi and Wu 2015). Prior to the conference, each participating country was asked to commit to reducing CO_2 emissions according to their respective capabilities, referred to as Intended Nationally Determined Contributions (INDCs). Although some studies by NGOs question the effectiveness of INDCs to limit global temperature rise to less than 2 °C, (Lomborg 2016, pp. 116–117) the bottom-up approach of INDCs is considered viable and promising for the problem of climate change (Slaughter 2015, p. 2). In the 2015 INDCs, China submits several concrete targets of its climate action until 2030, including detailed reduction targets and reiterating China's efforts to reduce CO_2 intensity by 60–65% compared to 2005, increase the share of non-fossil fuels in primary energy consumption to about 20%, and increase forest areas by about 4.5 billion cubic meters compared to 2005 (United Nations Climate Change 2015, p. 4). Nevertheless, China still refers to the principle of "common but differentiated responsibility", which underlies the principle of fairness and national contribution according to respective capabilities (Carter and Mol 2007, p. 16).

13.3.5 Emissions Trading

The State Council released a White Paper on Climate Change Policies and Measures in late November 2011, in which the government described a detailed plan for the gradual establishment of an emissions trading system (Han et al. 2012, pp. 12–13). The implementation is in line with China's INDCs under the 2016 Paris Agreement and the 13th Five-Year Plan commitment to limit greenhouse gas emissions. Then, in 2013 and 2014, the Chinese government launched seven pilot carbon emissions trading programs at the city or provincial level (Swartz 2016, p. 7). The goal of the emissions trading scheme is to increase control and gradually reduce CO_2 emissions in China. In 2014, the National Development and Reform Commission (NDRC) published a preliminary regulation on "Administration of Emissions Trading," which refers to the national trading of CO_2 emissions (Zhang et al. 2017, p. 10). With the experience gained from the pilot programs, China finally introduced a national emissions trading system in December 2017. This currently covers only the energy sector. The implementation is divided into three phases. In the first phase, market infrastructures should be developed starting in 2018. In the second phase, emissions trading should be simulated from 2019. From 2020, the deepening and expansion phase with the trading of allowances should start. In the short term, the seven pilots will run in parallel with the national market, covering more economic sectors than just the energy sector. In recent years, a functioning market for emissions trading has been established, but it needs to be developed in areas such as regulations, market infrastructure, reporting, and capacity (International Carbon Action Partnership 2019, pp. 1–3).

13.3.6 Example of the Beijing-Tianjin-Hebei Region

Air quality in the Pearl River Delta improved with the help of the 2013 Action Plan measures (Jiang et al. 2015b, pp. 10–11), so it was replaced by the "Fen Wei Plains" in the new Three-Year Action Plan. Air quality in the other two core regions is comparatively poor. The Beijing-Tianjin-Hebei region is particularly badly polluted. In 2015, it was the region with the worst air quality. Among the 161 Chinese cities, the top 20 with the most days of non-compliance were all concentrated in the Beijing-Tianjin-Hebei region (Clean Air Asia 2016, p. 7). This situation can be illustrated with the help of a case study on the effectiveness of laws and especially the past 2013 action plan regarding air quality in the region.

In the study by Wang et al. (2018, pp. 4–10), the changes in the five main pollutants (PM10, PM2.5, ozone, nitrogen dioxide, and sulfur dioxide) in Beijing, Tianjin, and the capital of Hebei province Shijiazhuang from early 2013 to November 2017 were investigated. The study showed that there are clear differences in air pollutants in winter and summer. The concentrations of PM, highly reactive gases, and sulfur dioxide are higher in winter and lower in summer, while ozone has a higher concentration in summer and lower in winter. This is largely due to meteorological conditions and the titration effect between highly reactive gases and ozone (Zhang et al. 2014, p. 6089). Progress in air pollution control was different in all three cities. For PM10, there was a significant decrease in all three cities over the five years considered, especially in 2016 and 2017. The number of days exceeding the national AQI level II was 56 in 2016 and 33 in 2017 for Beijing, 62 and 46 for Tianjin, respectively, and 147 and 132 for Shijiazhuang, respectively. Similar to the observations for PM10, PM2.5 decreased during the mentioned period. PM2.5 concentrations in Beijing, Tianjin, and Shijiazhuang decreased by 33, 34, and 45%, respectively. Beijing was able to achieve the target of an annual average of 60 $\mu g/m^3$ from the action plan (Feng 2018). Sulfur dioxides show a significant decrease in the three cities. Annual nitrogen dioxide concentration has also slightly decreased in all three cities over the 5 years. Ozone is the only pollutant that did not decrease during this period. In Beijing, Tianjin, and Shijiazhuang, the annual average of the daily maximum eight-hour concentration of ozone increased. As a secondary pollutant, ozone concentration correlates strongly, but not always positively, with the concentration of ozone precursor gases, especially highly reactive gases and volatile organic compounds (Liu et al. 2016a, p. 758).

Although sulfur dioxide, nitrogen dioxide, PM10, and PM2.5 concentrations declined in the three cities during this period, concentrations still exceed AQI Stage II and are well above WHO guidelines, especially in the winter months. PM2.5 remains the main pollutant and ozone is increasingly becoming the dominant pollutant (Anger et al. 2015, p. 263).

Emissions from both the transportation and industrial sectors remain too high, accounting for 12–25.2% and 11.4–24.6% of total PM2.5 pollution in Beijing-Tianjin-Hebei in 2014, respectively (Gao et al. 2018, p. 722), and thus can be identified as the main reasons for the failure to achieve national AQI Level II. Although the vehicle emission standards for gasoline and diesel light-duty vehicles in China have been raised to the national V or VI levels, which is equivalent to the Euro V or VI standard for light-duty vehicles, the total number of vehicles in the region is too large. Hebei has the largest steel industry in China and produces a quarter of China's total steel, so emissions from heavy industry particularly affect this region. Although compliance with emission standards gradually improved and the government-controlled steel production capacity, heavy emissions from the industrial sector still contributed the most to air pollution in Hebei and Tianjin (Wang et al. 2018, p. 11).

About 28–36% of PM2.5 in Beijing is due to interregional transport (Zhang et al. 2015, p. 11). Despite strict control measures, many industrial and vehicular emissions violated emission standards. Although overall air quality in the Beijing-Tianjin-Hebei region has improved as a result of stricter air pollution control guidelines under the Action Plan. However, these vary across regions and for different air pollutants. The national AQI level II continues to be significantly exceeded. According to the 2016 Environmental Report, 75% of cities did not meet the established standard in 2016 and 70.7% did so in 2017 (MEP 2017, p. 5; MEE 2018a, p. 8).

13.4 Environmental Awareness

Since the population in China is strictly controlled by the government, environmental protection as a mass movement is only possible to a limited extent. Nevertheless, civil activism has increased in recent years against government decisions that are deemed environmentally harmful (Hoffman and Sullivan 2015). Often, this is because of local governments' focus on economic growth rather than environmental protection. As a result, the Chinese public's tolerance for environmental problems has already decreased significantly. This is reflected in increases in public protests against polluting projects, both in urban and rural areas (Albert and Xu 2016). Moreover, the internet helps to spread information within the population, putting additional political pressure on the government. For example, in March 2015, the documentary film about China's air pollution, Under the Dome, spread very quickly on social media (Chai 2015). The Chinese government is responding to the pressure, but it is concerned that environmental activism could lead to a political movement (Economy 2014, p. 193).

13.5 Water Resources

If the water resources in the PRC are considered, the situation looks favourable at first glance. China has the fourth largest water resources in the world, is hardly dependent on external inflows, and has the third largest freshwater supply in the world on the Tibetan plateau. A closer look quickly reveals that China is experiencing a water crisis. Unequal water distribution between North and South, rural exodus, population growth, water pollution as well as the rapid economic upswing lead to a water shortage in the People's Republic, which will become even worse in the future. One solution to this problem is planned to divert water from the Himalayan region to the north, which repeatedly leads to conflicts with the southern riparian states. Between China and India, these conflicts are particularly serious. The US National Intelligence Council's Global Trends 2025 Report predicts that the diplomatic relations of countries in the Himalayan region will be particularly strained by water shortages in the future:

> With water becoming more scarce in several regions, cooperation over changing water resources is likely to be increasingly difficult within and between states, straining regional relations. Such regions include the Himalayan region, which feeds the major rivers of China, Pakistan, India, and Bangladesh. (US National Intelligence Council 2008, p. 66 f.)

This chapter discusses the water shortage and water resources in China and shows the consequences of Chinese water projects on the Indian border and the solutions to the Sino-Indian water conflict.

13.5.1 Existing Water Resources in China

In China, the problems of water scarcity were recognized early on: Premier Wen Jiabao declared in 1999 that water scarcity threatened "the very survival of the Chinese nation" (The Economist 2013). China hosts nearly 20% of the world's population with only 7% of global water resources (World Bank 2016, p. 56). In addition to economic and population growth, urbanization and rural migration will further exacerbate water shortages. By 2050, it is estimated that over 76% of China's population will live in cities, representing over one billion people (United Nations 2014, p. 21).

The Chinese Ministry of Water Resources predicted a "serious water crisis" in 2030, once the population reaches 1.6 billion and per capita water resources fall below the World Bank-defined level of water scarcity (Xinhua 2001). The economic effects of water scarcity are already becoming apparent. Between 2000 and 2003, water shortage and pollution were responsible for economic costs of about RMB 240 billion, which was equivalent to about USD 29 billion in 2003 (World Bank 2007, p. 81). In 2015, the combined cost of water and air pollution was estimated to be equivalent to 6% of China's gross domestic product (Yan 2015).

Although China is almost completely water-independent – dependence on external inflows is only 0.9% – the uneven distribution of water resources is causing major water shortages in the north of the country (FAO 2016). The north has only one-fifth of the water resources of the People's Republic; moreover, renewable water resources in the north are not nearly as abundant as in the south. The average rate of internal renewable freshwater resources per person per year in China is 2062 cubic meters, compared to only about 700 per person per year in the North, and the usable amount is much lower still (World Bank 2016, p. 56). For the megacities of Beijing and Tianjin, per capita, water availability is only 300 cubic meters per year (Xie 2009, p. 1). The World Bank warned of the consequences if China does not properly protect and use its water resources:

> [W]ithout a major concerted successful program to improve water resources management, the damage to the environment and to Chinese natural resources will become irreversible, resulting in huge negative impacts on the quality of life of the Chinese people and the Chinese economy in the future. (World Bank 2002, p. 4)

In northern and central China, 80% of the groundwater is already too polluted to be used as drinking water (Xia 2016). In southern China, on the other hand, especially on the Tibetan plateau, there are large freshwater reserves.

13.5.2 Water Resources on the Tibetan Plateau

The Tibetan Plateau is home to the third largest freshwater resource on earth. Since the North and South Poles are the two largest freshwater sources, the Tibetan Plateau is also known as the "third pole". The Food and Agriculture Organization of the United Nations (FAO) estimates that there are 5000 glaciers in the entire Himalayan region, storing a total of 3870 cubic kilometers of water (FAO 1999, p. 93). The melting of glaciers releases water, which makes its way downhill in the form of rivers. The meltwater feeds the largest rivers in Asia, forming a total of eleven mega deltas in which populous cities such as Guangzhou, Bangkok, Calcutta, Dhaka, and Karachi are located.

This flow of meltwater will continue to increase in the future, as the Tibetan plateau is particularly affected by global warming. The Chinese Tibetan Regional Meteorological Bureau warned in a report that the temperature on the plateau is rising faster than the world average (Xinhua 2007). Rising temperatures, in turn, are melting ice and snow faster, and accelerated glacier melt has already increased the amount of water in rivers by an average of 5.5% (Yao et al. 2007, p. 642). In the long term, the flow may again decrease dramatically as glaciers recede, because glaciers store only a limited amount of water. Brahma Chellaney, an expert who has worked extensively on the issue of water resources in Tibet, puts it this way:

> The Tibetian and Himalayan glaciers constitute, metaphorically, a water bank account built up over thousands of years. If the present trends of accelerated glacier thawing continue, this bank account would become empty. (Chellaney 2011, p. 112)

As the glaciers recede, precipitation in the region is becoming increasingly important for river discharge. Most important for precipitation is the monsoon, with monsoon precipitation expected to increase in the future as the Tibetan plateau warms (Briscoe and Malik 2006, p. 19). Air rises from the warmed plateau, creating a heat depression. Especially in summer, when moreover the land mass of Asia heats up behind the Himalayas, this creates an even stronger relative negative pressure compared to the high-pressure areas over the

Indian Ocean, drawing the water-rich air of the monsoon deep inland. In addition, due to climate change, rising temperatures result in increased evaporation of water over the ocean, causing the air masses to absorb more moisture overall. This increases the amount of water stored in the air in a gaseous state and thus the amount of precipitation. This further favors the flow of air into Southeast Asia. As the monsoon rains down on the Himalayas and the Tibetan Plateau, this will result in increased rainfall in a short period, which may be responsible for significant flooding. Thus, in the long term, affected countries will have to adjust to reduced river discharge outside the monsoon season and increased discharge during the monsoon (Wang et al. 2008, p. 913). Already in 2017, the worst floods in 30 years occurred (Davies 2017).

Because of these large freshwater resources, the Chinese government is taking several measures to harness the hydropower potential, or water, of the "third pole."

13.5.3 Measures Taken by the Chinese Government

To tackle water problems and make better use of its hydropower potential, China is relying on a large number of dams, reservoirs, and canals. In 2000, Wen Jiabao announced:

> In the twenty-first century, the construction of large dams will play a key role in exploiting China's water resources, controlling floods and droughts, and pushing the national economy and the country's modernization forward. (Mcelroy 2000)

13.5.3.1 Dams

China built more dams than the rest of the world combined within the last 50 years, including major projects such as the Three Gorges Dam (Chellaney 2011, p. 157). In total, China has over 80,000 reservoirs and dams (Chinese National Committee on Large Dams 2006). In the meantime, however, there are problems with many of the government's construction projects; for too long it has acted according to the old Mao doctrine of "think big, move fast, and worry about the consequences later" when building dams (Wines 2011a).

Problems arise due to the often poor quality of the dams. Construction sites are often not thoroughly examined in advance and construction work is carried out too quickly. The problems are so far-reaching that even state media report on them, for example in the China Daily:

> Improper construction procedures, disqualified workers, embezzlement of construction funds, and mismanagement of local water resource departments are threatening the safety of the dams. (Tan 2009)

In 2009, the Minister of Water Resources, Chen Lei, himself said the following regarding dam safety:

> China's dam safety is coming under heavy pressure and inspections show many of them are not in good condition. (Tan 2009)

Several large dams in Gansu Province, on the edge of the Tibetan Plateau, are on the verge of collapse just one or two years after they were built. Between 1999 and 2008, there were a total of 59 dam failures in China. More than 40,000 reservoirs and dams are in "potential danger" of breaking, more than 40% of the dams in total (Tan 2009). In an interview with China Economic Weekly, an offshoot of the state-run People's Daily, the director of the Water Reservoir Department, Xu Yuanming, said:

> These reservoirs are a major risk and will ruin farmland, railways, buildings and even cities when they collapse. (Moore 2011)

Another construction project that is not going according to plan is the South-North Water Transfer Project, one of the largest construction projects in Chinese history.

13.5.3.2 The South-North Water Transfer Project (SNWDP)

In the wake of the great water poverty in China's north, the government launched the largest water transfer project in the world. Mao Zedong himself provided the idea for the South-North Water Transfer Project in 1952. He said:

> The south has plenty of water, but the north is dry. If we could borrow some, that would be good. (Sohu 2002)

Water is to be piped from the south to the north along three routes with a total length of around 1200 kilometers. Beijing and Tianjin in particular are to benefit, as water shortages are worst here. The cost of the project was initially estimated at USD 62 billion, more than double the cost of building the Three Gorges Dam (Office of the South-to-North Water Diversion Project Commission of the State Council n.d.). However, in 2014, USD 79 billion has already been spent on it, making it the most expensive construction project in the world, and the cost is likely to be many times higher in the end (Chang 2014). The project is scheduled to end in 2050, at which time 45 billion cubic meters of water per year will be piped to the north (Xinhua 2009).

In addition to the sharp increase in costs, there are other problems with the SNWDP. The heavy pollution of the rivers in China also affects the water of the SNWDP. On the eastern route, the water is so polluted that 426 treatment plants have to be built; in a test run in July 2013, almost all the fish on a fish farm died (Xinhua 2013). Xinhua News Agency reported that measures against water pollution on some parts of the route take 44% of the budget (Xinhua 2013). There is even a risk that the water from the SNWDP will not be drinkable when it reaches the destination cities (Wong 2011).

While the first two canals, namely the eastern and middle routes, are already in operation, the western route is currently still being planned. This is also the route that is expected to carry the most water. Water from three tributaries of the Yangtze River, the Tongtian, the

Yalong, and the Dadu, is to be conveyed by it to the tributaries of the Yellow River and thus to the north (Office of the South-to-North Water Diversion Project Commission of the State Council n.d.). There are still plans for another type of construction of the western route, which is much discussed. They envisage diverting water from six headwaters of other rivers into the Yangtze and Yellow Rivers, totaling up to 200 billion cubic meters of water (Xinhua 2009; Southern Weekend 2006). Affected rivers include the Mekong, Brahmaputra, Salween, and Nu Rivers. This volume is almost five times that of the other two routes (Chellaney 2011, p. 191).

With regard to the construction of the western route, a change of opinion has become apparent in recent years. As recently as 2006, Liu Changming, a hydrologist at the Chinese Academy of Sciences, confirmed the construction of the western route. Although long tunnels and various dams are needed to channel the water through various mountain plateaus 450 kilometers further north, construction is to begin:

> Now the Western Route isn't just an abstract plan; it will go ahead. (The New York Times 2006)

However, criticism of the project is growing. Wang Hao, an academic at the Chinese Academy of Engineering, was quick to warn that the cost of the western route could be many times higher than planned. Wang:

> Given the complexity and magnitude of the project, water diversion on the Great Western Route would demand an investment of closer to 1 trillion yuan (US$125.4 billion) than 200 billion yuan. (Southern Weekend 2006)

Qiu Baoxing, former vice minister of the Ministry of Housing and Urban and Rural Development and current advisor to Premier Li Keqiang's State Council, publicly spoke out against the NSWDP in 2014, calling for more sustainable alternatives and an end to the project. "If we miss the opportunity to repair water ecology, we will pay dearly," Baoxing said.

> If we try to solve our water crisis by diverting water, then new ecological problems will emerge. This is not sustainable at all. (Wang 2014)

At a press conference that same year, Vice Minister of the Ministry of Water Resources Jiao Yong said of the progress of the western route that while it was still being studied, water conservation and environmental protection were a higher priority (Zhang 2015).

So the construction of the Western Route, though long-planned and much discussed, is by all accounts still a long way off, if not abandoned altogether, given the problems with the project so far and the growing criticism. Nevertheless, plans to take water for the western route from the Brahmaputra and other transboundary rivers are causing great unease among governments in other countries. In India, the fear is not only of taking water from the Brahmaputra but of diverting the entire river itself.

13.5.3.3 Diversion of the Brahmaputra River

In the following section, the course of the Brahmaputra is described first, followed by an account of the plans and discussions on the diversion to date. Then the previous dam constructions on the Brahmaputra are discussed and finally the consequences of a diversion.

The Course of the Brahmaputra

As one of the longest rivers on earth, the Brahmaputra flows through three states and many different cultural areas. This leads to the fact that it has various names, such as the Yarlung Tsangpo in Tibet, Siang in northern India, the Brahmaputra in India or Jamuna in Bangladesh. The Brahmaputra has its source in the middle Himalayas. It first flows 2000 kilometers through Tibet, at an altitude of over 3500 meters, making it the world's highest river system. After passing the town of Pei in Tibet, the river describes a bend to the northeast and then to the south, which is called the Great Bend. Here it flows through the narrow Dihang Gorge, also known as the Yarlung Tsangpo Great Canyon. With walls as high as 3000 meters and mountain peaks towering over the canyon by over 5000 meters, this is considered the deepest canyon in the world. The Brahmaputra describes a curve in this canyon, dropping some 3000 meters in a series of cascades over a length of more than 200 kilometers before crossing the border with India into Arunachal Pradesh. It then flows through Assam province, where it is fed by several other tributaries, such as the Lohit and the Dibang. In Assam itself, the Brahmaputra becomes a wide stream, the width of which can be as much as eight kilometers, and this is where it got its name. The Brahmaputra means "son of Brahma", one of the major gods of Hinduism. After flowing through India for several hundred kilometers, the Brahmaputra finally reaches Bangladesh. After the confluence with the Ganges, the holiest river in India, it then flows into the Bay of Bengal. The fact that the Brahmaputra flows not only through India, but also through Bangladesh, which is extremely water-scarce, lends the diversion and water withdrawal by China a threatening character.

Rerouting Plans

The idea of diverting the Brahmaputra has been circulating in the People's Liberation Army since the 1980s (Chellaney 2011, p. 135). In July 1986, at the first international conference of the Global Infrastructure Fund in Anchorage, this idea was discussed publicly and outside China (Jha 2014). For more than ten years, however, this project was then forgotten, as its implementation was considered technically impossible. In 2005, it received a new lease of life when a former officer of the People's Liberation Army, Li Ling, wrote the book Tibet's Waters Will Save China. In it, Li describes in detail how the waters of the Brahmaputra could be diverted into the Yellow River (Arpi 2011; Jha 2014). The book was blessed and funded by the Chinese government, and copies were purchased and circulated in official agencies and offices (Chellaney 2011, p. 154). The circulation of the book shows that this idea has become suitable for the masses and that the government no longer shies away from the public discussion of such plans.

Discussing the idea in public, in turn, does not mean that it meets with widespread approval everywhere. In 2005, Li Ling's book was published, but the then Chinese President Hu Jintao confirmed in the same year during a visit to New Delhi that there were no concrete plans to divert the river (Vasudeva 2011). In the same year, the then Chairman of the Ministry of Water Resources, Wang Shucheng, clearly stated at a seminar in Hong Kong that he did not think anything of diverting the Brahmaputra towards China:

> It is unnecessary, infeasible and unscientific to include the Yarlung Zangbo River in the western route of the massive project. (Xinhua 2009)

After his resignation, Wang Shucheng became even more outspoken. In an interview with a Chinese newspaper, he called the diversion project "gross nonsense" and warned of "unimaginable environmental consequences" (China Institute of Water Resources and Hydropower Research 2011). The interview was published on the website of the China Institute of Water Resources and Hydropower Research.

In the discussion about the Brahmaputra diversion, there is an additional discernible shift between generations. The old cadres like Guo Kai, the planner of SNWDP, usually still dream of solving all water projects with one mega project. Thus, Guo Kai himself said:

> With one engineering project we could solve all of northern China's water problems. (Simons 2006)

The younger generation, on the other hand, has enjoyed better education and attaches more importance to scientific approaches. The former minister of the Ministry of Water Resources, Qian Zhengying, together with the expert Zhang Guangdou and China's Engineering Academy, carried out the preparation of the Strategic Study on Sustainable Development of China's Water Resources in the twenty-first Century (Zhang 2015; cf. Southern Weekend 2006). Qian herself was born in the United States (Sullivan 2007, p. 414) and Zhang studied at Harvard University and the University of California (ChinaCulture.org 2017). They worked with 43 academics and 300 experts for a full year on the report, which was delivered to the State Council and various ministries in July 2000. The report cast doubt on the feasibility, viability, and necessity of the diversion, after which plans for it were halted (Zhang 2015; see Southern Weekend 2006). Zhao Yean, a senior member of the government's Yellow River Water Resources Committee, aptly summarized the generational conflict:

> Under Mao, scientists were often sidelined, but now the government has realized that it needs technical expertise to solve its problems.

However, despite good intentions, politicians are not scientists, Zhao commented on the problem in this context:

> [T]hey really don't understand the difficulties of building such a large water transfer project or the damage it would cause the environment. (Simons 2006)

While the diversion of the Brahmaputra was once seen as a miracle cure for all of China's water problems, it is now viewed much more critically by a new generation of Chinese politicians who are taking a more systematic and scientific approach to the issue. India and Bangladesh nevertheless worry about a diversion, especially seeing the massive dam construction on the Brahmaputra as a threat and potential preparation for larger projects.

Dam Construction on the Brahmaputra

India's concern over the diversion of the Brahmaputra stems primarily from Chinese dam construction on the Brahmaputra, which the Chinese government denied even after it began. Two dams, in particular, the Zangmu Dam and the Metog Dam, are causing controversy in this regard.

As early as 2009, China began to build a hydroelectric power plant on the upper part of the Brahmaputra, near Zangmu, at a cost of over one billion USD. Nevertheless, this construction was initially denied when asked by India. India's Foreign Minister Nirupama Rao said:

> What I want to say is that this matter has been taken up not just once but on several occasions with China and China has consistently denied that it is engaged in any such construction activity on the Brahmaputra. (The Times of India 2009)

Even when Prime Minister Manmohan Singh met his counterpart Wen Jiabao at the ASEAN-India and South Asia Summit in Thailand in October 2009, the construction of dams was denied (The Hindu 2009). In November, however, India's intelligence agency, the National Remote Sensing Agency, confirmed that China had begun dam construction at Zangmu. Despite this revelation, there was no comment from China. This can be interpreted as a sign of China's disgruntlement, as the Dalai Lama visited the disputed province of Arunachal Pradesh in November 2009 (India Today 2009).

It was not until April 2010, during a visit by the Indian Foreign Minister Somanahalli Mallaiah Krishna to Beijing, that China again took up the issue of dam construction and named the place where the dam was being built. Here, additional assurances were given that it was only run-of-river dams. These hardly store any water and are solely dependent on seasonal runoff, which means that there is no shortage of water downstream (Axe 2010). Then in June 2011, China responded to one of India's several requests to reveal more information about the project. Chinese Foreign Ministry spokesperson Hong Lei said that China would take a "responsible" stance and "fully consider" the interests of downstream countries (Krishnan 2011a). In addition, China assured us that it was only a small project that would not affect the downstream flow of the river (Outlook in India 2011). In January 2013, China's five-year energy plan was published, which talked about three more medium-sized dams to be built at Dagu, Jiacha, and Jiexu (Ho 2015, p. 188).

This led to serious diplomatic disagreements between Beijing and New Delhi, as India was not informed of these plans in advance and only learned of them through the Chinese press (Press Information Bureau Ministry of Water Resources 2015).

The Brahmaputra will remain a contentious issue between China and India. China is pursuing further plans involving the construction of more dams on the Brahmaputra but is deliberately not disclosing much information. For instance, it was reported in 2009 that China will soon sign off on even more plans to build dams on the Brahmaputra. The names of these dams are Dhongzhong, Guoduo, Xiangda, Ruxi, and Linchang (The Times of India 2009). The planning of more dams is no longer kept secret by the Chinese side. Zhang Boting, the deputy secretary general of the Chinese Society of Hydropower Engineers, gave an interview in the Indian newspaper The Hindu in 2011 in which he spoke extensively about Chinese dam construction plans. For example, he said, the Zangmu Dam is the only one of at least 28 proposed dams that has been approved so far. But due to the increasing number of power outages, combined with international pressure to reduce carbon emissions, China could no longer afford to let the Brahmaputra's potential go unused, he said. Zhang expressed confidence that it was only a matter of time before more hydropower projects were approved against the backdrop of the growing energy crisis (Krishnan 2011b).

The second dam that India is concerned about is the Metog Dam. This is set to become the largest dam in the world, harnessing the enormous hydropower potential of the Great Bend. The Metog dam is expected to be 160 meters high and generate 38 gigawatts of power, which is almost half the capacity of the British power grid or twice the capacity of the Three Gorges Dam (Krishnan 2011b). HydroChina Corporation published a map of planned dams in the area back in 2010. On this, the Metog Dam was already clearly marked (HydroChina n.d.). Zhang Boting commented:

> This dam could save 200m tonnes of carbon each year. We should not waste the opportunity of the biggest carbon emission reduction project. For the sake of the entire world, all the water resources that can be developed should be developed. (Watts 2010)

The state-owned hydropower company Gezhouba Group is also planning to build a dam at Metog. The company expects that the next five-year plan from 2020 onwards will massively promote hydropower in Tibet and has already published an agenda on its website, including construction plans for the dam construction at Metog (Deka and Krishnan 2015). China decided in 2006 to build a highway at a cost of over USD 87 million through Metog County (Chellaney 2011, p. 160). Although there is already a road connecting Metog to Bomi, 141 kilometers to the north, the highway is to be expanded and, if possible, an additional tunnel built to shorten the route (Xinhua 2004). It is unlikely that this effort will be undertaken for a county with a population of less than 11,000. After all, the site where the Metog Dam is to be built is less than 20 kilometers from Metog City (Chellaney 2011, pp. 160–161). Xu Xiaozhu, the deputy head of Metog County, said himself that economic development is the top priority for the province. In this context, the state media names the context of the highway and what is meant by economic development. For example, Xinhua News Agency quoted an official in charge of local tourism in the report. The newspaper wrote:

> Lahmo, a local tourist official, [...] also hopes the highway will allow investors to tap the rich water resources from Brahmaputra canyon. (Xinhua 2004)

The Indian government is particularly critical of the construction work on the Metog Dam. On the one hand, they show that China does not intend to stop the construction of the dam on the Brahmaputra; instead, the largest hydroelectric power plant in the world is to be built, which could disturb the course of the river with its 160-meter-high wall. On the other hand, these construction works could be a pretext and in reality, serve to prepare a diversion of the Brahmaputra towards China (Krishnan 2011a).

Consequences of a Detour

India itself has 17% of the world's population but only 4% of the world's water resources, making it very sensitive to any interference with its rivers (World Bank 2016, p. 102). A diversion of the Brahmaputra is a recurring topic in the Indian media and is often conjured up as a specter and disaster scenario (Krishnan 2011a). However, when the Brahmaputra is looked at more closely, it is reasonable to conclude that even if the Brahmaputra is diverted, the consequences for India are manageable. The most important argument that a diversion would have little consequence is the discharge of the Brahmaputra. A diversion would have to be done on the Chinese side, which would have little effect on river flow in India. For one thing, the Brahmaputra on China's side is fed primarily by glacial water, the importance of which will diminish in the medium to long term. Already, precipitation from the monsoon rains that fall on the Himalayas is much more important for India and the Brahmaputra (Bandyopadhyay et al. 2016, pp. 8–9). Second, the Brahmaputra takes much of its water from tributaries that originate in India itself (Singh et al. 2004, pp. 141–143). At the border crossing from China to India, it has an annual discharge of 31 billion cubic meters; at the crossing from India to Bangladesh, the annual river flow is 606 billion cubic meters (Bandyopadhyay et al. 2016, p. 9; Jiang et al. 2015a, p. 5857). Thus, the share of water from China is a fraction of the total flow volume.

Another reason why the diversion would not be so severe is the distribution of water in India itself. India hardly relies on the water of the Brahmaputra. While this accounts for 30% of the nation's water resources and 41% of the nation's hydropower reserves, it flows through areas with little or no water shortage. In 2001, only 20% of the irrigation potential was drawn down, as opposed to the national average of 56.4%. Only 3% of the hydropower potential was utilized at that time, as opposed to the national average of 16% (Bandyopadhyay et al. 2016, p. 13). It is also important to note that only 4% of the Brahmaputra's discharge is even usable due to its large volume, sudden monsoon rains, and rapid flow velocity (Ray 2014, p. 11).

India's fear that the dam construction on the Brahmaputra will result in the loss of valuable sediments is rather unjustified. While the Nuxia measuring station in Tibet receives around 350 mm of precipitation during the monsoon season, precipitation increases sharply when the Brahmaputra leaves the summit line of the Himalayas behind and flows through Arunachal Pradesh, where the annual precipitation is 4000 mm. Due to little rainfall, hardly any sediment is carried from the Chinese part of the Brahmaputra. At Nuxia, the annual rate of sediment washed away is 30 million tonnes, compared to the 735 million tonnes at Bahadurabad on the Bangladesh border, this amount is "insignificant" according to researchers (Bandyopadhyay et al. 2016, p. 57).

In summary, then, if the Brahmaputra were diverted, the damage to India would be severely limited. A fraction of the water of a barely used river flowing through a water-rich area would be lost, along with an insignificant amount of sediment (Bandyopadhyay et al. 2016, p. 57). The potential and water of the Brahmaputra are nevertheless important for India, where water shortages are even more severe than in China. India's water shortage will also be exacerbated in the future by its rapidly growing population combined with increasing affluence. By 2050, India's population is expected to approach 1.66 billion people, and annual water demand will reach 1.5 trillion cubic meters as early as 2030 (United Nations 2017, p. 5). The 2030 Water Resources Group estimates that with the 740 billion cubic meters of renewable freshwater per year, there will be a 50% deficit between demand and supply in India by 2030 (The 2030 Water Resources Group 2009, p. 9). The Indian government also assumes similar figures. It is estimated that India's annual national water demand will increase to 1.09 trillion by 2025 and 1.45 trillion cubic meters by 2050 (Government of India Planning Commission 2007, p. 1). The resulting water shortage will have serious consequences for India's growing economy and social costs. The former Indian Prime Minister, Manmohan Singh, said in his first State of the Nation Address in 2004:

> Water has emerged as a critical and contentious issue across the country, even access to safe drinking water remains a problem in many parts of the country. (Singh 2004)

Approaches

For China, the main problem is that the majority of Chinese agriculture is located in the water-scarce north. For example, more than half of the country's wheat and 40% of its cotton is grown in the three large, arid provinces of Henan, Hubei, and Shandong (Alpermann 2010, p. 170). In the North China Plain, 16 million hectares of fertile land produce 20% of the nation's food, mainly winter wheat and maize (Deng et al. 2006, p. 24). In northern China, 85% of the fertile land must be artificially irrigated (U. S. Water News Online 2001). Hence the idea of diverting the Brahmaputra to the north, precisely these dry cultivation areas are to become fertile with its water.

The China Geological Bureau reports that groundwater use has doubled since 1970. When rivers and rainfall can no longer irrigate the fields, farmers use pumps to get fresh water from the ground. However, water reserves under the North China Plain will be depleted within the next 30 years. An expert who has worked as a consultant for the World Bank and China's Ministry of Water Resources said:

> There's no uncertainty, the rate of decline is very clear, very well documented. They will run out of groundwater if the current rate continues. (Yardley 2007)

In general, China and India face the problem that many farmers are smallholders. They have only a low income and lack the money for modern technology and cultivation techniques (Chellaney 2011, p. 96). More than 70% of irrigated cropland in northern and northwestern China lacks water-saving provisions (Jin and Young 2001, pp. 215–218). In addition, agriculture and industry put a strain on water resources. Fertilization of fields is extremely damaging to groundwater, and the use of chemical fertilizers has quadrupled since the reform period. The World Bank warned of damage to groundwater and to the soil itself (World Bank 2001, p. 58).

Furthermore, cultivation in the water-rich south offers a good alternative to irrigation. Instead of diverting water to the north in problematic, expensive mega-projects, the fields, and farmers could move to the south. While this solution is not without problems, Western and Chinese analysts estimate that 30–40 million farmers will have to relocate by 2025 due to environmental degradation. So it would behoove the Chinese leadership to prevent this trend with a resettlement program. The success story of desalination plants in Israel could become a model for China and India. As one of the most water-scarce countries in the world, Israel meets almost half of its water needs through desalination and thus has more water than it needs (Jacobsen 2016).

13.5.3.4 Desalination Plants

Desalination plants have several advantages for China. Firstly, large cities near the coast are particularly affected by water shortages. The urbanization rate is estimated to rise to 60–65% by 2030, and the demand for freshwater will increase by around 63% by 2030 (Peng 2011). The proximity to the sea provides a good starting point for desalination plants, as they require seawater. This eliminates the need to transport water over long distances, which is expensive and complex. Secondly, these plants can provide water security. Completely independent of precipitation or other influencing factors, fresh water can be produced constantly. The sea as the source of origin is inexhaustible, unlike the glaciers of Tibet, for example. Furthermore, the water from the plants is of very good quality and free of pollutants or other impairments. The water from the SNWDP, on the other hand, has major problems with pollution (Wong 2011).

For this reason, China plans to significantly increase the production of fresh water with the help of desalination plants. From 2006 to 2010, capacity increased by almost 70% per year. By 2015, the amount of water produced should be between 2.2 and 2.6 million cubic meters, and by 2020 as much as three million liters per day. However, this target is likely to be missed; in 2015, the daily amount of produced water was only one million cubic meters (Sun 2016). One reason for this is the drastic underfunding of projects in this area. The government itself does not invest very much. Instead, the "Strict Guidelines of Water Management" were given in the 12th Five-Year Plan. These set binding limits on water abstraction, so that water-intensive industrial factories are no longer allowed to use urban water for production, but must secure their water supply (Zuo et al. 2015). As a result, most plants are built by industrial end-users. 131 desalination plants so far supply fresh water, of which more than 60% is used for industrial purposes (WaterWorld 2017). Seventy to 80% of project costs are borne by end users, and the government finances at most 10% of projects with direct investment (Wines 2011b). On the one hand, this results in the construction of many small and medium-sized plants that could achieve much higher output with little financing. For another, this means that only financially successful and experienced companies build such plants; less large or inexperienced firms buck this trend (Wines 2011b). There is little incentive for local governments or state-owned enterprises to build desalination plants.

> When there is a drought, local officials and enterprises all come to see us and say, 'We want to desalinate seawater,'" said Wang Zhi, the director of Tianjin University's Key Laboratory of Membrane Science and Desalination Technology, "but if there is sufficient rainfall the next year, they will drop the idea and invest their money in other things first". (Sun 2016)

Furthermore, the construction of desalination plants is not profitable at current water prices. The price of a cubic meter of tap water in China is often less than USD 0.5 (Sun 2016), in Beijing USD 0.66 (Kuo 2014), and a cubic meter of desalinated water costs between USD 0.75 and 1.2 to produce (Sun 2016; cf. Peng 2011). Nevertheless, there are companies that are aware of their role as state-owned enterprises and invest massively in desalination plants. The government-controlled conglomerate State Development and Investment Corporation (SDIC) invested four billion USD in the construction of one of the most modern plants in Tianjin, based on Israeli technology. Although the water is sold for only half as much as it costs to produce, SDIC plans to quadruple capacity even further. The power plant's general manager, Guo Qigang, said in an interview:

> Someone has to lose money. We're a state-owned corporation, and it's our social responsibility. (Wines 2011b)

For China, the investment is nevertheless worthwhile, because the desalinated fresh water is almost as expensive as the water from the SNWDP. This currently provides water that costs USD 1.2 to 1.5 per cubic meter (Peng 2011). In addition, the still rising costs of SNWDP have to be considered, whereas with desalination the costs decrease significantly with increasing volume.

The Chinese government should therefore invest in desalination plants in a structured and transparent manner, as the previous strategy of letting companies build the plants has failed. Israel, as a role model, has the most cost-efficient, largest, and most energy-efficient plants in the world. 75% of the tap water there is obtained through desalination, and this is soon to be 100% (Küffner 2018). In China, the construction of plants as large as those in Israel by companies alone is not possible, but only large plants can properly exploit economies of scale. This is where the state would therefore have to invest more. Instead, China invested over USD 64 billion in the SNWDP, which so far only supplies polluted water at an increased price, while also being highly dependent on environmental factors (Wong 2011). This opinion is also shared by Qiu Baoxing, the minister who spoke out against the NSWDP:

> Recycled water could replace diverted water. Most Chinese cities are capable of finding more water if we develop water desalination technology and collect more rainwater. (Wang 2014)

The use of wastewater mentioned by Qiu Baoxing will be discussed in more detail in the next section.

13.5.3.5 Wastewater Use

In Asia, wastewater is not managed carefully, tens of billions of USD worth of treated water is wasted every year in Asian cities, and 90% of wastewater is not treated in any way. Instead, groundwater is tapped and wasted (Thapan 2010). In China, the situation looks better, but there is a strong urban-rural divide in wastewater treatment. In the cities, 77.5% of wastewater is treated, and this figure is expected to rise to 85% in the future (Hu et al. 2014). In rural areas, on the other hand, only 6% of wastewater is treated (Li 2013). In addition, only half of the available capacity for wastewater treatment is used, and in nearly 15% of all cases, treatment is done inadequately (Zhang et al. 2016). Therefore, in 2015, China decided to invest USD 87.5 billion in wastewater treatment and recycling in the 13th Five-Year Plan (Soh 2018).

Israel is also a leader and role model in wastewater recycling and water consumption. In Israel, almost all wastewater is recycled, and even slightly polluted water is reused in industry and agriculture (OECD 2017). In 2010, 21% of the water used was recycled wastewater and 8% was untreated wastewater; in the future, it is precisely the amount of treated wastewater is expected to increase (State of Israel Water Authority Planning Division 2012, p. 14). To use more wastewater, it is offered more cheaply to industry and farmers than fresh water. In addition, each farmer is allocated a certain amount of fresh water. He can exchange part of it for wastewater, giving a bonus of 20% more. This reduces the consumption of fresh water and encourages the use of wastewater. In addition, the Israeli state pays 60% of all costs necessary to provide the wastewater (Rejwan 2011, pp. 24–25). As a result of these measures, 38% of the water used in Israeli agriculture is treated and recycled wastewater (Rejwan 2011, p. 12). Untreated wastewater is even used for special cereals, and the overall share of untreated water in agricultural consumption is 14% (State of Israel Water Authority Planning Division 2012, p. 54).

To reduce household water consumption and thus the volume of wastewater itself, Israel relied on several measures simultaneously. In 2008, a major media campaign was launched to draw attention to the water shortage and call on the population to save water.

The campaign exceeded the expectation of 8%, and in 2009 the average water consumption per household dropped by 10%. The reduction in water consumption continues to this day, as a result of which the government has so far seen no need to restart the campaign (Rejwan 2011, p. 23). In a second step, the water price for households increased by 40% in 2010. On the one hand, this is intended to further reduce water consumption, and on the other hand, the money is used to finance Israel's desalination plants. As a result, water consumption fell by a further 10% (Rejwan 2011, p. 23).

13.5.4 Water Conflict Between China and India

Many newspaper articles about the Chinese projects on the Brahmaputra give the impression that a water war between China and India is imminent, with the imminent diversion of the Brahmaputra as the trigger. The Indian author and expert Brahma Chellaney, for example, spoke of a "covert war"; newspaper articles with him as an interview partner bear titles such as "China's new war front" or "The coming water wars" (Chellaney 2013). In an article by the international press agency UPI, the source of the Brahmaputra is even dubbed the "most dangerous place in the world" (Walker 2007).

Judging by the facts, a war between China and India seems to be only a matter of time for now. China and India are two countries with major water problems that will get much worse in the near future. China is not disclosing anything about its dam construction plans, leaving the Indian government uninformed about existing projects. Plans and advocates for diverting the Brahmaputra have been around for a long time, and construction near the Great Bend shows that China has ambitions there. Moreover, the relationship between the two countries is at its worst in a long time, strained mainly by border disputes. In 2017, unarmed fighting between Chinese and Indian soldiers and threats of war broke out on the disputed Doklam Plateau after construction work by China (Global Times 2017). The conflict ended after 73 days, however, China stopped sharing information on Brahmaputra outflows shortly after. It is likely that the border conflict and strained diplomatic relations, as a result, are responsible (Jayanth and Sutirtho 2017). The crisis on the Doklam plateau and the sudden halt in the exchange of data on the Brahmaputra seem like the harbingers of a larger, violent conflict that a diversion of the Brahmaputra could trigger. If the construction of a simple road in a remote corner of the Himalayas already leads to threats of war and violent clashes between Chinese and Indian troops, what kind of conflict would the diversion of one of the largest and most sacred rivers in water-scarce India trigger?

A closer look at the Brahmaputra issue reveals that a diversion of the river is unlikely and would be less problematic than many claims. The Brahmaputra flows through one of the most water-rich regions of India and its water is hardly used. The amount of water coming from China is only a fraction of the total flow, and most of the Brahmaputra's water comes from tributaries within India. Whether China diverts the Brahmaputra or simply builds massive dams, India's water resources will hardly be affected. Further, it is inefficient for China to divert the Brahmaputra. The cost of the other routes of the SNWDP has multiplied and so far only relatively expensive polluted water can be delivered. Meanwhile, it is even questionable whether the western route will be built at all without a diversion of the Brahmaputra. Already, many key Chinese politicians have spoken out against such projects, and the government is showing early signs of changing its mind. A diversion of the Brahmaputra would also be accompanied by international ostracism as well as a strong reaction from India. It is hardly worth putting up with these problems for a river that will lose much of its water on the Chinese side in the long run. Thus, diverting the Brahmaputra may be the dream of old CCP cadres who want to solve China's problems in one fell swoop through gigantic construction projects, but with a new, more educated, and rational generation of leaders, this dream is unlikely to become a reality.

Nevertheless, the question arises as to why the diversion of the Brahmaputra is so strongly discussed, especially on the Indian side, and repeatedly conjured up as a "spectre". It is probably because a potential diversion of the Brahmaputra can be used well as a propaganda tool against China. While India, with its existing water treaties, is seen as a model of cooperation and selflessness, the discussed diversion of the Brahmaputra remains a perfect template to illustrate Chinese egoism and megalomania. Against the backdrop of border conflicts, the diversion of the Brahmaputra is therefore readily taken up by India.

That the danger of a water war is quite low, despite water shortages and diplomatic tensions, is proven by the example of Israel. Israel's rate of renewable water resources per capita is one-twentieth that of China and one-tenth that of India. Yet Israel manages to have water in abundance through modern agriculture and desalination plants. If Israel can manage to overcome its water shortage from such a position, China and India will be able to solve their water problems peacefully. Further, the economic relations between the two countries are far too important to be damaged in the long run by a full-blown war. China is India's largest trading partner, and India, soon to be the world's most populous country, has huge market potential.

So while the current relationship between China and India is strained by the water shortage, it is precisely this that can make for a better relationship in the future. Water shortage is a common problem between these two nations, which can and should be combated jointly through cooperation and projects. The Global Trends 2025 Report mentioned at the beginning of this article sees not only risks but also opportunities in the challenges of water shortage:

> Such dire scenarios are not inevitable even with worse-than-anticipated climate change impacts, however. Economic development, the spread of new technologies, and robust new mechanisms for multilateral cooperation to deal with climate change may foster greater global cooperation. (US National Intelligence Council 2008, p. 68)

A sentence from Israel's Water Master Plan aptly sums up the danger of a "water war", it states:

> A desalination plant is cheaper than one day of a battle. (State of Israel Water Authority Planning Division 2012, p. 19)

13.6 Waste Management

Waste management generally involves the prevention, recovery, and disposal of waste (Bilitewski et al. 2000, p. 13). In recent decades, Chinese waste management has improved considerably. However, many people generate a lot of waste, so this issue remains virulent. Different types of waste can be distinguished: Municipal waste is categorized as waste from private households, public areas, and institutions. Industrial waste, hazardous waste or e-waste are disposed of separately (Lianghu et al. 2014, p. 95). In addition, there are distinctions in waste from, for example, e-waste, which is considered hazardous waste. Paper can be recycled and organic waste can be composted. In all areas, the PRC is experimenting – in some cases very innovatively.

With the 13th Five-Year Plan, the government is pursuing several goals in waste management: strengthening the circular economy, promoting recycling, doubling the amount of waste treated, and building more waste treatment facilities (NDRC 2016, pp. 124–125). The strengthened circular economy will prevent more waste in the future by reintroducing materials fully into the cycle. A total of RMB 251.84 billion is to be invested in China's waste management, including RMB 169.93 billion in the construction of facilities (Nelles et al. 2017b, p. 35). In addition, waste separation and recycling of household waste, especially in 46 selected cities, will be promoted to achieve a separation rate of nearly 90%, which is why up to RMB 26 billion will be invested in waste collection and transportation (Nelles et al. 2017b, p. 35). Restaurant waste is categorized by the People's Republic as a separate type of waste. Proper separation and treatment play a major role in this waste category, and RMB 13.6 billion is to be invested in projects for this purpose. When it comes to recycling restaurant waste, China wants to aim for a rate of 35% that can be returned to the circular economy (Nelles et al. 2017a, p. 221). Another point is to promote the construction of hazard-free municipal waste treatment plants, the capacity of which is to be expanded to 510 thousand tons per day. Therefore, 169.93 billion RMB will be raised for the construction of the plants (Nelles et al. 2017b, p. 35). In urban areas, plastics, metals, and papers are already collected and recycled separately (Nelles et al. 2017a, p. 216). Municipal waste is recycled in waste-to-energy plants or sent to secure landfills, with about 65% of waste sent to landfills and 32% incinerated in 2013 (Johnson 2017, p. 306).

China has long maintained an out-of-date approach to hazardous wastes, such as e-waste, by not storing these wastes properly and allowing them to enter the environment, which can be devastating to people and the environment (Hicks et al. 2005, p. 461). Pollution of soil and air has been linked to waste management several times. China was also the largest waste importer of e-waste, plastic, paper, and industrial waste until 2018 (EUWID 2018). In 2012, China imported an estimated nine million tonnes of plastic in total (Greenpeace 2017). According to People's Daily, China also imported more than 70% of the world's electronic waste (Hongyu 2018). In 2017, waste import bans were placed on more than 24 types of waste. Rapid economic growth and waste imports from around the world are contributing to the waste problem in China (Holdinghausen and Lee 2018, p. 86).

13.6.1 Institutions

Currently, two ministries manage China's waste management, the Ministry of Ecology and Environment (MEE) is responsible for general waste prevention and control (MEE 2019a). On the other hand, the Ministry of Housing and Urban Development (MOHURD) is entrusted with statutory waste management, waste disposal and landfilling of municipal waste, and setting limits for waste and wastewater treatment emissions (Nelles et al. 2017b, p. 20). In addition to the central government institutions, the relevant regional and local authorities in China are responsible for waste management (Nelles et al. 2017b, p. 18).

The mission of the MEE is to improve the environment in China, which involves regulating water, soil, and air quality. The MEE is composed of 23 departments, including the Solid Waste and Chemicals Department and the Radioactive Sources and Safety Regulation Department (MEE 2019b). The Solid Waste and Chemicals Division is responsible for the monitoring and management of solid waste, chemicals, and heavy metals (MEE 2019c). This includes the regulation and monitoring of solid waste imports as well as exports, and the management of toxic chemicals. The Radioactive Sources and Safety Regulations Division handles all matters relating to radioactive waste, its lawful storage as well as disposal (MEE 2019d).

In the MOHURD, two departments are responsible for municipal waste management; the Department of Town and Village Planning and the Department of Urban Planning (Nelles et al. 2017b, p. 18). In 2016, the MOHURD published a Compulsory Waste Classification System, with which the People's Republic aims to invest in waste separation (Nelles et al. 2017b, p. 13).

Two other ministries contribute to waste management in China: the Ministry of Industry and Information Technology (MIIT) and the Ministry of Agriculture (MOA). The MIIT includes the Department of Energy Conservation and Comprehensive Utilization, which are responsible for the management of commercial waste. In MOA, there are two departments involved in the management of waste, the Livestock Department, which is responsible for livestock manure, and the Crop Production Department, which is responsible for straw and other agricultural resources (Nelles et al. 2017b, p. 18).

13.6.2 Waste Disposal

China was the world's largest producer of waste in 2016, with 200 million tons of municipal waste and 3.3 billion tons of industrial waste (Holdinghausen and Lee 2018, p. 85), ahead of the United States (Zhu et al. 2009, p. 1227). Currently, China's waste management is characterized by three features, namely the scale of waste generation, coping strategies, and a public debate on waste management (Johnson 2017, pp. 303–304).

China's municipal waste includes kitchen waste, paper, rubber, textiles, wood and bamboo, dust, ceramics, glass, metal, and ash, and also includes market waste, street sweepings, waste from public areas, park waste, and sewage sludge (Cheng and Hu 2010, p. 3819). In China, municipal waste is currently primarily incinerated or landfilled (Johnson 2017, p. 306). The waste composition in the north and south of the country differs due to the different climatic conditions as well as the living conditions and dietary habits of the population. In kitchen waste, the concentration of water is increased due to the high organic content (Zhang et al. 2010, p. 1623). In the North, the moisture content of municipal waste is 30–50%, while in the South it is 50–60% (Lianghu et al. 2014, p. 96). As a result, the decomposition process of MSW, during landfilling, produces biogenic waste fractions. These lead to climate-relevant landfill gas emissions (Havukainen et al. 2017, p. 1). Already in the past, China has shown increasing interest in learning from the experiences of countries that operate advanced waste management and is very involved in know-how transfer with Germany (Dorn et al. 2009, pp. 451–452).

Incineration and landfilling are among the "safe" treatment methods in China (Havukainen et al. 2017, p. 2). The overall treatment capacities have increased immensely compared to previous years, in 2008 the treatable amount was only around 250,000 tons per day, in 2013 the treatable amount was already almost 500,000 tons per day because more waste was disposed of in incinerators (Johnson 2017, p. 305). Further waste incineration plants are to be built (Germany Trade and Invest 2018).

Waste separation is currently still in its infancy in China. It has already been introduced in the cities of Shanghai, Beijing, Tianjin, and Chongqing. A further 27 cities and municipalities are to follow (Germany Trade and Invest 2018).

13.6.2.1 Landfilling

Landfilling is currently the dominant disposal method in China. Previously, waste was disposed of in simple, unsecured landfills (Dorn et al. 2009, p. 453). In 1991, the first landfill in China was constructed and put into operation, the Hangzhou Tianziling MSW. However, it was not until 1997 that the first landfill with a liner was built to meet international standards, the Shenzhen Xiaoping MSW Landfill (Zhang et al. 2010, p. 1627). Between 2001 and 2010, large landfills were eventually built in cities and coastal regions. Landfills that did not meet international standards were closed. Gradually, landfills are also being built in medium-sized and small cities (Nelles et al. 2017b, p. 9).

As of 2010, 498 secured landfills have been constructed in the People's Republic, and about 290,000 tons of municipal waste per day are currently landfilled there (Lianghu et al. 2014, p. 103). The largest landfill in China is located in Chongqing and measures 420 hectares (SASAC 2018). Accordingly, the majority of waste is landfilled, but not only in secured, legal landfills. A large proportion of these landfills do not meet international standards (Zhang et al. 2010, p. 1629). The landfills are technically inferior, as evidenced by the lack of filtration systems, poor groundwater protection mechanisms, and the release of emissions (Chen 2015). The biogenic waste fractions also lead to the generation of leachate, some of which is released untreated from the landfills into the sewage system, polluting groundwater (Han et al. 2016, p. 1255). These are the current major difficulties of China's waste management in terms of managing municipal waste. The goal of the Chinese government is to replace landfills with other methods, such as waste incineration, in a timely manner. Cities such as Shanghai, Shenzhen, and Beijing plan to stop landfilling waste altogether by 2020 (Schonberg 2017).

13.6.2.2 Waste Incineration Plants

Waste incineration is counted as thermal waste treatment. In this process, the combustible fractions in the waste are incinerated to achieve a reduction in volume and to use the energy contained in the waste. In 2010, only 16.2% of waste was incinerated in China (Lianghu et al. 2014, p. 103), but by 2013, the figure was 32% (Johnson 2017, p. 306). Waste incineration is used for electricity generation and heating. In 2004, China had only 54 waste incineration plants with a total capacity of 16,907 tons per day; by 2013, it had over 166 plants with a total capacity of 158,488 tons per day (Johnson 2017, p. 306). By 2020, China will need an additional 800 waste-to-energy plants to cope with rising waste volumes (Erling 2016). In Beijing, the largest waste incinerator in the world, Lujiashan, was built in 2014. This waste incinerator can dispose of up to 3000 tons of waste per day, producing 420 million kilowatt hours (kWh) of electricity (Johnson 2017, p. 302).

The incineration of waste also produces fly ash as a by-product (Cheng and Hu 2010, p. 3817). These fly ashes contain a high concentration of toxic heavy metals and dioxins (Steinhardt and Wu 2016, p. 62). In China, fly ash is categorized as hazardous waste (Wang et al. 2009, p. 955). Public opposition to the construction of waste incinerators is therefore rising among the population (Johnson 2017, p. 303).

13.6.2.3 Composting

In composting, organic material is broken down by soil organisms under oxygen, resulting in compost. In the early days of waste management, China still made extensive use of the traditional composting method (Wei et al. 2000, p. 286). In the 1990s, the People's Republic invested in composting facilities in the cities of Changzhou, Beijing, and Shanghai, but in the production of compost, these facilities show only moderate success (Lianghu et al. 2014, p. 107). The compost is also of low quality due to poor waste separation (Johnson 2017, p. 306).

In 2001, there were a full 140 composting plants operating in China with a total capacity of 25,461 tons per day; by 2010, there were only eleven, bringing the total capacity down to 5480 tons per day (Lianghu et al. 2014, p. 107). Since municipal waste is not separated and it is difficult to segregate household waste and organic materials afterward, composting is not profitable. On agricultural land, farmers try to bury their waste. They assume that the organic fractions will rot, but non-organic fractions, such as plastic, are also disposed of in the fields (Dorn et al. 2009, p. 451).

13.6.2.4 Cockroach Farms

Some of the kitchen waste in municipal solid waste in China is fed to cockroaches and disposed of in this way. A former employee of a pharmaceutical company had the idea of setting up a cockroach farm in Jinan in Shandong province. It houses a total of 300 million American cockroaches, which can recycle up to 15 million tons of household waste per day. The life cycle of the American cockroach is 700 days. It is suitable as an omnivore for effective waste disposal. The cockroaches are sold as feed to pig farms and fisheries at the end of their life cycle (Leng 2018).

The largest cockroach farm in China is located in Xichang, Sichuan province, where six billion cockroaches are bred per year (Leng 2018). The farm is run by a company called "Gooddoctor" and produces medicines for ulcers and abdominal diseases from the ingredients of the cockroaches. After their life cycle ends, the cockroaches are first cleaned with a steam jet, washed, dried, and finally, their nutrients are extracted in large tanks (Suen and Woo 2018). There are plans to build more cockroach farms in China (Leng 2018).

13.6.3 Waste Import Ban

In the 1990s, China exported numerous cheap consumer goods to the rest of the world by container ships, but imported very little, resulting in high export surpluses. The empty containers were loaded with waste from industrialized countries on their way back to China, and so import companies were founded to supply China with waste from the West, specializing in various types of waste. For example, the import company "Nine Dragons" specialized in waste paper imports from the USA (Holdinghausen and Lee 2018, p. 86).

However, in 2015, the People's Republic started to inspect waste imports from abroad more strictly and tighten import regulations (Rehbock 2018, p. 53). On 1 June 2017, the Chinese government imposed a temporary ban on permits for waste imports and at the same time established 60 inspection teams to inspect all Chinese waste import companies (BIR 2017). In 2018, the project continued under the name of Green Fence and imposed a strict import ban on 24 types of waste, including plastic, e-waste, metallic slag, and unsorted paper. The MEE has added another 16 solid waste types to the list called "Catalogue of Solid Wastes Prohibited for Import" on December 31, 2018, and December 31, 2019, respectively. From July 1, 2019, eight more types of solid wastes whose import was not restricted before have been removed from the list called "Catalogue of Solid Wastes Not Restricted for Import as Raw Materials" and placed on the list called "Catalogue of Solid Wastes Restricted for Import as Raw Materials". This includes, for example, waste and scrap of cast iron, alloy steel, and tinned iron (Bundesverband Sekundärrohstoffe und Entsorgung 2018). Consequently, more and more types of waste are being excluded from import by China.

Furthermore, China's General Administration of Customs (GAC) has issued lists of approved pre-shipment inspection agencies, including 21 agencies responsible for the legitimate inspection and monitoring of waste imports. The waste is divided into six types:

(1) metal waste, (2) plastic waste, (3) recycled paper and cardboard waste, (4) metal alloy waste, (5) mixed metal waste, (6) wood and cork waste. For example, China Inspection and Certification Group Germany Co, Ltd. is responsible for inspecting waste imports from Germany (Bundesverband Sekundärrohstoffe und Entsorgung 2019).

For the People's Republic, the waste imports were a good source of raw materials and money. It is clear, however, that the imported waste included a great deal of unusable material that could not be recycled due to its high pollution levels. Particularly in the case of plastic waste, attention must be paid to qualitative purity. Toxic substances are also frequently discovered in imported waste, which is particularly dangerous for people and the environment. Furthermore, China's waste quantities are increasing year by year, so it becomes necessary to organize its waste and plastic management first (Holdinghausen and Lee 2018, pp. 85–86).

Experts in Germany see China's import ban as an opportunity to invest in the production of high-quality recyclates and improve recycling facilities, thus promoting the circular economy (Roth et al. 2018, p. 87). However, this is currently only a future prospect because instead of enacting uniform regulations in Germany, waste is exported to other countries such as Malaysia, India, or Vietnam. Waste also continues to be transferred to China via countries such as Malaysia despite the import ban (Holdinghausen and Lee 2018, p. 85). According to NABU, Germany produced 6.2 million tons of plastic waste in 2017, less than 50% of packaging waste can be recycled and almost 51% of it is incinerated (NABU 2017). The crucial problem is that developed countries with high incomes send their plastic waste to less developed countries, which ultimately still struggle with their waste management system and are not able to deal with the high amounts of waste (Brooks et al. 2018, p. 2).

China's import ban has far-reaching consequences, especially for the plastics industry. Either the exporting countries must improve their waste and plastic management, reduce the quantities or switch to other importing countries, which have increased the quantities of imported waste since China's waste import ban. According to NABU, care must be taken not to incinerate even higher quantities of plastics, as this produces toxic by-products that are harmful to the environment (NABU 2017).

13.7 Conclusion

Environmental policy is extremely important for China's future because a better environment contributes to a sustainable model of economic growth. Since 2013, the Chinese government has been increasingly committed to environmental protection at the national and international levels, as demonstrated by national laws and the Paris Climate Change Agreement.

Nevertheless, the implementation of policy measures remains difficult. In this respect, China's decentralized authoritarian system seems to be detrimental to the environment, as measures cannot be directly enforced by the central government (Economy 2014, p. 185).

Another issue is that while enforcement of control measures has been integrated into the performance and promotion evaluation system for civil servants, no punishments follow for government officials in case of non-enforcement or failure of these measures (Khan and Chang 2018, p. 14). In addition, the population could be involved by increasing general environmental awareness to support the government's policies.

The PRC suffers from an acute water shortage, especially in the north. In the 2000s, Chinese scientists seemed to prefer to solve this shortage with water from the Himalayan mountains, but in the meantime, there has been a change in thinking so that these giga-projects are no longer pursued with the original vehemence. The water supply of the north via desalination plants at the sea seems to have a promising future. Especially since nuclear power plants are being built in these regions to supply the necessary electricity. A dispute over water resources between the PRC and India, therefore, seems unlikely.

China has started to practice modern and advanced waste management. Although waste disposal methods meet today's standards, it is nevertheless difficult to dispose of the increasing amounts of waste. Currently, the focus of waste management is on the expansion of waste incineration plants, and rapid progress in this area has already been realized. New developments such as maintaining cockroach farms for waste disposal offer the potential to invest in more sustainable methods of waste management in the future and to cope with increasing waste volumes. The PRC has enacted an import ban on waste, which better protects the Chinese environment. It would be up to the industrialized nations to take care of their waste management instead of looking for other exporting countries.

References

Albert, Elanor, and Beina Xu. 2016. *China's environmental crisis. Council on foreign relations.* https://www.cfr.org/backgrounder/chinas-environmental-crisis. Accessed on 28.08.2019.

Alpermann, Björn. 2010. *China's cotton industry. Economic transformation and state capacity.* London: Routledge.

Anger, Annela, Olivier Dessens, Fengming Xi, Terry Barker, and Wu. Rui. 2015. China's air pollution reduction efforts may result in an increase in surface ozone levels in highly polluted areas. *Ambio* 45 (2): 254–265.

Arpi, Claude. 2011. *Hydropower Projects on the Brahmaputra.* http://claudearpi.blogspot.de/2011/09/hydropower-projects-on-brahmaputra.html. Accessed on 28.08.2019.

Axe, David. 2010. *War is boring: China dam project stokes regional tensions.* https://www.globalpolicy.org/the-dark-side-of-natural-resources-st/water-in-conflict/49047-war-is-boring-china-dam-project-stokes-regional-tensions.html. Accessed on 28.08.2019.

Bandyopadhyay, Jayanta, Nilanjan Ghosh, and Chandan Mahanta. 2016. *IRBM for Brahmaputra sub-basin: Water governance, environmental security and human Well-being.* New Delhi: Observer Research Foundation. http://brahmaputrariversymposium.org/wp-content/uploads/2017/09/Monograph_IRBM-for-Brahmaputra_Z-Final.pdf. Accessed on 28.08.2019.

Beyer, Stefanie. 2006. Environmental law and policy in the People's republic of China. *Chinese Journal of International Law* 5 (1): 185–211.

Bilitewski, Bernd, Georg Härdtle, and Klaus Marek. 2000. *Abfallwirtschaft – Handbuch für Praxis und Lehre.* Berlin: Springer.

BIR. 2017. *Important: Update on China.* https://bir.org/news-press/latest-news/important-update-on-china/. Accessed on 10.06.2019.

BP p.l.c. 2019. *BP statistical review of world energy*, 68. Aufl. https://www.bp.com/content/dam/bp/business-sites/en/global/corporate/pdfs/energy-economics/statistical-review/bp-stats-review-2019-full-report.pdf. Accessed on 16.06.2019.

Briscoe, John, and Ravinder P.S. Malik. 2006. *India's water economy. Bracing for a turbulent future.* New Delhi: Oxford University Press.

Brooks, Amy L., Shunli Wang, and Jenna R. Jambeck. 2018. The Chinese import ban and its implications on global plastic waste trade. *Science Advances* 4 (6): 1–7.

Browne, Andrew. 2015. Another kind of climate change: China warms to superpower role. *The Wall Street Journal.* https://www.wsj.com/articles/another-kind-of-climate-change-china-warms-to-superpower-role-1450154527. Accessed on 28.08.2019.

Bundesverband Sekundärrohstoffe und Entsorgung. 2018. *China verkündete weitere 32 Einfuhrverbote für feste Abfälle.* https://www.bvse.de/recycling/recycling-nachrichten/3022-china-verkuendete-weitere-32-einfuhrverbote-fuer-feste-abfaelle.html. Accessed on 10.06.2019.

———. 2019. *China erweitert Abfallimportverbote und Liste zugelassener Pre-Shipment-Agencies.* https://www.bvse.de/recycling/recycling-nachrichten/3976-china-erweitert-abfallimportverbote-und-liste-zugelassener-pre-shipment-agencies.html. Accessed on 10.06.2019.

Carter, Neil, and Arthur Mol. 2007. *Environmental governance in China, 16.* London: Routledge.

Chai, Jing. 2015. *Chai Jing's review: Under the Dome – Investigating China's Smog.* Youtube. https://www.youtube.com/watch?v=T6X2uwlQGQM. Accessed on 19.06.2019.

Chang, Gordon. 2014. China's water crisis made worse by policy failures. *World Affairs*, 08. Januar 2014. http://www.worldaffairsjournal.org/blog/gordon-g-chang/china%E2%80%99s-water-crisis-made-worse-policy-failures. Accessed on 02.01.2018.

Chellaney, Brahma. 2011. *Water. Asia's new battleground.* Washington, DC: Georgetown University Press.

———. 2013. China's new war front. *The Times of India*, 23. April 2013. http://timesofindia.indiatimes.com/edit-page/Chinas-new-war-front-Natural-resource-as-a-political-tool/articleshow/19683339.cms. Accessed on 28.08.2019.

Chen, Yifang. 2015. *Land of 1,000 Landfills.* https://slate.com/human-interest/2015/06/unregulated-landfill-epidemic-chinas-capital-is-cleaning-up-more-than-1000-unregulated-dumps.html. Accessed on 30.08.2019.

Cheng, Hefa, and Yuanan Hu. 2010. Municipal solid waste (MSW) as a renewable source of energy: Current and future practices in China. *Bioresource Technology* 101: 3816–3824.

China Council for International Cooperation on Environment and Development. 2014. *CCICED special policy study report: Performance evaluation on the action plan of air pollution prevention and control and regional coordination mechanism.* http://www.cciced.net/cciceden/POLICY/rr/prr/2014/201411/P020160928409089848299.pdf. Accessed on 10.06.2019.

China Institute of Water Resources and Hydropower Research. 2011. *Shuǐlì bù qián bùzhǎng yánzhōng de shuǐdiàn shìfēi. Wāngshùchéng: „Bùnéng zǒng yòng zāinàn xìng hòuguǒ jiàoyù rénmen".* http://www.iwhr.com/zgskyww/hyxw/webinfo/2011/07/1309484269094627.htm. Accessed on 02.01.2018.

ChinaCulture.org. 2017. *Expert on water conservancy and hydroelectric engineering: Zhang Guangdou.* http://en.chinaculture.org/library/2008-02/01/content_127157.htm. Accessed on 29.08.2019.

Chinese National Committee on Large Dams. 2006. *State of the art, dam construction in China.* http://web.archive.org/web/20141101040556/http://www.iwhr.com/english/newsview.asp?NewsID=16831. Accessed on 29.08.2019.

Clean Air Asia. 2016. *Air pollution prevention and control progress in Chinese cities*. https://cleanairasia.org/wp-content/uploads/2016/08/China-Air-2016-Report-Full.pdf. Accessed on 08.06.2019.

Davies, Richard. 2017. India – Third wave of flooding hits Assam, 2 million affected. *Floodlist*, August 14. floodlist.com/asia/india-assam-floods-august-2017. Accessed on 29.08.2019.

Deka, Kaushik, and Ananth Krishnan. 2015. Bend it like Beijing. *India Today*, 10. November. https://www.indiatoday.in/magazine/the-big-story/story/20151123-bend-it-like-beijing-820842-2015-11-10. Accessed on 29.08.2019.

Deng, Xiping, Lun Shan, Heping Zhang, and Neil C. Turner. 2006. Improving agricultural water use efficiency in arid and semiarid areas of China. *Agricultural Water Management* 80 (1): 23–40.

Dimitrov, Radoslav S. 2010. Inside UN climate change negotiations: The Copenhagen conference. *Review of Policy Research* 27 (6): 795–821.

———. 2016. The Paris agreement on climate change: Behind closed doors. *Global Environmental Politics* 16 (3): 1–11.

Dorn, Thomas, Mohamad Al-Ahmad, Gert Morscheck, Abdallah Nassour, and Michael Nelles. 2009. Internationale Abfallwirtschaft als Zukunftsaufgabe – aktuelle Entwicklungen im arabischen Raum und der VR China. *Fachzeitschrift Müll und Abfall* 19 (9): 448–454.

Economy, Elizabeth. 2014. Environmental governance in China: State control to crisis management. *Dædalus, the Journal of the American Academy of Arts & Sciences, Spring* 143 (2): 184–197.

Erling, Johnny. 2016. *Was deutschen Müll für die Chinesen so spannend macht*. https://www.welt.de/wirtschaft/article159655412/Was-deutschen-Muell-fuer-die-Chinesen-so-spannend-macht.html. Accessed on 29.08.2019.

EUWID. 2018. *China kündigt Importverbote für weitere Abfälle an*. https://www.euwid-recycling.de/news/international/einzelansicht/Artikel/china-kuendigt-importverbote-fuer-weitere-abfaelle-an.html. Accessed on 16.06.2019.

FAO. 1999. *Irrigation in Southern and Eastern Asia in figures*. FAO water reports 18. http://www.fao.org/3/i2809e/i2809e.pdf. Accessed on 29.08.2019.

———. 2016. *Aquastat main database, total renewable water resources, dependency ratio by country*. http://www.fao.org/nr/water/aquastat/data/query/index.html?lang=en. Accessed on 02.01.2018.

Feng, Hao. 2018. China releases 2020 action plan for air pollution. *China Dialogue*. https://www.chinadialogue.net/article/show/single/en/10711-China-releases-2-2-action-plan-for-air-pollution. Accessed on 29.08.2019.

Feng, Lu, and Wenjie Liao. 2015. Legislation, plans, and policies for prevention and control of air pollution in China: Achievements, challenges, and improvements. *Journal of Cleaner Production* 112 (2): 1549–1558.

Flitton, Daniel. 2015. Paris UN climate conference 2015: A global deal made in China (and the US). *The Sydney Morning Herald*. https://www.smh.com.au/environment/climate-change/paris-un-climate-conference-2015-a-global-deal-made-in-china-and-the-us-20151213-glmfo3.html. Accessed on 29.08.2019.

Florence, Eric, and Pierre Defraigne. 2012. Towards a new development paradigm in twenty-first century China: Economy. In *In Society and politics*. London: Routledge.

Gao, Jiajia, Kun Wang, Yong Wang, Shuhan Liu, Chuanyong Zhu, Jiming Hao, Huanjia Liu, Shenbing Hua, and Hezhong Tian. 2018. Temporal-spatial characteristics and source apportionment of PM2.5 as well as its associated chemical species in the Beijing-Tianjin-Hebei region of China. *Environmental Pollution* 233: 714–724.

GBD MAPS Working Group. 2016. *Burden of disease attributable to coal-burning and other major sources of air pollution in China*, Special report 20. Health Effects Institute, Boston. https://www.healtheffects.org/publication/

burden-disease-attributable-coal-burning-and-other-air-pollution-sources-china. Accessed on 29.08.2019.

Germany Trade and Invest. 2018. *Branche kompakt: China will bis 2020 über 38 Mrd. US-Dollar in die Abfallwirtschaft investieren.* https://www.gtai.de/GTAI/Navigation/DE/Trade/Maerkte/Branchen/Branche-kompakt/branche-kompakt-recycling-und-entsorgungswirtschaft,t=branche-kompakt-china-will-bis-2020-ueber-38-Mrd.-usdollar-in-die-abfallwirtschaft-investieren,did=1859194.html. Accessed on 28.08.2019.

Global Times. 2017. India will suffer worse losses than 1962 if it incites border clash. *Global Times*, 04. Juli2017. http://www.globaltimes.cn/content/1054925.shtml. Accessed on 02.01.2018.

Government of India Planning Commission. 2007. *Report of the expert group on "ground water management and ownership"* S. 1–3. http://www.planningcommission.nic.in/reports/genrep/rep_grndwat.pdf. Accessed on 02.01.2018.

Greenpeace. 2017. *China's ban on imports of 24 types of waste is a wakeup call to the world.* http://m.greenpeace.org/eastasia/high/press/releases/toxics/2017/Chinas-ban-on-imports-of-24-types-of-waste-is-a-wake-up-call-to-the-world%2D%2D-Greenpeace/. Accessed on 12.06.2019.

Han, Guoyi, Marie Olsson, Karl Halldin, and David Lunsford. 2012. *China's carbon emission trading an overview of current development.* Stockholm: Forum for Reforms, Entrepreneurship and Sustainability, FORES.

Han, Zhiyong, Haining Ma, Guozhong Shi, Li He, Luoyu Wei, and Qingqing Shi. 2016. A review of groundwater contamination near municipal solid waste landfill sites in China. *Science of the Total Environment* 569–570: 1255–1264.

Havukainen, Jouni, Mingxiu Zhan, Jun Dong, Miia Liikanen, Ivan Deviatkin, Xiaodong Li, and Mika Horttanainen. 2017. Environmental impact assessment of municipal solid waste management incorporating mechanical treatment of waste and incineration in Hangzhou, China. *Journal of Cleaner Production* 141: 453–461.

Health Effects Institute. 2019. *State of global air 2019, special report.* Health Effects Institute, Boston. https://www.stateofglobalair.org/sites/default/files/soga_2019_report.pdf. Accessed on 29.08.2019.

Hicks, Charlotte, Rolf Dietmar, and Martin Eugster. 2005. The recycling and disposal of electrical and electronic waste in China – Legislative and market responses. *Environmental Impact Assessment Review* 25: 459–471.

Ho, Selina. 2015. A river flows through it: A Chinese perspective. In *China – India relations. Cooperation and conflict*, ed. Kanti Bajpai, Jing Huang, and Kishore Mahbubani, 182–197. London: Routledge.

Hoffman, Samatha, and Jonathan Sullivan. 2015. Environmental protests expose weakness in China's leadership. *Forbes.* https://www.forbes.com/sites/forbesasia/2015/06/22/environmental-protests-expose-weakness-in-chinas-leadership. Accessed on 16.06.2019.

Holdinghausen, Heike, and Felix Lee. 2018. Der Müll der anderen. In *Chinas Aufstieg – Mit Kapital, Kontrolle und Konfuzius*, ed. Seven Hansen and Barbara Bauer, 84–87. Berlin: TAZ Verlag.

Hongyu, Bianji. 2018. *China firmly says not o foreign waste.* http://en.people.cn/n3/2018/0328/c90000-9442858.html. Accessed on 20.05.2019.

Hu, Feng, Debra Tan, and Inna Lazareva. 2014. *8 Facts on China's wastewater.* http://chinawaterrisk.org/resources/analysis-reviews/8-facts-on-china-wastewater/. Accessed on 29.08.2019.

Huang, Cary. 2015a. Press freedom needed to win China's choking air pollution battle. *South China Morning Post.* http://www.scmp.com/news/china/society/article/1887084/press-freedom-needed-win-choking-air-pollution-battle. Accessed on 08.06.2019.

Huang, Keira. 2015b. Smog in northeast China at nearly 50 times World Health Organization safe limits. *South China Morning Post.* https://www.scmp.com/news/china/society/article/1877282/smog-northeast-china-nearly-50-times-world-health-organisation. Accessed on 08.06.2019.

HydroChina. n.d. HydroChina dam map. http://web.archive.org/web/20130507134607/http://www.hydrochina.com.cn/zgsd/images/ziyuan_b.gif. Accessed on 02.01.2018.

India Today. 2009. Satellite images confirm construction of Chinese dam on Brahmaputra. *India Today*, 04. November 2009. http://indiatoday.intoday.in/story/Satellite+images+confirm+construction+of+Chinese+dam+on+Brahmaputra/1/69296.html. Accessed on 02.01.2018.

Intergovernmental Panel on Climate Change (IPCC). 2013. *Klimawandel 2013 Naturwissenschaftliche Grundlagen*. https://www.ipcc.ch/site/assets/uploads/2018/03/IPCC_WG1_AR5_Headlines_deutsch.pdf. Accessed on 08.06.2019.

International Carbon Action Partnership. 2019. *China National ETS*. https://icapcarbonaction.com/en/?option=com_etsmap&task=export&format=pdf&layout=list&systems[]=55. Accessed on 08.06.2019.

Jabeen, Azra, Huangxi Sheng, and Muhammad Aamir Aamir. 2015. Environmental stability still in danger, loopholes of new Environmental Protection Law (EPL) in China. *US-China Law Review* 12: 951–964.

Jacobsen, Rowan. 2016. Israel proves the desalination era is here. *Scientific American*, 29. Juli 2016. https://www.scientificamerican.com/article/israel-proves-the-desalination-era-is-here/. Accessed on 29.08.2019.

Jayanth, Jacob, and Patranobis Sutirtho. 2017. Doklam standoff ends: India pulls troops, no word on Beijing's road along border. *Hindustan Times*, 28. August 2017. http://www.hindustantimes.com/india-news/doklam-standoff-india-says-has-agreed-to-disengagement-of-troops-after-talks-with-china/story-xybEpTjdsyFbobAjrf8yEN.html. Accessed on 29.08.2019.

Jha, Prem Shankar. 2014. Why India and China should leave the Yarlung Tsangpo alone. *China Dialogue*, 05. März 2014. https://www.chinadialogue.net/article/show/single/en/6753-Why-India-and-China-should-leave-the-Yarlung-Tsangpo-alone. Accessed on 29.08.2019.

Jiang, Chong, Linbo Zhang, Daiqing Li, and Fen Li. 2015a. Water discharge and sediment load changes in China. Change patterns, causes, and implications. *Water* 7(10): 5849–5875. http://www.mdpi.com/2073-4441/7/10/5849/pdf. Accessed on 29.08.2019.

Jiang, Xujia, Chaopeng Hong, Yixuan Zheng, Bo Zheng, Dabo Guan, Andy Gouldson, Qiang Zhang, and Kebin He. 2015b. To what extent can China's near-term air pollution control policy protect air quality and human health? A case study of the Pearl River Delta region. *Environmental Research Letters* 10(10). https://iopscience.iop.org/article/10.1088/1748-9326/10/10/104006. Accessed on 29.08.2019.

Jin, Leshan, and Warren Young. 2001. Water use in agriculture in China. Importance, challenges, and implications for policy. *Water Policy* 3 (3): 215–228.

Johnson, Thomas. 2017. Municipal solid waste management. In *Handbook of environmental policy in China*, ed. Eva Sternfeld, 302–313. London: Routledge.

Kaiman, Jonathan. 2014. China strengthens environmental laws. *The Guardian*. https://www.theguardian.com/environment/2014/apr/25/china-strengthens-environmental-laws-polluting-factories. Accessed on 08.06.2019.

Kan, Haidong, Renjie Chen, and Shilu Tong. 2012. Ambient air pollution, climate change, and population health in China. *Environmental International* 42: 10–19.

Khan, Mehran Idris, and Yen-Chiang Chang. 2018. Environmental challenges and current practices in China – A thorough analysis. *Sustainability* 10 (7): 2547.

Krishnan, Ananth. 2011a. Brahmaputra waters will not be diverted, indicates China. *The Hindu*, 01. Juni 2011.http://www.thehindu.com/news/international/Brahmaputra-waters-will-not-be-diverted-indicates-China/article13835249.ece#. Accessed on 29.08.2019.

———. 2011b. Push for new dams across Brahmaputra as China faces drought. *The Hindu*, 10. Juni 2011.http://www.thehindu.com/news/

push-for-new-dams-across-brahmaputra-as-china-faces-drought/article2093981.ece. Accessed on 02.01.2018.

Küffner, Georg. 2018. Die Welt hat Durst: So wird Trinkwasser produziert. *Frankfurter Allgemeine Zeitung*. 7. Juli. https://www.faz.net/aktuell/technik-motor/technik/salzwasser-trinkbar-machen-durchentsalzungsanlagen-15669505.html. Accessed on 29.08.2019.

Kuo, Lily. 2014. China has launched the largest water-pipeline project in history. *The Atlantic*, 07. Mai 2014. https://www.theatlantic.com/international/archive/2014/03/china-has-launched-the-largest-water-pipeline-project-in-history/284300/. Accessed on 29.08.2019.

Leng, Sidney. 2018. *Chinese farmer unleashes swarm of hungry cockroaches to chew through mountain of food scraps*. https://www.scmp.com/news/china/society/article/2143886/chinese-farmer-unleashes-swarm-hungry-cockroaches-chew-through. Accessed on 29.08.2019.

Li, Zhenxiang. 2013. Zhōngguó gèdì kāizhǎn cūnzhuāng qīngjié yùndòng. *China News*, 16. Dezember 2013. http://www.chinanews.com/gn/2013/12-16/5620868.shtml. Accessed on 29.08.2019.

Lian, Ruby, and Melanie Burton. 2017. China orders aluminum, steel cuts in war on smog. *Thomson Reuters*. https://www.reuters.com/article/us-china-pollution/china-orders-aluminum-steel-cuts-in-waron-smog-idUSKBN1683G6. Accessed on 19.06.2019.

Lianghu, Su, Huang Sheng, Dongjie Niu, Chai Xiaoli, Nie Yongfeng, Zhao Youcai. 2014. Municipal solid waste management in China. In Municipal Solid Waste Management in Asia and the Pacific Islands, 95–112. Singapur: Springer.

Lin, Xinyan, and Mark Elder. 2014. Major developments in China's national air pollution policies in the early 12th Five-Year Plan period. Institute for Global Environmental Strategies (IGES): 79–81.

Liu, Jianqiang. 2014. China's new environmental law looks good on paper. *China Dialogue*. https://www.chinadialogue.net/blog/6937-China-s-new-environmental-law-looks-good-on-paper/en. Accessed on 29.08.2019.

Liu, Baoshuang, Danni Liang, Jiamei Yang, Qili Dai, Xiaohui Bi, Yinchang Feng, Jie Yuan, Zhimei Xiao, Yufen Zhang, and Xu. Hong. 2016a. Characterization and source apportionment of volatile organic compounds based on 1-year of observational data in Tianjin, China. *Environmental Pollution* 218: 757–769.

Liu, Gengyuan, Zhifeng Yang, Bin Chen, Yan Zhang, Su Meirong, and Sergio Ulgiati. 2016b. Prevention and control policy analysis for energy-related regional pollution management in China. *Applied Energy* 166 (C): 292–300.

Liu, Miaomiao, Yining Huang, Zongwei Ma, Zhou Jin, Xingyu Y. Liu, Haikun Wang, Yang Liu, Jinnan Wang, Matti Jantunen, Jianzhao Bi, and Patrick L. Kinney. 2017. Spatial and temporal trends in the mortality burden of air pollution in China: 2004–2012. *Environmental International* 98: 75–81.

Lomborg, Bjorn. 2016. Impact of current climate proposals. *Global Policy* 7(1). https://onlinelibrary.wiley.com/doi/full/10.1111/1758-5899.12295. Accessed on 29.08.2019.

Mcelroy, Damien. 2000. China planning nuclear blasts to build giant hydro project. *The Telegraph*, 22.10.2000. http://www.telegraph.co.uk/news/worldnews/asia/china/1371345/China-planning-nuclear-blasts-to-build-giant-hydro-project.html. Zuletzt geprüft am 29.08.2019. Accessed on 02.01.2018

MEE. 2017. *Jīng-Jīn-Jì jí zhōubiān dìqū 2017–2018 niánqiū dōngjì dàqì wūrǎn zònghé zhìlǐ gōngjiān xíngdòng fāng'àn* (Action plan to comprehensive control autumn and winter air pollution in Beijing-Tianjin-Hebei and surrounding regions 2017–2018). http://www.mee.gov.cn/gkml/hbb/bwj/201708/W020170824378273815892.pdf. Accessed on 30.08.2019.

———. 2018a. *2017 Report on the state of the ecology and environment in China*. http://english.mee.gov.cn/Resources/Reports/soe/SOEE2017/201808/P020180801597738742758.pdf. Accessed on 10.06.2019.

———. 2018b. Ministry of ecology and environment inaugurated. *News Release*. http://english.mee.gov.cn/News_service/news_release/201804/t20180419_434955.shtml. Accessed on 16.06.2019.

———. 2019a. *Mandates*. http://english.mee.gov.cn/About_MEE/Mandates/. Accessed on 10.06.2019.

———. 2019b. *Departments*. http://english.mee.gov.cn/About_MEE/Internal_Departments/. Accessed on 10.06.2019.

———. 2019c. *Department of solid wastes and chemicals*. http://english.mee.gov.cn/About_MEE/Internal_Departments/200910/t20091015_162418.shtml. Accessed on 10.06.2019.

———. 2019d. *Department of radiation source safety regulation*. http://english.mee.gov.cn/About_MEE/Internal_Departments/201605/t20160526_346914.shtml. Accessed on 10.06.2019.

MEP. 2017. *2016 Report on the state of the environment in China*. http://english.mee.gov.cn/Resources/Reports/soe/ReportSOE/201709/P020170929573904364594.pdf. Accessed on 10.06.2019.

Moore, Malcom. 2011. More than 40,000 Chinese dams at risk of breach. *The Telegraph*, 26. August 2011. http://www.telegraph.co.uk/news/worldnews/asia/china/8723964/More-than-40000-Chinese-dams-at-risk-of-breach.html. Accessed on 30.08.2019.

NABU. 2017. *Kunststoffabfälle in Deutschland*. https://www.nabu.de/umwelt-und-ressourcen/abfall-und-recycling/22033.html. Accessed on 10.08.2019.

NDRC. 2016. *Zhōnghuá rénmín gònghéguó guómín jīngjì hé shèhuì fāzhǎn dì shísān gè wǔ nián guīhuà gāngyào*. The 13th Five-Year-Plan for Economic and Social Development of the PRC (2016–2020). http://www.ndrc.gov.cn/gzdt/201603/P020160318576353824805.pdf. Accessed on 16.06.2019.

Nelles, Michael, Astrid Lemke, Gert Morscheck, Abdallah Nassour, Andrea Schüch, and Ying Zhou. 2017a. Entsorgung von biogenen Abfallfraktionen in der VR China. *Müll und Abfall* 17 (5): 216–224.

Nelles, Michael, Abdallah Nassour, Aymann El Naas, Astrid Lemke, Gert Morschek, Andrea Schüch, Pinjing He, Liming Shao, and Hua Zhang. 2017b. *Studie – Verwertung von biogenen Fraktionen aus Siedlungsabfällen in der VR China*. https://www.vdma.org/documents/266241/18172648/BioChina%20deutsch_1498038474187.pdf/20251b0d-11be-46dd-8cb4-fad5636b2b7b. Accessed on 30.08.2019.

OECD. 2017. *Waste water treatment*. https://data.oecd.org/water/waste-water-treatment.htm. Accessed on 02.01.2018.

Office of the South-to-North Water Diversion Project Commission of the State Council. n.d. *South-to-North Water Diversion. Eastern Route Project (ERP)*. http://www.nsbd.gov.cn/zx/english/erp.htm. Accessed on 02.01.2018.

Outlook in India. 2011. No threat from Chinese dam plan on Brahmaputra. *Outlook in India*, 04. April 2011. https://www.outlookindia.com/newswire/story/no-threat-from-chinese-dam-plan-on-brahmaputra/717662. Accessed on 02.01.2018.

Peng, Jennie. 2011. *Market report: Developing desalination in China*. http://www.waterworld.com/articles/wwi/print/volume-25/issue-6/regional-spotlight-asia-pacific/market-report-developing-desalination.html. Accessed on 30.08.2019.

Press Information Bureau Ministry of Water Resources. 2015. *Dam on Brahmaputra River by China*. http://pib.nic.in/newsite/PrintRelease.aspx?relid=132646. Accessed on 02.01.2018.

Qi, Ye, and Tong Wu. 2015. China's 'yes' to new role in climate battle. *China Daily Europe*. http://europe.chinadaily.com.cn/epaper/2015-12/04/content_22625313.htm?from=singlemessage&isappinstalled=0. Accessed on 19.06.2019.

Ray, Pranab Kumar. 2014. *Rivers of Conflict or Rivers of Peace. Water Sharing between India and China*. In: ORF Seminar Series 1(13) : 4–12. http://www.indiaenvironmentportal.org.in/files/file/Rivers%20of%20Conflict%20or%20Rivers%20of%20Peace.pdf. Accessed on 30.08.2019.

Rehbock, Eric. 2018. Zukunftsinvestition Kunststoffrecycling. *Fachzeitschrift Müll und Abfall* 2: 53.

Rejwan, Ariel. 2011. *The State of Israel: National Water Efficiency Report*. http://www.water.gov.il/Hebrew/ProfessionalInfoAndData/2012/24-The-State-of-Israel-National-Water-Efficiency-Report.pdf. Accessed on 30.08.2019.

Roth, Sascha, Thomas Fischer, and Hartmut Hoffmann. 2018. Stellungnahme aus den Umweltverbänden. *Fachzeitschrift Müll und Abfall* 2 (18): 87–88.

Ryall, Julian, and Audrey Yoo. 2013. Japan, South Korea concerned that China's smog will affect them. *South China Morning Post*. https://www.scmp.com/news/china/article/1348605/japan-south-korea-concerned-chinas-smog-will-affect-them. Accessed on 16.06.2019.

SASAC. 2018. *Das landesweit größte Projekt zur ökologischen Sanierung von Deponien*. http://wap.sasac.gov.cn/n2588025/n2588124/c8927765/content.html. Accessed on 08.06.2019.

Schonberg, Alison. 2017. *China's landfills are closing: Where will the waste go?* https://www.coresponsibility.com/shanghai-landfills-closures/. Accessed on 05.06.2019.

Simons, Craig. 2006. In China, a water plan smacks of Mao. *Cox*, 10. September 2006. http://web.archive.org/web/20070911233235/http://www.coxwashington.com/hp/content/reporters/stories/2006/09/10/BC_CHINA_WATER10_COX.html. Accessed on 30.08.2019.

Singh, Manmohan. 2004. *The text of prime minister's speech*. http://www.rediff.com/news/2004/jun/24pm4.htm. Accessed on 30.08.2019.

Singh, Vijay, Nayan Sharma, C. Ojha, and P. Shekhar. 2004. Hydrology. In *The Brahmaputra basin water resources*, 139–195. Berlin: Springer Science & Business Media.

Slaughter, Anne Marie. 2015. *The Paris approach to global governance. Project syndicate*. https://scholar.princeton.edu/sites/default/files/slaughter/files/projectsyndicate12.28.2015.pdf. Accessed on 10.06.2019.

Soh, Tin Siao. 2018. *China's 13th five year plan: What role will wastewater play?*. https://www.waterworld.com/international/wastewater/article/16201297/chinas-13th-five-year-plan-what-role-will-wastewater-play. Accessed on 30.08.2019.

Sohu. 2002. *Nán shuǐ běi tiáo zīliào: Máozédōng „jiè shuǐ" shuō hé guīhuà biānzhì guòchéng*. [auf Chinesisch]. *Sohu.com*, 27. Dezember 2002. http://news.sohu.com/22/31/news205293122.shtml. Accessed on 02.01.2018.

Southern Weekend. 2006. Controversial plan to tap Tibetan waters. *China.org.cn*, 02. Juni 2006. http://web.archive.org/web/20150925063248/http://www.china.org.cn/english/MATERIAL/177295.htm. Accessed on 02.01.2018.

Standardization Administration of the PRC (SAC) (Zhōnghuá rénmín gònghéguó guójiā biāozhǔn). 2012. *GB 3095 – 2012 Ambient air quality standards* (Huánjìng kōngqì zhí liàng biāozhǔn). http://210.72.1.216:8080/gzaqi/Document/gjzlbz.pdf. Accessed on 19.06.2019.

State Council of the PRC. 2013. *Zhōushēngxián zài 2013 nián quánguó huánjìng bǎohù gōngzuò huìyì shàng de jiǎnghuà* (Speech by MEP Minister Zhou Shengxian at 2013 National Work Meeting on Environmental Protection). http://www.gov.cn/gzdt/2013-02/04/content_2326581.htm. Accessed on 19.06.2019.

———. 2018a. *Thirteenth National People's Congress, Guówùyuàn jīgòu gǎigé fāng'àn* (State Council Institutional Restructuring Program). http://www.gov.cn/xinwen/2018-03/17/content_5275116.htm. Accessed on 16.06.2019.

———. 2018b. *Guówùyuàn guānyú yìnfā dǎ yíng lántiān bǎowèi zhàn sān nián xíngdòng jìhuà de tōngzhī, guó fā* (2018) 22 hào (Three-year action plan for winning the Blue Sky War (Guo Fa (2018) No. 22)). http://www.gov.cn/zhengce/content/2018-07/03/content_5303158.htm. Accessed on 16.06.2019.

———. 2018c. *Guānyú guówùyuàn jīgòu gǎigé fāng'àn de shuōmíng* (Explanation on the institutional reform plan of the State Council). http://www.gov.cn/guowuyuan/2018-03/14/content_5273856.htm. Accessed on 05.06.2019.

State of Israel Water Authority Planning Division. 2012. *Long-term master plan for the National Water Sector*. http://www.water.gov.il/Hebrew/Planning-and-Development/Planning/MasterPlan/DocLib4/MasterPlan-en-v.4.pdf. Accessed on 02.01.2018.

Steinhardt, Christoph, and Fengshi Wu. 2016. In the name of the public: Environmental protest and the changing landscape of popular contention in China. *The China Journal* 75 (1): 61–82.

Suen, Thomas, and Ryan Woo. 2018. Bug business: Cockroaches corralled by the millions in China to crunch waste. https://www.reuters.com/article/us-china-cockroaches-idUSKBN1O90PX. Accessed on 16.06.2019.

Sullivan, Lawrence. 2007. *Historical dictionary of the People's Republic of China*. Vol. 414. Lanham: Rowman and Littlefield.

Sun, Yiting. 2016. China's massive effort to purify seawater is drying up. *MIT Technology Review*, 11. Juli 2016. https://www.technologyreview.com/s/601861/chinas-massive-effort-to-purify-seawater-is-drying-up/. Accessed on 02.01.2018.

Swartz, Jeff. 2016. *China's national emissions trading system: Implications for carbon markets and trade*. Global Economic Policy and Institutions. https://www.ieta.org/resources/China/Chinas_National_ETS_Implications_for_Carbon_Markets_and_Trade_ICTSD_March2016_Jeff_Swartz.pdf. Accessed on 30.08.2019.

Tambo, Ernest, Duoqian Wang, and Xiaonong Zhou. 2016. Tackling air pollution and extreme climate changes in China: Implementing the Paris climate change agreement. *Environmental International* 95: 152–156.

Tan, Yingzi. 2009. Yellow River dams verge on collapse. *China Daily*, 19. Juni 2009. http://www.chinadaily.com.cn/china/2009-06/19/content_8301942.htm. Accessed on 30.08.2019.

Thapan, Arjun. 2010. *ADB and partners conference on water: Crisis and choices*. https://www.adb.org/news/speeches/water-crisis-and-choices-adb-and-partners-conference. Accessed on 30.08.2019.

The 2030 Water Resources Group. 2009. *Charting our water future, economic frameworks to inform decision-making*. http://www.2030wrg.org/wp-content/uploads/2014/07/Charting-Our-Water-Future-Final.pdf. Accessed on 30.08.2019.

The Economist. 2013. Costly drops. Removing salt from seawater might help slake some of northern China's thirst, but it comes at a high price. *The Economist*, 09. Februar 2013. https://www.economist.com/news/china/21571437-removing-salt-seawater-might-help-slake-some-northern-chinas-thirst-it-comes-high. Accessed on 02.08.2018.

The Hindu. 2009. China has denied building dam across Brahmaputra. *The Hindu*, 05. November 2009. http://www.thehindu.com/news/national/ldquoChina-has-denied-building-dam-across-Brahmaputrardquo/article16890313.ece. Accessed on 02.01.2018.

The New York Times. 2006. China taps Tibetan waters. *The New York Times*, 01. August 2006. http://www.nytimes.com/2006/08/01/business/worldbusiness/01iht-river.2352899.html. Accessed on 02.01.2018.

The Times of India. 2009. China denies building dams on Brahmaputra: Foreign secretary. *The Times of India*, 05. November 2009. http://timesofindia.indiatimes.com/india/China-denies-building-dams-on-Brahmaputra-Foreign-secretary/articleshow/5197237.cms. Accessed on 02.01.2019.

Thinktank-Resources. 2016. *The water crisis: A major challenge for China in the 21st century*. Paris. http://www.thinktank-resources.com/en/events/morning-conference/water-crisismajor-challenge-china-21st-century. Accessed on 05.06.2019.

U. S. Water News Online. 2001. World Bank presses China to take action to prevent water shortages. *U. S. Water News*, 01. Juni 2001. http://web.archive.org/web/20130908220143/www.uswaternews.com/archives/arcglobal/1worban6.html. Accessed on 02.01.2018.

United Nations. 2014. *World urbanization prospects: The 2014 revision, highlights*. http://esa.un.org/unpd/wup/Publications/Files/WUP2014-Highlights.pdf. Accessed on 02.01.2018.

———. 2017. *World population prospects, Key findings & advance tables*. https://esa.un.org/unpd/wpp/. Accessed on 02.01.2018.

United Nations Climate Change. 2015. *INDC*. https://www4.unfccc.int/sites/submissions/indc/Submission%20Pages/submissions.aspx. Accessed on 08.06.2019.

US National Intelligence Council. 2008. *Global trends 2025. A transformed world*. https://www.dni.gov/files/documents/Newsroom/Reports%20and%20Pubs/2025_Global_Trends_Final_Report.pdf. Accessed on 02.01.2018.

Vasudeva, P.K. 2011. The great dam of China will endanger millions of Indian lives. *Daily News & Analysis*, 23. Juni 2011. http://www.dnaindia.com/world/comment-the-great-dam-of-china-will-endanger-millions-of-indian-lives-1558052. Accessed on 30.08.2019.

Walker, Martin. 2007. Walker's world: The most dangerous place. *UPI*, 04. Mai 2007. https://www.upi.com/business_news/security_industry/2007/05/14/walkers-world-the-most-dangerous-place/UPI-87261179157455. Accessed on 30.08.2019.

Wang, Yue. 2014. Chinese minister speaks out against South – North water diversion project. *Forbes*, 20.02.2014. https://www.forbes.com/sites/ywang/2014/02/20/chinese-minister-speaks-out-against-south-north-water-diversion-project/#29a9c6477d83. Accessed on 30.08.2019.

Wang, Jinpeng. 2018. Reform of China's environmental governance: The creation of a Ministry of Ecology and Environment. *Chinese Journal of Environmental Law* 2 (1): 112–117. Brill Nijhoff.

Wang, Bin, Qing Bao, Brian Hoskins, Guoxioang Wu, and Yimin Liu. 2008. Tibetan Plateau warming and precipitation changes in East Asia. *Geophysical Research Letter* 35(14). https://agupubs.onlinelibrary.wiley.com/doi/full/10.1029/2008GL034330. Accessed on 30.08.2019, S. 909–914.

Wang, Qin, Jianhua Yan, Tu Xin, Yong Chi, Xiaodong Li, Lu Shengyong, and Kefa Cen. 2009. Thermal treatment of municipal solid waste incinerator fly ash using DC double arc argon plasma. *Fuel* 88 (5): 955–958.

Wang, Guoqian, Xuequan Wang, Wu Bo, and Lu. Qi. 2012. Desertification and its mitigation strategy in China. *Journal Resources Ecology* 3: 97–104.

Wang, Li, Fengying Zhang, Eva Pilot, Yu Jie, Chengjing Nie, Jennifer Holdaway, Linsheng Yang, Yonghua Li, Wuyi Wang, and Sotiris Vardoulakis. 2018. Taking action on air pollution control in the Beijing-Tianjin-Hebei (BTH) region: Progress, challenges and opportunities. *International Journal of Environmental Research and Public Health* 15: 306.

WaterWorld. 2017. China's desalination project tally tops 130. *Water & Wastewater International*, 20.07.2017. http://www.waterworld.com/articles/wwi/2017/07/china-s-desalination-project-tally-tops-130.html. Accessed on 02.01.2018.

Watts, Jonathan. 2010. Chinese engineers propose world's biggest hydro-electric project in Tibet. *The Guardian*, 24.05.2010. https://www.theguardian.com/environment/2010/may/24/chinese-hydroengineers-propose-tibet-dam. Accessed on 30.08.2018.

Wei, Yuangsong, Yaobo Fan, Minjian Wang, and Jusi Wang. 2000. Composting and compost application in China. *Resources, Conservation and Recycling* 30: 277–300.

Wines, Michael. 2011a. China admits problems with three Gorges Dam. *New York Times*, 19.05.2011. http://www.nytimes.com/2011/05/20/world/asia/20gorges.html. Accessed on 30.08.2019.

———. 2011b. China takes a loss to get ahead in the business of fresh water. *New York Times*, 26.10.2016. http://www.nytimes.com/2011/10/26/world/asia/china-takes-loss-to-get-ahead-in-desalination-industry.html. Accessed on 02.01.2018.

Wong, Edward. 2011. Plan for China's water crisis spurs concern. *New York Times*, 01.06.2011. http://www.nytimes.com/2011/06/02/world/asia/02water.html. Accessed on 30.08.2019.

World Bank. 2001. *China: Air, land, and water – Environmental priorities for a new millennium*. http://siteresources.worldbank.org/INTEAPREGTOPENVIRONMENT/Resources/china-environment1.pdf. Accessed on 02.01.2018.

———. 2002. *China. Country water resources assistance strategy.* https://openknowledge.worldbank.org/bitstream/10986/15526/1/484990ESW0CN0w10Box338912B01PUBLIC1.pdf. Accessed on 02.01.2018.

———. 2007. *Cost of pollution in China: Economic estimates of physical damage.* https://siteresources.worldbank.org/INTEAPREGTOPENVIRONMENT/Resources/China_Cost_of_Pollution.pdf. Accessed on 02.01.2018.

———. 2016. *The little green data book 2016.* https://data.worldbank.org/products/data-books/little-green-data-book. Accessed on 02.01.2018, S. 31–118.

World Bank and Institute for Health Metrics and Evaluation. 2016. *The cost of air pollution: Strengthening the economic case for action.* http://documents.worldbank.org/curated/en/781521473177013155/pdf/108141-REVISED-Cost-of-PollutionWebCORRECTEDfile.pdf. Accessed on 18.11.2019.

Xia, Chen. 2016. *80 % underground water undrinkable.* http://www.china.org.cn/environment/2016-04/11/content_38218704.htm. Accessed on 30.08.2019.

Xie, Jian. 2009. *Addressing China's water scarcity. Recommendations for selected water resource management issues.* http://elibrary.worldbank.org/content/book/9780821376454. Accessed on 30.08.2019.

Xin, Dingding. 2018. China announces cabinet reshuffle plan to streamline government work. *China Daily.* http://usa.chinadaily.com.cn/a/201803/13/WS5aa7224ca3106e7dcc1412f6.html. Accessed on 05.06.2019.

Xinhua. 2001. Water shortage to hit danger limit in 2030. *Xinhua*, 16.11.2001. http://www.china.org.cn/english/BAT/22262.htm. Accessed on 02.01.2018.

———. 2004. Road to link Tibet's last roadless county. *Xinhua*, 01.12.2004. http://web.archive.org/web/20081203201802/http://en.tibettour.com.cn/geography/200412006816102549.htm. Accessed on 02.01.2018.

———. 2007. Weather keeps heating up in Tibet. *Xinhua*, 23.07.2007. http://en.people.cn/90001/90782/6221656.html. Accessed on 30.08.2019.

———. 2009. Nation won't divert Yarlung Tsangpo River to thirsty north. *Xinhua*, 26.05.2009. http://china.org.cn/environment/news/2009-05/26/content_17838473.htm. Accessed on 02.01.2018.

———. 2013. China water diversion faces pollution control challenge. *Xinhua*, 27.07.2013. http://china.org.cn/environment/2013-07/27/content_29545902.htm. Accessed on 02.01.2018.

Yan, Sophia. 2015. The cost of pollution in China. *CNN*, 08.12.2015. http://money.cnn.com/2015/12/08/news/economy/china-pollution-business/index.html. Accessed on 30.08.2019.

Yao, Tandong, Pu Jianchen, Lu Anxin, Youqing Wang, and Yu. Wusheng. 2007. Recent glacial retreat and its impact on hydrological processes on the Tibetan Plateau, China, and surrounding regions. *Arctic, Antarctic, and Alpine Research* 39 (4): 642–650.

Yardley, Jim. 2007. Beneath booming cities, China's future is drying up. *New York Times*, 28.09.2007. http://www.nytimes.com/2007/09/28/world/asia/28water.html. Accessed on 30.08.2019.

Zentralkomitee. 2015. *Shēnhuà dǎng hé guójiā jīgòu gǎigé fāng'àn* (The plan on deepening reform of party and state institutions). http://www.gov.cn/zhengce/2018-03/21/content_5276191.htm#1. Accessed on 05.06.2019.

Zhang, Hongzhou. 2015. China – India: Revisiting the 'water wars' narrative. *The Diplomat*, 30.06.2015. http://thediplomat.com/2015/06/china-india-revisiting-the-water-wars-narrative/. Accessed on 29.08.2019.

Zhang, Dongqing, Soon Keat Tan, and Richard M. Gersberg. 2010. Municipal solid waste management in China: Status, problems and challenges. *Journal of Environmental Management* 91: 1623–1633.

Zhang, Qian, Bin Yuan, and Minzhao Shao. 2014. Variations of ground-level O3 and its precursors in Beijing in summertime between 2005 and 2011. *Atmospheric Chemistry and Physics* 14: 6089–6101.

Zhang, Hefeng, Shuxiao Wang, Jiming Hao, Xinming Wang, Shunlan Wang, Fahe Chai, and Mei Li. 2015. Air pollution and control action in Beijing. *Journal of Cleaner Production* 112 (2): 1519–1527.

Zhang, Qionghua, Wennan Yang, Huu Hao Ngoc, Wenshan Guo, Pengkang Jin, Mawuli Dzakpasu, Shengjiong Yang, Qian Wang, Xiaoshang Wang, and Dong Ao. 2016. Current status of urban wastewater treatment plants in China. *Environment International* 92 (1): 11–22.

Zhang, Mengya, Yong Liu, and Su. Yunpeng. 2017. Comparison of carbon emission trading schemes in the European Union and China. *Climate* 5 (3): 70.

Zhao, Suping, Yu Ye, Daiying Yin, Jianjun He, Na Liu, Qu Jianjun, and Jianhua Xiao. 2016. Annual and diurnal variations of gaseous and particulate pollutants in 31 provincial capital cities based on in situ air quality monitoring data from China National Environmental Monitoring Center. *Environmental International* 86: 92–106.

Zhu, Liu. 2015. China's carbon emissions report 2015. Harvard Kennedy School Belfer Center for Science and International Affairs. http://belfercenter.ksg.harvard.edu/files/carbon-emissionsreport-2015-final.pdf. Accessed on 08.06.2019.

Zhu, Minghua, Xiumin Fan, Alberto Rovettac, Qichang He, Federico Vicentinic, Bingkai Liu, Alessandro Giustic, and Liu Yi. 2009. Municipal solid waste management in Pudong New Area, China. *Waste Management* 29 (3): 1227–1233.

Zuo, Qiting, Runfang Jin, Junxia Ma, and Guotao Cui. 2015. China pursues a strict water resources management system. *Environmental Earth Sciences* 72 (6): 2219–2222.

Energy Policy

14

Nils Wartenberg, Fabian Stein, and Barbara Darimont

The immense economic growth in the PRC is leading to a strong increase in energy demand so that China has advanced to become the world's largest consumer of energy. However, the growth in energy demand has already slowed down as economic growth has also reduced. In the coming decades, India will overtake China in the growth of energy demand (BP 2019, p. 69). So far, coal has been the main source of energy to meet the growing demand. In 2010, this still accounted for 66% of total energy production, while renewable energy sources comprised just over 10%. According to the 13th Five-Year Plan, this share is to be increased to 20% (Sadler 2017, p. 7). The increasing health burden on the population due to smog and due to the exceeding of carbon dioxide targets set by the government in large parts of China makes new energy sources come to the fore. In addition to renewable energy sources, such as wind power, solar power, hydropower, and geothermal power, nuclear power is one of the most important new energy sources (Zhou and Zhang 2010, pp. 4282–4283). In China, the state is the largest energy consumer, so increasing energy efficiency in state-owned enterprises takes priority over other measures.

Coal remains the main supplier of energy despite the development of renewable energy. However, a complete turnaround is unrealistic in the next few years. Therefore, the Chinese government propagates to make the existing coal-fired power plants cleaner. The construction and operation of coal-fired power plants in the densely populated east of the country

N. Wartenberg • F. Stein
Mannheim, Germany
e-mail: fabian.stein@mein.gmx

B. Darimont (✉)
East Asia Institute of Ludwigshafen University of Business and Society,
Ludwigshafen am Rhein, Germany
e-mail: darimont@oai.de

B. Darimont (ed.), *Economic Policy of the People's Republic of China*,
https://doi.org/10.1007/978-3-658-38467-8_14

should be limited and instead promoted in central or northeast China to minimize the impact on the population (NDRC 2016, pp. 25, 84, 88).

In the energy mix of the 13th Five-Year Plan of the People's Republic, the focus, in addition to the expansion of renewable energies, is on the construction of a nuclear energy belt along China's coastline. The plan lists concrete projects such as the completion of reactors still under construction and the start of construction of several nuclear power plants near the coast. Furthermore, nuclear power plants are planned in the interior of the country, close to the major rivers. The capacity of the nuclear power plants is to be increased from 30 gigawatts to over 58 gigawatts by 2020. This represents a doubling of capacity within 5 years (Central Committee of the CCP 2016, pp. 86–87).

In March 2018, the Chinese government was restructured, not creating a separate Ministry of Energy, but retaining the National Energy Administration (NEA), which was established by the National Development and Reform Commission in 2010 (Chinese Government n.d.). Thus, the energy sector still does not have an authority of ministerial rank.

14.1 Nuclear Energy

Nuclear power has only been used as an energy source in China since the 1990s. According to the Chinese government, nuclear power is to be expanded more intensively in the future because it is seen by the Chinese government as a less environmentally damaging option for energy production.

The most important player in the Chinese nuclear energy sector is the China National Nuclear Corporation (CNNC), which reports directly to the central government. The president and vice-president of CNNC are both directly appointed by the State Council and jointly supervise all civil and military projects in China, yet it is officially a state-owned enterprise and not a national authority (CNNC 2016). CNNC is the largest operator of nuclear power plants in China. CNNC also includes the Shanghai Nuclear Engineering Research and Design Institute (SNERDI) and the China Nuclear International Uranium Corporation (Sino-U). SNERDI is responsible for the development of some Chinese reactors, and Sino-U for the search and subsequent mining of uranium abroad.

China General Nuclear Power Group (CGN) is another state-owned enterprise, it emerged from the former China Guangdong Nuclear Power Group, which in turn is under the State-owned Assets Supervision and Control Commission (SASAC) of the State Council. CGN also operates several nuclear power plants in China (CGN 2019).

14.1.1 Nuclear Power Plants

China began developing its national nuclear power capacity in the 1980s (Zhou et al. 2011, p. 772). The first nuclear reactor on Chinese territory was the Qinshan-1 type CNP-300, which was completed in 1991, and was mostly designed and built by the Shanghai Nuclear Engineering Research and Design Institute. Only the reactor pressure vessel was manufactured abroad by the Japanese company Mitsubishi. After that, reactors were initially imported from abroad:

- The Daya-Bay-1 (1993) and Daya-Bay-2 (1994) reactors and the identical Lingao-1 and Lingao-2 (2002) reactors were developed in France and were manufactured by Framatome.
- The reactors Qinshan-3-1 (2002) and Qinshan-3-2 (2003) of the type CANDU 6 were developed by the Canadian Atomic Energy of Canada Ltd.
- The reactors Tianwan-1 (2006) and Tianwan-2 (2007) of the VVER type were built by Atomstroiexport from Russia.

In addition to reactors from abroad, China started building its reactors at an early stage. In addition to the 1994 Qinshan-1 reactor, China currently has 36 other reactors of its development in operation (International Atomic Energy Agency 2019a). China is also working on new developments, such as fast breeder reactor technology (China Experimental Fast Reactor) (PRIS 2019). In addition, high-temperature HTR-PM reactors have been built in Shandong for commercial use (Chen et al. 2018, pp. 82–83). With 50 gigawatts of electrical power installed, 2000 tons of spent fuel are expected to be generated annually (World Nuclear Association 2019).

In 2018, the PRC had 42.8 gigawatts of nuclear energy-based electric capacity installed, representing 4.2% of the total electric capacity installed in the PRC. This ranks the PRC third in the world in terms of installed nuclear energy-based electric capacity, behind the United States and France (International Atomic Energy Agency 2019b).

14.1.2 Resources

China has its uranium resources but also relies on the import of uranium. There are differing statements on the weighting of the individual areas. The Chinese government states that by 2020 the distribution should consist of one-third domestic production, one-third foreign production, and one-third purchases on the open market. However, independent market analysts estimate the distribution to be much less balanced, with a higher focus on sourcing uranium from abroad through production or trade. Domestic production is estimated to be much lower (Faul 2011). China has already imported significantly more uranium than needed in its reactors in recent years, for example importing about four times the amount of uranium needed in 2010 and 2011 (Ding and Liu 2012). In this way, China would like to secure the increasing consumption of uranium for nuclear power plants in the medium term.

14.1.2.1 People's Republic of China

In China, 1885 tonnes of uranium were produced in 2017 (World Nuclear Association 2018). The demand of 8289 tons (as of 2017) is thus covered by domestic production by only 22% (World Nuclear Association 2019). This means that China has to procure uranium through other channels.

The PRC's uranium reserves are approximately 260,000 to 290,000 tonnes. The deposits are mainly located in Inner Mongolia with a volume of 79,000 tons, Jiangxi with 57,000 tons, Guangdong with 44,000 tons, and Xinjiang with 43,000 tons. Other deposits are in Guangxi, Hunan, Hebei, Yunnan, Shanxi, Zhejiang, and Liaoning with a cumulative deposit of another 43,000 tons of uranium (OECD Nuclear Energy Agency and International Atomic Energy Agency 2014, p. 201). According to some PRC forecasts, the estimated amount of potential uranium deposits is much higher. Up to two million additional tonnes of uranium are suspected, particularly in Inner Mongolia, Xinjiang, and the Songliao Basin in northeast China (OECD Nuclear Energy Agency and International Atomic Energy Agency 2014, pp. 199–201).

A disadvantage of uranium ores mined in the PRC is a relatively low uranium content of 0.02–0.05%. Foreign uranium ores contain up to 0.1% uranium. The processing of the domestic ore is, therefore, more complex and involves the production of larger quantities of environmentally hazardous tailings. Uranium must be extracted from the ore using acid (World Nuclear Association 2018).

14.1.2.2 Kazakhstan

Currently, 70% of China's uranium imports come from Kazakhstan (Patton Schell 2014, p. 49). On 21 February 2011, the Chinese Nuclear Energy Industry Corporation signed a contract with Kazakhstan's Kazatomprom to supply a total of 30,000 tonnes of uranium between 2011 and 2020 (Cong 2009, p. 29).

In addition, on October 31, 2008, the former foreign ministers of China and Kazakhstan, Yang Jiechi and Marat Tazhin expanded the 2004 China-Kazakhstan Cooperation Agreement. This extension of the protocol includes, among other things, cooperation between the two state-owned enterprises Kazatomprom and China Guangdong Nuclear Power Co, which in later years was transformed into China General National Nuclear Power Co. The cooperation includes uranium mining, nuclear energy production, long-term supply of uranium, nuclear power generation, and construction of nuclear power plants.

In October 2007, CGN and CNNC entered into an agreement with Kazatomprom by holding 49% of the shares and Kazatomprom 51% in a joint venture of a Kazakh uranium mine. In return, Kazatomprom has received stakes in Chinese nuclear power plants and nuclear fuel reprocessing facilities. Despite the intensive cooperation between the two countries in the nuclear power sector, China does not have a safeguard agreement with Kazakhstan, unlike Australia and Canada, which prohibits the use of uranium for military purposes, among other things (Patton Schell 2014, pp. 52–53).

14.1.2.3 Australia

On 26 April 2006, China and Australia signed the Sino-Australia Uranium Agreement (SAUA). The agreement allows Australian uranium producers to export up to 20,000 tonnes to China for energy production. The agreement excludes the use of uranium for military purposes. This is ensured by the International Atomic Energy Agency (IAEA). However, shortly after the conclusion of the agreement, criticism was raised in Australia because the mining and use of uranium could lead to environmental damage in both Australia and China (Wang 2009, p. 2490). In addition, it was noted that the mining and export of uranium contributed little to the Australian economy. Neither the export volume is particularly large, nor are jobs created to any significant extent (Wu et al. 2008, p. 413). Australia does not own any nuclear power plants and does not plan to enter the nuclear power generation market in the coming years (Wang 2009, p. 2491). Furthermore, many uranium deposits are located on the territories of the Aborigines, the indigenous people of Australia, who strictly reject any mining or use of uranium. This is respected by the Australian government. Mining of uranium in original territories is not permitted (Katona 2002, p. 38). Therefore, exports from Australia are not expected to reach the 20,000 tonnes allowed by the treaty in the future.

14.1.2.4 Canada

Along with Australia and Kazakhstan, Canada is the third major supplier of uranium to Chinese nuclear power plants. In 2010, the world's largest producer of uranium Cameco, based in Canada, signed a contract to supply 23 million pounds of uranium, equivalent to 10,432 tonnes, to China Nuclear Energy Industry Corporation, a subsidiary of CNNC, by 2020. That same year, a contract to supply 29 million pounds of uranium, the equivalent of 13,154 metric tons, through 2025 was signed with China Guangdong Nuclear Power Co. (now China General Nuclear Power Co.) under a contract (Nickel 2013). After restrictions on Canada's part prevented the uranium from being shipped directly to China, it was diverted through countries such as Namibia and Kazakhstan. Former Canadian Prime Minister Stephen Harper made an agreement with China that significantly eased the restrictions on exporting uranium to China. This not only facilitated uranium exports through existing treaties but also facilitated further cooperation between China and Canada's nuclear power sector (Patton Schell 2014, p. 51). The treaties are subject to safeguard agreements, such as the exclusive use of uranium for civilian purposes (Clark and McCarthy 2012).

14.1.3 Reprocessing

The high use of nuclear power produces significant quantities of spent fuel that must be either reprocessed or disposed of. Large-scale spent fuel reprocessing plants are planned and are expected to be operational by 2030. Several plants exist today, but the output of these plants is still very small. In the so-called "closed fuel cycle" or closed fuel cycle, the portion of fissile material that can still be used is extracted in a reprocessing plant and can be used again as fuel in reactors as a MOX element (mixed oxide element). In 2008, an experimental MOX fuel element production facility was built. Since 2018, 40 tons of MOX fuel elements are produced per year in China (World Nuclear Association 2019).

14.1.4 Waste Disposal

The disposal and final storage of nuclear waste is still an unsolved problem worldwide, which China has to deal with. With the increasing number of nuclear power plants, the safe storage of radioactive waste is becoming a problem in China. Based on the target of 50 gigawatts by 2020, the generation of radioactive waste will reach 1300 tons per year. Cumulatively, the amount will reach 14,000 tons by 2020 (World Nuclear Association 2019). Radioactive waste is differentiated based on an official classification by the International Atomic Energy Agency. A total of six classes are differentiated so that the materials must be treated differently during disposal, interim storage or final storage.

China currently has three repositories for low and intermediate-level radioactive waste, in Gansu near the city of Yumen with a capacity of 20,000 m^3 and a planned expansion to 200,000 m^3, in Guangdong near the Daya Bay nuclear power plant with a capacity of 8800 m^3 and a planned expansion to 80,000 m^3, and in Sichuan in the Feifang Mountains with a capacity of 20,000 m^3 and a planned expansion to 180,000 m^3. In addition, two other repositories of low to intermediate-level radioactive waste are planned (World Nuclear Association 2019).

For the storage of high-level radioactive waste, a temporary landfill was built in Gansu in the northeast of the city of Lanzhou with a capacity of 550 tons. However, the largest share is stored on the site of the nuclear power plants. Other sites are being evaluated for their suitability as repositories. There is also the possibility of using a vitrification plant for liquid radioactive waste, which has been under construction since 2014 and is used by the military, for waste from civilian nuclear power plants in the future (World Nuclear Association 2019). The prerequisite is an existing spent fuel reprocessing facility. The vitrification plant, which binds liquid radioactive waste using vitrification and thus makes it suitable for final storage, is a facility of the Karlsruhe Institute of Technology (Karlsruher Institut für Technologie 2014).

14.1.5 Safety of Nuclear Installations

The safety of nuclear power plants, which are often located near urban centers with high energy demand, is a challenge for China. The International Atomic Energy Agency, which is responsible for regulating the safe use of nuclear power in the international community, drafted a document in 1994 that was signed by its members, including China. The so-called "Convention on Nuclear Safety" addresses the issues of siting, design, construction, operation, availability of sufficient financial resources and human capital, safety assessment and verification, quality assurance, and emergency response (International Atomic Energy Agency 2009).

14.1.5.1 Locations

When building nuclear power plants, attention must be paid to geographical hazard potentials such as earthquakes, floods or meteorological influences, but also the risk of political unrest, proximity to populated areas or use of water and land areas in the vicinity (International Atomic Energy Agency 2009). The first nuclear power plant sites in Zhejiang and Guangdong provinces were selected in the 1980s based on good access to cooling water, relative distance from populated areas, and protected location in natural bays. Subsequent nuclear power plants were also built near the coast in Jiangsu, Shandong, Liaoning, Fujian, Hainan, and Guangxi provinces. Since the early 2000s, more nuclear power plants have been built inland in Hubei, Hunan, Sichuan, Henan, and Jiangxi provinces, but without the positive location factors. Here it is reasonable to assume that economic motives have come to the fore. The side effects of nuclear power plants include job creation in the region, infrastructure development, and increased GDP (Lieberthal 2011, p. 23). The problem for China is not to let the necessary safety aspects be displaced by the economic advantages when choosing a location (Aldrich 2008, p. 5).

14.1.5.2 Provision of Specialist Staff

Skilled workers, such as engineers, etc., are needed to build and operate nuclear power plants according to international standards (International Atomic Energy Agency 2009). Estimates suggest that the Chinese nuclear power industry will need an additional 12,000 to 13,000 engineers and physicists by 2020 (Zhu 2012). CNNC and CGN are trying to improve training through investment. This is done both through in-company training and by encouraging companies to collaborate with universities (Xu 2014, p. 26). The long training periods make recruitment difficult. It takes 7 years to become licensed as a plant operator and another 3 years of training to take a managerial position (Zhu 2012).

Before the construction of the first nuclear power plants, the career of an atomic and nuclear physicist in the PRC often led to nuclear weapons engineering. Nuclear weapons programs in China are located in remote desert areas. Parents were unwilling to send their only child far away due to the one-child policy, instead favoring secure employment nearby and thus ensuring their retirement (Xu 2014, p. 26). This led to the closure of nuclear and atomic physics programs in most universities in China. By 2005, only four universities with such a program remained. The number of students dropped to just 800 per year (Duan 2010). By comparison, the United States had as many as 3900 students at 32 universities in the 1970s, when it had comparable growth in the nuclear energy sector to China (Johnson 2017, p. 2). Through financial support, China is trying to rebuild nuclear and nuclear physics programs at universities. Another problem is the lack of qualified teachers in the field, as teacher training declined drastically. For some time, China has been trying to bring this expertise back into the country through training in France and Canada, but this is happening slowly and not at a sufficient rate to meet the demand (Xu 2014, p. 26).

In addition to the skilled workers in the construction and operation of nuclear power plants, the employees in the nuclear regulatory authorities are an important component of safety in energy production. There is a shortage of qualified professionals in this area. In most developed countries, about 30 to 40 regulatory staff are responsible for one reactor. In China, the ratio was 1 to 11 in 2012 (Xu 2014, p. 26). From 2012 to 2014, the number of staff increased from 300 to 1000 (Green 2014). Another problem is that due to inadequate training, safety regulations are not understood and thus not implemented (Zhou et al. 2011, p. 780).

14.1.5.3 Quality Assurance

Quality assurance is achieved through the adoption and implementation of regulations, laws, and standards as issued by a national regulatory authority. The first laws and rules were adopted by China by the International Atomic Energy Agency and the USA and laid down in the two documents "Safety Regulations for Nuclear Power Reactors" of 1991 and "Rules on Nuclear Power Safety" of 1996. Due to the rapid expansion of the nuclear power sector that followed from 2003, gaps appeared, which were supposed to be filled by the documents "Nuclear Power Safety Plan" and "Medium and Long-Term Development for Nuclear Power" issued in October 2012. However, the documents were adopted by five different institutions with different competencies and objectives, so they are not concrete enough to ensure safety at nuclear power plants (Xu 2014, pp. 23–27).

The high diversity of reactor types in China due to the import of technology from France, Canada, and Russia, as well as additional proprietary reactor types, presents a hurdle to the development of safety standards. A comparatively high effort is necessary to ensure the safety of all plants (Zhou and Zhang 2010, p. 4285). In addition, there are critical voices among experts who portray China's authorities as unprepared and overstretched, which can also play a role in the export of reactor technology (Thomas 2017, p. 690).

14.1.6 Export of Chinese Nuclear Power Technology

The opportunities for Chinese reactors on the world market are ambivalent. On the one hand, China has good opportunities in developing and emerging countries due to attractive financing options and comparatively low prices. In addition, they can exploit economies of scale, as they manufacture relatively many reactors in China and thus have cost advantages. Furthermore, they receive state support. China has already signed contracts with Pakistan and Argentina to build Chinese-designed reactors. China's manufacturers participate in all major tenders worldwide (Pike and Koop 2019; Thomas 2017, p. 687). On the other hand, it is very difficult to penetrate the much more lucrative European market. Although China is in negotiation with several European countries (Czech Republic, Slovakia, Hungary, Romania, Bulgaria, and Poland), the negotiations did not yield any results due to the countries' tight financial situation. Finland and France have shown interest in buying Chinese reactors, but currently, there are only concrete offers from the UK and Turkey. Both countries have sufficient funds and have an energy demand that makes the sale of a larger number of reactors possible (Schneider and Frogatt 2015).

Especially the sale of nuclear reactors to the UK would lead to the international recognition of Chinese technology on the world market. In 2013, an agreement was concluded between the French EDF and the Chinese CGN to build a total of two nuclear power plants with four potential reactors at Hinkley Point and Sizewell. The Chinese share was initially 40% and was reduced to 33.5% in 2015 during renegotiations (EDF Energy 2015). Currently, the project is still in the approval phase. After the International Atomic Energy Agency officially approved the Chinese reactor in 2014, the project is now awaiting approval from the EU and approval in the UK. The process can take more than 4 years and a decision is not expected before 2021 (Pike and Koop 2019; Schneider and Frogatt 2015, p. 96). Since then, approval in the UK has been delayed several times (Thomas 2017, p. 688). Exact reasons have not been given, but national security concerns are suspected to be the main reason (Timothy 2015).

14.2 Renewable Energies

Renewable energies are comparatively clean and sustainable sources that cause fewer environmental problems than coal, for example. For this reason, the Chinese government is massively promoting renewable energies. In 2005, a law was passed requiring producers of energy in their energy mix to include renewable energy at a certain percentage. Electricity from renewable energy is also subsidized (National People's Congress 2005). In addition to the wind, hydro, and solar as energy sources, China also uses biomass and geothermal energy (Ahmed et al. 2016, p. 217). Although investment in renewable energy has fallen compared to the record year of 2017, companies and the government invested USD 100 billion in China's renewable energy sector (Efstathiou 2019).

The PRC is currently the country that invests the most money in renewable energy research. Their investments in this field amounted to USD 100 billion in 2018, which is one-third of the USD 300 billion invested in renewable energy worldwide (Henze 2019). The patents filed in the renewable energy sector show that China is massively promoting exploration. In 2018, Chinese developers filed 150,000 new patents related to renewable energy with Chinese patent authorities, representing 29% of global patent filings that year (Dudley 2019).

14.2.1 13th Five-Year Plan

In the 13th Five-Year Plan of the PRC (2016–2020), the Chinese government explains its ideas about the development of renewable energy and gives the exact key figures it wants to achieve. It states that the Chinese government's goal is to increase the share of renewable energy in total energy demand from approximately 12% to over 14% (NDRC 2016, p. 19). The focus is to be on the wind, thermal and solar energy. Hydropower is excluded from these percentages (NDRC 2016, p. 19). According to the five-year plan, the aim is to achieve slower growth in energy consumption and greater energy efficiency (NDRC 2016, p. 34).

The expansion of China's energy grid was set out in the 13th Five-Year Plan. The plan provides for the renewal of the Chinese energy grid (NDRC 2016, p. 85), as many lines are outdated and thus the performance of the power grid does not meet the requirements. The power grid is at its load limit, with many provinces experiencing power outages that sometimes affect entire districts for hours. The Chinese power grid requires high investment in maintenance due to its length of 543,000 km (Wei et al. 2017, p. 163). Part of the solution to the problems, according to the 13th Five-Year Plan, should be a so-called smart grid, i.e. an intelligent power grid (National Development and Reform Commission 2016, p. 88). A smart grid gives the possibility to control electricity tariffs and consumers according to availability. In interaction with renewable energies and conventional power plants, the effectiveness of the electricity network can thus be significantly increased without causing additional costs for operators or end users.

In addition to the overarching 13th Five-Year Plan, a plan for the development of solar energy in China was published in December 2016. Among other things, this detailed plan stated that China aims to produce solar cells in the field of mono-crystalline silicon technology with an efficiency of at least 23% by 2020, as opposed to poly-crystalline silicon technologies, which have an efficiency of about 20%. However, the final version of the plan does not mention mono-crystalline technology, but rather "advanced crystalline" silicon technology (Ball et al. 2017, p. 92).

14.2.2 Hydro Energy in China

The country's geography, with several major river systems, supports the development of hydroelectric power. China has the longest river in all of Asia, the Yangtze River, which runs 6304 km across the entire country, making it the third longest river in the world. China's second largest river, the Yellow River, is also one of the ten largest rivers in the world, stretching over 5464 km. Through its hydro energy resources, China was able to build over 352 gigawatts of hydro energy capacity in 2018. A total of 1,232,900 gigawatts/h of electricity was produced (International Hydropower Association 2019a). However, the growth of the industry has declined significantly within the last few years and therefore experts expect a moderate growth rate of 3.5–4% until the year 2020 (International Hydropower Association 2019b). China is doing everything it can to double its hydropower capacities.

Hydro energy is the renewable energy source with the longest history in the PRC. Already under the leadership of Mao Zedong, the Chinese government began to build the first hydroelectric projects (Wu 2013, p. 24). Probably the most famous hydroelectric power plant from that time is the Sanmenxia Dam, which was built in 1960 not far from the city of Xi'an (Behrens 2016). The structure was declared one of the symbols of the communist economic boom, as its planned 1310 gigawatts/h of electricity produced per year would have far exceeded comparable dams in the country (Global Energy Observatory 2018). However, the project proved to be uneconomical. The engineers who built the Sanmenxia dam had not considered the nature of the riverbed, and the power plant's turbines quickly became clogged with silt and sediment. The power plant delivered less than 100 gigawatts/h per year, which was less than one-tenth of its planned capacity (Behrens 2016).

The Three Gorges Dam is the largest dam in China, with a capacity of 22.5 gigawatts. The largest dam has become a national symbol of China's interest in hydro energy. The Three Gorges Dam is located on the Yangtze River in central China. Currently, the Three Gorges Dam produces 10% of all electricity from hydroelectric power (Dodson 2012, p. 125). But the construction of the dam has been criticized for its high cost, which amounted to USD 41.8 billion (RMB 254.2 billion) by project completion. This makes it the most expensive hydroelectric project in China (Graham-Harrison 2009). The reason for the far higher than initially budgeted costs was the relocation of 1.4 million people who, according to government statements, were relocated to surrounding villages as the mega-project needed the space (BBC 2012). Nevertheless, other giant projects followed, such as the Xiluoda Dam (12,600 megawatts), the Xiangjiaba Dam (6000 megawatts), and the Longtan Dam (6300 megawatts) (World Energy Council 2019).

The use of hydroelectric power plants is dependent on the available water resources. China is facing chronic water shortages in many provinces, especially in the north, north-west, and northeast. The great drought of 2011 led to emergency conditions at many hydroelectric power plants, especially as river levels were too low for the power plants to operate (World Energy Council 2019). To avoid this problem, the Chinese government has decided to move the majority of future hydroelectric projects to the more water-rich south of the country. This development is currently under international criticism as many neighboring nations are beginning to worry about their water resources and rivers. One of the most discussed projects is China's plan to build several dams on the Mekong River (Dodson 2012, p. 125).

The Mekong River, called Lancangjiang in Chinese territory, is the twelfth longest river in the world. The body of water flows through the PR of China, Myanmar, Laos, Thailand, and Cambodia and ultimately empties into the sea near Vietnam. The countries and a total of 60 million people living near the river depend on it as a water-giving lifeline. A large part of the population there lives from fishing or supplies the rice fields with water from the river. Within the last 10 years, three giant dams have already been completed. The largest is the Xiaowan Dam, completed in 2013, which at 292 meters is the largest dam after the Three Gorges Dam (Mekong River Commission 2017, p. 10).

Due to the increasing demand for energy in the region with economic development, further projects of this scale are being planned. However, the governments of neighbouring countries complain because the damming of the water leads to extremely reduced levels in the countries further downstream (Schumacher-Voelker 2007, p. 152). There is a concern there that in the event of a military conflict or political tensions, China would cut off the water to the countries, which would have devastating effects on the population (Clark 2014).

14.2.3 Wind Energy in China

The second most important renewable energy source in China by installed capacity is wind power. In 2016, China had 149 gigawatts of installed wind power capacity. Wind power is divided into two types: On-shore wind power on land and off-shore wind power on the sea or the coast. China has 253 gigawatts of on-shore wind power potential and off-shore wind power potential is over 750 gigawatts (Schumacher-Voelker 2007 p. 142).

It is expected that China will have developed wind power capacities of 593 gigawatts by 2040. This development would mean a doubling of the share of wind power in the overall energy mix. Already, Chinese investment in wind power accounts for 28% of global investment (IEA 2017). This investment enabled China to overtake the US in total installed capacity in 2010. To meet government targets, China manufactures and connects 1½ wind turbines to the grid every day due to government investment (Dodson 2012, pp. 116–117).

One company that has particularly driven development, for example, is Goldwind, based in Ürümqi, the provincial capital of Xinjiang. The location in the westernmost province of Xinjiang is particularly favorable because the average wind speeds are the highest, the wind is relatively steady, the available area to build wind turbines is the largest, and the land prices are comparatively low. Furthermore, the highest average wind speeds are found in Inner Mongolia, making the province predestined for the construction of wind turbines (Schumacher-Voelker 2007, p. 143). All in all, especially the northern and northwestern provinces show a large growth in wind power. The total installed capacity of wind power in China is 32% in the north and 30% in the northwest. This is remarkable compared to other provinces, as, for example, 13% of total capacity has been installed in the east so far (as of 2017), 6% in the southwest, 4% in central China, and 3% in the southwest. Therefore, onshore wind power is mainly built in the north of China. Off-shore wind power is particularly important in southwest China, as most of China's 14,500 km coastline is located there.

Chinese wind turbine manufacturers were initially criticized by Western manufacturers for their handling of intellectual property. Initially, foreign manufacturers dominated the Chinese wind power market, such as the Danish wind turbine manufacturer Vestas or the Spanish manufacturer Gamesa (Dodson 2012, p. 117). Within only a few years, local manufacturers such as Goldwind or Dongfang took over the Chinese market and displaced foreign manufacturers using cheaper prices and better relations with authorities and customers (Schumacher-Voelker 2007, p. 130). The prices of Chinese wind turbines increased in recent years and lost their attractiveness, so they were subsidized by the Chinese government (Cusick 2016).

Investment in wind farms has been declining for several years because Chinese operators often do not include maintenance costs in their price calculations. Furthermore, there is a lack of technical expertise to compete with countries such as Denmark and Germany (Dodson 2012, p. 119). Furthermore, there are quality problems. Furthermore, land prices on the coasts of China are rising, making offshore wind farms less profitable (Schumacher-Voelker 2007, p. 147). Therefore, fewer large wind farms are built, but rather medium and small wind turbines (International Trade Administration 2016, p. 36). Especially the long distances of wind farms in the north and northwest ensure that electricity from wind power remains relatively expensive and cannot compete with cheaper electricity from fossil fuels without support from the state (Zhang et al. 2016, p.335).

14.2.4 Solar Energy

Solar energy includes solar thermal and photovoltaics. The terms solar thermal and photovoltaic are to be distinguished, in solar thermal solar collectors produce heat from sunlight, and in photovoltaic solar radiation is directly converted into electrical energy (Umwelt Bundesamt 2015).

In 2018, the PRC had 130 gigawatts of installed solar power, making it the global leader (Boyle 2018). In 2017, 32% of global photovoltaic capacity was installed in China (Frauenhofer ISE 2019, p. 15). In the last two years, large solar energy capacities, 43 gigawatts in 2018 and 53 gigawatts in 2017 have been installed (Shen and Stapczynski 2019). In 2018, 177.5 terrawatts/h of electricity were generated (Liu 2019).

Furthermore, the PRC is the world's largest producer of solar technology and the largest manufacturers of solar panels come from China (Dickson 2016, pp. 35–36). The value chain also includes the production of solar silicon. GCL-Poli is one of the largest producers of solar silicon and its market share in the solar silicon sector is estimated at 30% of the global market (Roselund 2016).

Solar water heating systems are another form of solar energy. This form of heat generation is cheap to produce and does not require an electricity grid, so it is used extensively in China.

14.2.4.1 Geographical Distribution of Solar Energy Capacity

A large part of China's solar energy capacity is located in the western provinces of Tibet, Qinghai, and Xinjiang, as this is where the solar irradiation per year is highest (Schumacher-Voelker 2007, p. 131). The construction of solar energy parks on the high plateaus in Tibet and Qinghai has, apart from the favourable land prices, the advantage that the population in remote high altitudes can be supplied with solar energy much more cheaply than with other energy sources because there are no fuel transport costs and the lower efficiency of power plants at high altitudes is avoided (Pletcher 2019).

One of the disadvantages is the extremely long distance between the western provinces where the plants are located and the main electricity consumers on the east coast of the country. For example, there is almost 3000 km between the city of Ürümqi in the province of Xinjiang and the Chinese economic center of Shanghai. In the Chinese west, the infrastructure is underdeveloped; moreover, the long distance of transporting electricity involves significant losses. In 2017, Xinjiang province was only able to feed 61% of the solar power it generated into the grid, and 39% of that was lost. In the neighboring province of Gansu, 19% of the solar electricity generated was lost before or during transport due to less than optimal transmission lines (Parnell 2018).

Building large-scale solar plants in big cities like Shanghai or Beijing is unprofitable because land prices are very high (Zhang 1998, p. 52). Therefore, the promotion of small-scale solar panels that can be installed on rooftops is emphasized in big cities (Obertreis 2018). The building-integrated photovoltaic panels have the advantage that they can be installed directly on the roofs of buildings in cities or in some cases even integrated into the façade, thus producing electricity directly to the consumer (Reijenga 2005). Especially entrepreneurs who build new or renovate buildings should be persuaded to integrate solar panels or install systems for solar water heating. Small photovoltaic systems contributed about one-third of newly installed solar capacity in 2017 (Economist 2018).

14.2.4.2 Photovoltaic Market

The development of the Chinese photovoltaic industry can be used as an example to illustrate how Chinese industrial policy has functioned over the last 20 years. In the 2000s, the solar industry was primarily driven by the aim of exporting solar technology (Ball et al. 2017, p. 17). The PRC wanted to become the world leader in photovoltaic technology and the entire solar panel value chain. In the mid-2000s, solar energy became more in demand in European countries (Zhang and He 2013, p. 394). Stimulated by the increasing demand, the Chinese industry mass-produced to dominate the market through cheap prices. In addition, the production of photovoltaics was automated, so as a result, the prices of solar panels fell. In addition, the Chinese government subsidizes its manufacturers of solar systems to become the technology leader. As a result, numerous foreign manufacturers – including many German companies – had to file for bankruptcy. Companies such as Siemens and Bosch withdrew from the photovoltaic business with losses (Guyton 2012).

With the global economic crisis in 2008, Chinese manufacturers themselves came under pressure. Many countries cut subsidies for solar systems. High overcapacities accumulated in China. In this phase, the Chinese state-supported its companies massively through subsidies and began to equip its own country with solar plants (Ball et al. 2017, p. 148; Reints 2019). Starting in 2011, anti-dumping measures were imposed on the PRC by foreign countries, putting pressure on the Chinese solar industry. The Chinese government responded with further subsidies for photovoltaic systems. This led to China massively expanding its solar energy in the following years. While China had only 800 megawatts of installed solar panels at the end of 2010, it had 76,500 megawatts by the end of 2016 (Ball et al. 2017, p. 19; Reints 2019). No other country in the world has expanded its solar energy by nearly 100 times in 5 years. However, the country has high overcapacity of solar panels.

In the 13th Five-Year Plan, the Chinese government holds out the prospect of cutting subsidies for solar panel manufacturers to achieve a fair price on the international market (National Development and Reform Commission 2016, p. 39). This decision by the government resulted in smaller suppliers being forced out of the market, as they could not compete against large corporations without government assistance. After many smaller

Chinese solar panel manufacturers went bankrupt, they were taken over by large corporations. This allowed powerful conglomerates to form. Among the top ten solar panel manufacturers in 2018, there were six Chinese companies, including JinkoSolar, JA Solar Trina Solar, and LONGi Solar ranked first to fourth. This illustrates the market power that China has in the solar panel market (Colville 2019). Due to the advancement in knowledge achieved, it was possible to reduce the price per kWh of solar power to RMB 0.316. This price is still below the 0.325 RMB/kWh that electricity generated from coal currently costs (Parnell 2018). Solar panel prices in China dropped by 80% between the years 2009 and 2017 (Stacey 2018). The price reduction was also caused by the increased production of silicon in China, the main component of solar cells (Asian Metal 2019).

Due to international criticism, China wants to cut the subsidies which were one of the reasons for the competitive prices of Chinese manufacturers. This would avoid the punitive tariffs that several nations, including India and the United States, have imposed on imported Chinese solar panels. Another reason for removing the subsidies is that the fund created specifically for this purpose has already been completely depleted and even overdrawn by $15 billion (Baraniuk 2018). To protect the national solar panel industry, the US government imposed a 30% import tax on Chinese solar panels (Executive Office of the President of the United States 2018, p. 4). Given the United States' ongoing trade conflict with China, it remains to be seen how the market will continue to develop and whether these tariffs will be raised even further. The EU, on the other hand, ended the import ban on solar modules from China in 2018, which was introduced to protect the European solar module market from cheap Chinese products (Witsch 2018). European manufacturers were unable to compete with the much cheaper Chinese products and were forced out of the domestic market.

In June 2018, the Chinese government reduced subsidies for solar power as well as quotas for solar projects to reduce overcapacity in the solar industry. The annual quotas for solar installation had already been met in the first 5 months of 2018 (Hook and Hornby 2018). The government wants to make subsidies more efficient, but this poses problems because the solar industry is speculating on a continuation of subsidies in the coming years and is very slow to reduce capacity (Ball et al. 2017, p. 27).

14.3 Conclusion

In summary, the People's Republic is investing in several new energy sectors. Nuclear energy is being massively expanded, with up to 100 nuclear power plants planned by 2030, which would give China more power plants than any other nation in the world. Nuclear energy is an alternative to coal for the Chinese government. Strikingly, the PRC is working with more than five different types of nuclear reactors, all of which require specific expertise in operation, but skilled workers, in particular, are scarce. The future development of nuclear power technology in China is impossible to predict. Further development in the event of a serious nuclear accident cannot be predicted in any way.

In recent years, China has built up massive capacities in hydro energy, wind power, and photovoltaics. It is proving to be problematic that the respective plants take on gigantic dimensions, but then high losses occur as a result of the large geographical distances and an electricity grid that does not meet the requirements.

As the example of the solar industry shows, China has tried in recent years to flood the international market with new technologies by massively supporting the relevant industry with high state investments. The aim is to emerge from this price competition as a global player. On the one hand, this has led to protests from the international trading community, with high tariffs on the relevant goods as a result. In addition, the high subsidies have led to overcapacities in the respective industry. Reducing these overcapacities is a challenge for the Chinese government under President Xi. In the solar industry, falling prices could lead to other Asian countries being equipped with photovoltaics. The expansion of the Silk Road may be another way for Chinese industry to reduce overcapacity.

References

Ahmed, Saeed, Anzar Mahmood, Ahmad Hasan, Guftaar A. S. Sidhu, and Muhammad F. U. Butt. 2016. A comparative review of China, India and Pakistan renewable energy sectors and sharing opportunities. Renewable and Sustainable Energy Reviews 57(5): 216–225.

Aldrich, Daniel P. 2008. *Site fights: Divisive facilities and civil society in Japan and the West*. Ithaca: Cornell University Press.

Asian Metal. 2019. *Silicon Metal 5-5-3 delivered China RMB/mt. Asian Metal.* New York. http://www.asianmetal.com/SiliconPrice/Silicon.html. Accessed on 12.06.2019.

Ball, Jeffrey, Dan Reicher, Xiaojin Sun, and Caitlin Pollock. 2017. *The new solar system: China's evolving solar industry and its implications for competitive solar power in the United States and the world.* Stanford Universität, Stanford. https://www-cdn.law.stanford.edu/wp-content/uploads/2017/03/2017-03-20-Stanford-China-Report.pdf. Accessed on 04.08.2019.

Baraniuk, Chris. 2018. *How China's giant solar farms are transforming world energy*. London: BBC. http://www.bbc.com/future/story/20180822-why-china-is-transforming-the-worlds-solarenergy. Accessed on 14.06.2019.

BBC. 2012. China's Three Gorges Dam may displace another 100,000. *BBC*, London. https://www.bbc.com/news/world-asia-china-17754256. Accessed on 19.06.2019.

Behrens, Christoph. 2016. Maos Monstrum am Gelben Fluss. *Süddeutsche Zeitung*, München. https://www.sueddeutsche.de/wissen/china-maos-monstrum-1.3312059. Accessed on 09.06.2019.

Boyle, Matthew. 2018. Which country uses the most solar power? *Finder*, London. https://www.finder.com/uk/nation-most-solar-power. Accessed on 12.06.2019.

BP. 2019. *Energy outlook 2019.* https://www.bp.com/content/dam/bp/business-sites/en/global/corporate/pdfs/energy-economics/energy-outlook/bp-energy-outlook-2019.pdf. Accessed on 26.08.2019.

Central Committee of the CCP. 2016. *Central Committee of the Communist Party of China: The 13th five-year plan for economic and social development of the People's Republic of China (2016–2020)*, 1–219. Beijing: Central Compilation & Translation Press. http://en.ndrc.gov.cn/newsrelease/201612/P020161207645765233498.pdf. Accessed on 26.08.2019.

CGN. 2019. CGN – *About us.* http://en.cgnpc.com.cn/encgn/c100028/Profile.shtml. Accessed on 26.08.2019.

Chen, Yanxin, Guillaume Martin, Christine Chabert, Romain Eschbach, Hui He, and Guo-an Ye. 2018. Prospects in China for nuclear development up to 2050. *Progress in Nuclear Energy* 103: 81–90.

Chinese Government. n.d. *The organizational structure of the State Council.* http://english1.english.gov.cn/links.htm. Accessed on 04.08.2019.

Clark, Campbell, and Shawn McCarthy. 2012. Harper relaxes accountability rules for China's use of uranium. *The Globe and Mail*, Toronto. https://www.theglobeandmail.com/news/politics/harper-relaxes-accountability-rules-for-chinas-use-of-uranium/article4171375/. Accessed on 11.06.2019.

Clark, Pilita. 2014. Troubled waters: The Mekong River crisis. *Financial Times*, London. https://www.ft.com/content/1add7210-0d3d-11e4-bcb2-00144feabdc0. Accessed on 10.06.2019.

CNNC. 2016. CNNC – *About us.* http://en.cnnc.com.cn/2016-02/01/c_49164.htm. Accessed on 16.06.2019.

Colville, Finlay. 2019. Top 10 solar module suppliers in 2018. *PV tech*, London. https://www.pv-tech.org/editors-blog/top-10-solar-module-suppliers-in-2018. Accessed on 06.06.2019.

Cong, Weike. 2009. *Nuclear industry in China – CNNC presentation at the International Atomic Energy Agency.* http://www-pub.iaea.org/mtcd/meetings/PDFplus/2009/cn175/URAM2009/Session%201/8_33_Cong_China.pdf. Accessed on 28.06.2019.

Cusick, Daniel. 2016. China blows past the U.S. in wind power. *Scientific American*, New York. https://www.scientificamerican.com/article/china-blows-past-the-u-s-in-wind-power/?redirect=1. Accessed on 24.05.2019.

Dickson, Cora. 2016. Top markets report renewable energy. *International Trade Administration*, Washington, DC. https://www.trade.gov/topmarkets/pdf/Renewable_Energy_Top_Markets_Report.pdf. Accessed on 04.08.2019.

Ding, Qingfen, and Yiyu Liu. 2012. Nation plans to import more uranium. *China Daily*, Peking. http://www.chinadaily.com.cn/cndy/2012-03/13/content_14818316.htm. Accessed on 19.06.2019.

Dodson, Bill. 2012. *China fast forward. The technologies, green industries and innovations driving the Mainland's future*. Hoboken: Wiley.

Duan, X. (2010): How to improve the quality of education for nuclear sciences and nuclear enineering. In: China Energy News, 2010 (15.05.2010).

Dudley, Dominic. 2019. China is set to become the world's renewable energy superpower, according to new report. *Forbes*, New York. https://www.forbes.com/sites/dominicdudley/2019/01/11/china-renewable-energy-superpower/#51801dc2745a. Accessed on 20.04.2019.

Economist. 2018. China is rapidly developing its clean-energy technology. *The Economist*,15. Mai 2018. https://www.economist.com/special-report/2018/03/15/china-is-rapidly-developing-its-clean-energy-technology. Accessed on 04.08.2019.

EDF Energy. 2015. *Press release: Agreements in place for construction of Hinkley Point C nuclear power station*, London. https://www.edfenergy.com/energy/nuclear-new-build-projects/hinkley-point-c/news-views/agreements-in-place. Accessed on 12.06.2019.

Efstathiou, Jim. 2019. *Global clean energy funding dips 8% as China cools solar boom.* New York: Bloomberg. https://www.bloomberg.com/news/articles/2019-01-16/global-clean-energy-funding-dips-8-as-china-cools-solar-boom. Accessed on 06.06.2019.

Executive Office of the President of the United States. 2018. *Section 201 cases: Imported large residential washing machines and imported solar cells and modules.* https://ustr.gov/about-us/policy-offices/press-office/fact-sheets/2018/january/section-201-cases-imported-large. Accessed on 07.06.2019.

Faul, Jeffrey F. 2011. *Chinas uranium procurement strategies.* Hg. v. NUKEM Inc. World nuclear Fuel Market Conference, Seville. https://www.yumpu.com/en/document/view/13997230/chinas-uranium-procurement-strategies-wnfm-june-6-nukem. Accessed on 26.08.2019.

Frauenhofer ISE. 2019. *Photovoltaics report.* https://www.ise.fraunhofer.de/content/dam/ise/de/documents/publications/studies/Photovoltaics-Report.pdf. Accessed on 04.08.2019.

Global Energy Observatory. 2018. Sanmenxia Hydroelectric Power Plant China. *Global Energy Observatory*, Los Alamos. http://globalenergyobservatory.org/geoid/44153. Accessed on 09.06.2019.

Graham-Harrison, Emma. 2009. *China says Three Gorges Dam cost $37 billion.* London: Reuters. https://www.reuters.com/article/idUSPEK84588. Accessed on 09.06.2019.

Green, Jim. 2014. China's nuclear power plans: Safety and security challenges. *Nuclear Monitor Issue*, 796. Amsterdam: World Information Service on Energy. https://www.wiseinternational.org/nuclear-monitor/796/chinas-nuclear-power-plans-safety-and-security-challenges#. Accessed on 11.06.2019.

Guyton, Patrick. 2012. Baustelle Siemens. *Der Tagesspiegel*, 24.11.2012, Berlin. https://www.tagesspiegel.de/wirtschaft/baustelle-siemens/7432038.html. Accessed on 04.08.2019.

Henze, Veronika. 2019. *Clean energy investment exceeded $300 billion once again in 2018.* London/New York: Bloomberg NEF. https://about.bnef.com/blog/clean-energy-investment-exceeded-300-billion-2018/. Accessed on 10.06.2019.

Hook, Leslie, and Lucy Hornby. 2018. China's solar desire dims. *Financial times*, 08.06.2018. London. https://www.ft.com/content/985341f4-6a57-11e8-8cf3-0c230fa67aec. Accessed on 04.08.2019.

International Energy Agency. 2017. World energy outlook 2017: China. Paris: International Energy Agency. https://www.iea.org/weo/china/. Accessed on 18.01.2020.

International Hydropower Association. 2019a. *China. International Hydropower Association*, London. https://www.hydropower.org/country-profiles/china. Accessed on 25.05.2019.

———. 2019b. *Hydropower status report. International Hydropower Association*, London. https://www.hydropower.org/statusreport. Accessed on 25.05.2019.

International Trade Administration. 2016. *2016 Top markets report: Renewable energy.* Washington, DC: International Trade Administration. https://www.trade.gov/topmarkets/pdf/Renewable_Energy_Top_Markets_Report.pdf. Accessed on 06.06.2019.

International Atomic Energy Agency. 2009. *Classification of radioactive waste. General safety guide no. GSG-1.* https://www-pub.iaea.org/MTCD/publications/PDF/Pub1419_web.pdf. Accessed on 11.06.2019.

———. 2019a. *Power reactor information system – Country details (People's republic of China).* https://pris.iaea.org/PRIS/CountryStatistics/CountryDetails.aspx?current=CN. Accessed on 26.05.2019.

———. 2019b. *Nuclear share of electricity generation in 2018.* https://pris.iaea.org/PRIS/WorldStatistics/NuclearShareofElectricityGeneration.aspx. Accessed on 28.06.2019.

Johnson, Don. 2017. *Nuclear engineering enrollments & degrees survey data, 50-year trend assessment, 1966–2015, 1–5.* Washington, DC: U.S. Department of Energy. https://orise.orau.gov/stem/reports/ne-assessment-2017.pdf. Accessed on 26.08.2019.

Karlsruher Institut für Technologie. 08.10.2014. *Verglasungsanlage für China.* http://www.kit.edu/kit/15762.php. Accessed on 11.06.2019.

Katona, Jaqui. 2002. Cultural protection in frontier Australia. *Flinders Journal of Law Reform* 6: 29–39. Adelaide: Flinders University School of Law.

Lieberthal, Kenneth G. 2011. *Managing the China challenge.* Washington, DC: Brookings Institution Press.

Liu, Yuanyuan. 2019. *China's renewable energy installed capacity grew 12 percent across all sources in 2018.* Los Angeles: Renewable Energy World. https://www.renewableenergyworld.com/articles/2019/03/chinas-renewable-energy-installed-capacity-grew-12-percent-across-all-sources-in-2018.html. Accessed on 12.06.2019.

Mekong River Commission. 2017. *Mitigation of the impacts of dams on fisheries*. Vientiane: Mekong River Commission. https://www.researchgate.net/figure/Xiaowan-Dam-on-the-Lancang-Upper-Mekong-River-in-China-Completed-in-2007-Xiaowan-is_fig2_323300979. Accessed on 10.06.2019.

National Development and Reform Commission. 2016. The 13th five-year plan for economic and social development of the People's Republic of China (2016–2020). https://en.ndrc.gov.cn/newsrelease_8232/201612/P020191101481868235378.pdf.

National People's Congress. 2005. Zhōnghuá rénmín gònghéguó kě zàishēng néngyuán fǎ. http://www.gov.cn/ziliao/flfg/2005-06/21/content_8275.htm. Accessed on 03.08.2019.

Nickel, Rod. 2013. *Canada's Cameco begins uranium sales to China*. London: Reuters. https://www.reuters.com/article/canada-uranium-china/update-1-canadas-cameco-begins-uranium-sales-to-china-idUSL1N0C5AD120130313. Accessed on 11.06.2019.

Obertreis, Rolf. 2018. China mit Rekordinvestitionen in Solarenergie. *Der Tagesspiegel*, Berlin. https://www.tagesspiegel.de/wirtschaft/energiewende-china-mit-rekordinvestitionen-in-solarenergie/21146442.html. Accessed on 06.06.2019.

OECD Nuclear Energy Agency, and International Atomic Energy Agency. 2014. Uranium 2014: Resources, production and demand. In *NEA (7209)*, 1–504. Issy-les-Moulineaux: OECD Nuclear Energy Agency.

Parnell, John. 2018. New Chinese solar plant undercuts cost of coal power. *Forbes*, New York. https://www.forbes.com/sites/johnparnell/2018/12/30/new-chinese-solar-plant-undercuts-cost-of-coal-power/#730da011182a. Accessed on 12.06.2019.

Patton Schell, Tamara. 2014. Governing uranium in China. In *DIIS report 2014:03*, 1–64. https://www.diis.dk/files/media/publications/import/extra/rp2014-03_uranium-china_patton-schell_cve_1703_web.pdf. Accessed on 28.06.2019.

Pike, Lili, and Fermin Koop. 2019. China eyes Argentina in global nuclear roll out. *Hg. v. China Dialouge*. https://www.chinadialogue.net/article/show/single/en/11293-China-eyes-Argentina-in-global-nuclear-roll-out. Accessed on 11.06.2019.

Pletcher, Kenneth. 2019. Plateau of Tibet. In *Encyclopaedia Britannica*. Chicago. https://www.britannica.com/place/Plateau-of-Tibet. Accessed on 12.06.2019.

PRIS, Ed. 2019. *Country details. China, Peoples Republic of.* https://pris.iaea.org/PRIS/CountryStatistics/CountryDetails.aspx?current=CN. Accessed on 19.06.2019.

Reijenga, Tjerk. 2005. *BIPV opportunities in China. KOW DDC sustainable architects + urban planners*. Shanghai. https://www.rvo.nl/sites/default/files/bijlagen/Opportunities%20in%20China_%20BIPV.pdf. Accessed on 12.06.2019.

Reints, Renae. 2019. The price of Chinese solar panels are about to go up. *Fortune*, New York. http://fortune.com/2019/01/24/chinese-made-solar-panels-recover/. Accessed on 06.06.2019.

Roselund, Christian. 2016. *GCL Poly reports flat growth & healthy profit, appoints new CEO*. Berlin. https://www.pv-magazine.com/2016/03/23/gcl-poly-reports-flat-growth-healthy-profit-appoints-new-ceo_100023859/. Accessed on 26.08.2019.

Sadler, Christine. 2017. China. *Klima- und Energiepolitik – Im Spannungsfeld zwischen internationalen Zusagen und nationaler Entwicklungsstrategie*. Berlin: Heinrich-Böll-Stiftung e.V. https://www.boell.de/sites/default/files/e-paper_g20_china_dt_-_baf.pdf. Accessed on 03.08.2019.

Schneider, Mycle, and Antony Frogatt. 2015. *World nuclear industry – Status report 2015*. Paris/London: Mycle Schneider Consulting Project.

Schumacher-Voelker, Emma, Ed. 2007. Business focus China energy. *A comprehensive overview of the Chinese energy sector.* Karlsruhe: gic Deutschland.

Shen, Feifei, and Stapczynski. 2019. *Top solar producer says China's new plan will sustain growth*. New York City: Bloomberg. https://www.bloomberg.com/news/articles/2019-03-07/china-to-sustain-solar-growth-with-new-plan-top-producer-says. Accessed on 12.06.2019.

Stacey, Kiran. 2018. China and India lead the surge to solar energy. *Financial Times*, London. https://www.ft.com/content/a42e23be-8900-11e8-affd-da9960227309. Accessed on 12.06.2019.

Thomas, Steve. 2017. China's nuclear export drive: Trojan horse or Marshall plan? *Energy Policy* 101: 683–691.

Timothy, Nick. 2015. *The government is selling our national security to China.* http://www.conservativehome.com/thecolumnists/2015/10/nick-timothy-the-government-is-selling-our-national-security-to-china-html. Accessed on 19.06.2019.

Umwelt Bundesamt. 2015. *Solarenergie.* https://www.umweltbundesamt.de/themen/klima-energie/erneuerbare-energien/solarenergie. Accessed on 10.06.2019.

Wang, Qiang. 2009. China needing a cautious approach to nuclear power strategy. *Energy Policy* 37 (7): 2487–2491.

Wei, Wendong, Xudong Wu, Xiaofang Wu, Qiangmin Xi, Xi Ji, and Guoping Li. 2017. Regional study on investment for transmission infrastructure in China based on the state grid data. *Frontiers of Earth Science* 1 (1): 162–183.

Witsch, Kathrin. 2018. *EU hebt Schutzzölle für Solarmodule auf – deutsche Hersteller müssen sich Wettbewerb stellen.* Düsseldorf: Handelsblatt. https://www.handelsblatt.com/unternehmen/energie/solarbranche-eu-hebt-schutzzoelle-fuer-solarmodule-auf-deutsche-hersteller-muessen-sich-wettbewerb-stellen/22992682.html?ticket=ST-1710056-jzivGgbfw1EZIwdWz4Y2-ap5. Accessed on 06.06.2019.

World Energy Council. 2019. *Hydropower in China.* https://www.worldenergy.org/data/resources/country/china/hydropower/. Accessed on 25.05.2019.

World Nuclear Association. 2018. *Supply of uranium.* http://www.world-nuclear.org/information-library/nuclear-fuel-cycle/uranium-resources/supply-of-uranium.aspx (Zuletzt aktualisiert im Dezember 2018). Accessed on 26.05.2019.

———. 2019. *China's nuclear fuel cycle.* http://www.world-nuclear.org/information-library/country-profiles/countries-a-f/china-nuclear-fuel-cycle.aspx (Zuletzt aktualisiertim Januar 2019). Accessed on 11.06.2019.

Wu, Jiaping, Stephen T. Garnett, and Tony Barnes. 2008. Beyond an energy deal: Impacts of the Sino-Australia uranium agreement. *Energy Policy* 36 (1): 413–433.

Wu, Kang. 2013. Energy economy in China. In *Policy imperative, market dynamics, and regional developments*. Singapore/Hackensack: World Scientific Pub. Co.

Xu, Yi C. 2014. The struggle for safe nuclear expansion in China. *Energy Policy* 73: 21–29.

Zhang, Sufang, and Yongxiu He. 2013. Analysis on the development and policy of solar PV power in China. *Renewable and Sustainable Energy Reviews* 21: 393–401.

Zhang, Yuning, Ningning Tang, Yuguang Niu, and Du. Xioaze. 2016. Wind energy rejection in China: Current status, reasons and perspectives. *Renewable and Sustainable Energy Review* 66: 322–344.

Zhang, Zhongxiang. 1998. The economics of energy policy in China. *Implications for global climate change*. Cheltenham: Edward Elgar Publishing.

Zhou, Sheng, and Xiliang Zhang. 2010. Nuclear energy development in China: A study of opportunities and challenges. *Energy* 35 (11): 4282–4288.

Zhou, Yun, Christhian Rengifo, Peipei Chen, and Jonathan Hinze. 2011. Is China ready for its nuclear expansion? *Energy Policy* 39 (2): 771–781.

Zhu, Xu. 2012. China needs to speed up training for nuclear energy specialists. *China Energy News* 16 (07): 2012.

15 Foreign Trade

Johanna Hehl, Barbara Darimont, and Louis Margraf

The PRC is now the most important foreign trade partner for many countries, such as Germany. This fact in itself gives China very great political power, which it has learned to use in recent years. The PRC under President Xi Jinping has repeatedly emphasized that, after years of restraint, it now wants to take up this position in world politics and actively help shape the global environment. With this desire, or to put it another way, with the "Chinese Dream", the PRC and the USA have become direct rivals. This development will determine the political structure and thus foreign trade over the next 10 or 20 years.

From a German and European perspective, the question arises as to which strategy makes sense in dealing with the PRC. The economic interdependencies are complex, and foreign trade is hardly conceivable at present without the PRC, but the Chinese economic model is nevertheless a systemic challenge because with it an authoritarian state is gaining influence. This may not be a problem in itself, but the Chinese Silk Road initiative can be described as a gigantic project in which the PRC is pushing foreign trade and expanding its sphere of influence to a wide variety of countries. This is causing anxiety in many countries because the conditions under which this influence is being exerted are unclear.

J. Hehl
Northumbrai University, Newcastle, UK

B. Darimont (✉)
East Asia Institute of Ludwigshafen University of Business and Society,
Ludwigshafen am Rhein, Germany
e-mail: darimont@oai.de

L. Margraf
Renmin University, Beijing, PR China

B. Darimont (ed.), *Economic Policy of the People's Republic of China*,
https://doi.org/10.1007/978-3-658-38467-8_15

15.1 Chinese Foreign Trade Policy

With the opening up, China's activities in international trade grew. China was anything but a major trading partner at the time. In 1980, the PRC's foreign trade volume was just USD 37.6 billion. Only 20 years later, the volume of foreign trade had already risen to USD 474 billion. Since 2009, China has been considered the world's leading exporter and is now the second largest importing nation after the USA. Within 32 years, China has developed from an insignificant economy into the world's largest trading nation. Since 2013, the People's Republic has been the largest trading nation in nominal terms, replacing the USA, which topped the list for years.

China's accession to the World Trade Organization (WTO) in 2001 has led to its integration into global trade. China has thereby shown that it is prepared to play an active role in shaping world trade. Since its admission to the WTO, the role of the People's Republic has evolved from a rather passive observer role, a "norm-taker" to an active "norm-maker" role (Noesselt 2018, pp. 189–192).

15.1.1 China in the WTO

As early as 1986, China submitted its application to join the General Agreement on Tariffs and Trade (GATT), the forerunner of the WTO. A committee was established, which over the next 15 years deliberated on the application for admission to the GATT and, from 1995 to the WTO, as well as negotiating the terms of admission for the People's Republic (United States Trade Representative (USTR) 2018, p. 1). However, it was not until November 11, 2001, that China became the 143rd member of the WTO. WTO's Director-General Mike Moor on the accession (World Trade Organization 2001b):

> This is a historic moment for the WTO, China and the international economic community.

15.1.1.1 Accession in 2001

With the admission of the People's Republic to the WTO, the USA and Europe in particular hoped that the acceptance of the WTO guidelines would be accompanied by the liberalization of the Chinese market and thus make free economic activity possible (Unger 2019). For China, admission to the WTO meant rapid integration into international trade and the global economy. Among other things, the People's Republic thereby secured access to important sales markets and technological know-how (Hilpert 2014, p. 12). The accession negotiations comprised three parts. First, China had to submit a detailed description of its foreign trade strategy and market situation to the relevant working group. These were regularly updated during the 15 years of negotiations. Furthermore, China had to negotiate bilateral trade agreements with each country involved in the accession consultations. The best outcome in each case, in terms of market access, tariffs, etc., was included in the accession treaties, i.e. applicable to all WTO members. The third part of the negotiations, which took place at the same time as the bilateral negotiations, consisted of multilateral negotiations with members of the working group. Further access requirements and concessions were negotiated (United States Trade Representative (USTR) 2018, p. 26). The results of all negotiations were incorporated into the Working Party Report and the accession treaties (China's Protocol of Accession), thus forming the contractual basis for China's accession to the WTO. For example, the PRC had to guarantee the almost complete elimination of import quotas. China was contractually obliged to open its market. The accession protocol also stipulates that prices may be regulated solely by the market itself and that China must, among other things, liberalize the market and privatize its state-owned enterprises (Hilpert 2014, p. 12). On 10 November 2001, the Ministerial Conference of the WTO finally confirmed the accession treaties and with the confirmation from the Chinese side, the People's Republic became a member of the WTO (World Trade Organization 2001b).

15.1.1.2 Concessions and Commitments

When entering the WTO, the respective country must not only accept the generally applicable rules of the WTO, such as market transparency and fair trade without discrimination but also existing multilateral agreements. In China's case, this meant 20 trade agreements. Although the People's Republic was still considered a developing country until the time of accession, many concessions were demanded of China to be admitted to the WTO. These are still considered unprecedented in the history of the WTO and GATT. Upon entry, China was required to make more commitments than the founding countries and yet had fewer rights in return. For example, China gave up developing country status, which would have guaranteed the PRC some protections and advantages over industrialized nations (Hilpert 2014, p. 10). The PRC had to make commitments, such as holistic transparency with regard to trade- and market-based legislative changes and policy decisions. This means, as stated in Art. 2c of the accession treaties, that any WTO member can request a report in this regard at any time. Furthermore, it is imposed on the People's Republic that state-owned enterprises may not be given any further preferential treatment. Rather, according to Article 3, all companies, whether foreign or Chinese, must be treated equally. Time limits have been set for the concessions agreed in the contract. The liberalization of free trade called for in Art. 5 had to be implemented within the following 3 years (World Trade Organization 2001a, pp. 3–10).

15.1.1.3 Implementation of the Agreements

Accession brought about many changes in the law for China and changes in the Chinese market. A transition period of 15 years was agreed upon. During this time, the concessions and obligations that the PRC entered into with its entry into the WTO had to be implemented. With the expiration of the 15 years in 2016, the last transitional phase has expired and all obligations and the associated legal changes and regulations should have been implemented.

In the period just before China's accession to the WTO and in the early years, significant reform of trade strategy and the market could be observed. Laws and regulations were enacted to meet the commitments negotiated in the accession treaties. However, to date, not all concessions have been fulfilled by the Chinese side. From the perspective of Chinese trading partners, there is a great need for improvement in implementation. In contrast, Hu Jintao, the then Chinese president, declared in 2011 on the occasion of China's 10-year WTO membership that all agreements had been successfully implemented. He further said that China has always acted according to the principle of equality in the past 10 years and has continued to focus on new reforms and market opening to expand cooperation that benefits everyone (Beijing Review 2011). This is also emphasized by Yuan Yuan, Director of Trade Policy – Review and Notification Department for WTO Affairs of the Ministry of Commerce of the PRC (MOFCOM). In her speech to the third "China Round Table 2015", she said that due to China's WTO membership, over 2300 laws and regulations have been passed at the country level in the last 14 years to create legal certainty and ensure market transparency. She also stressed that China had revised its Foreign Trade Law, implemented in 1994, as agreed. This was done 6 months before the deadline agreed with the accession treaties. The resulting liberalization of the trade law allowed market participants to act freely and independently (Yuan 2015, pp. 4–5). In fact, China fulfilled many accession obligations on time. All tariff reduction requirements were fully implemented, meeting deadlines. Import and export licenses were abolished and thus foreign companies have been able to import and export to China since 2005. Legal requirements on direct investment have also been significantly relaxed. Reports by the WTO (2018) and the US Department of Commerce (2018) are nevertheless rather critical, as accession treaties have not been fulfilled in some key positions. For example, there is still a lack of transparency in areas such as customs clearance or sanitary and phytosanitary (SPS) measures. In addition, China is accused that some of the agreements have only been formally implemented, but that there are major gaps in the implementation and enforcement of the laws, especially in the handling and protection of intellectual property (TRIMS). New laws have also been passed that continue to favour Chinese companies (Hilpert 2014, pp. 38–39).

Among the many special agreements, China signed when it entered the WTO was the understanding that it would not be possible for China to obtain market economy status before the end of the 15 years. But since 2016, the transition period has been over. Since 2016, the EU Parliament and the US have now been discussing whether China meets the requirements of a market economy.

15.1.2 Is China a Market Economy?

China is struggling to be recognized as a market economy. China's market is not a classic market economy, but countries like Russia and Saudi Arabia gained market economy status when they joined the WTO, and their market systems are not necessarily classic market economies either. With its market economy status, China would have various advantages. Western countries fear that it would cause a drop in the price of steel production. In the USA and Europe, it is feared that cheaper competition from China would make steel producers in the USA and Europe no longer competitive and thus endanger jobs (Dreger 2016).

To be able to assess whether China's market system corresponds to that of a market economy, it is necessary to define what is understood by a classical market economy today. In general, a distinction is made between two basic economic systems: The market economy and the centrally administered economy. The WTO has no regulation on the definition of a market economy. The definition of a market economy is determined by the individual states and varies at some points. The following definition represents the German view.

In a market economy, market processes are controlled by the price mechanism. In contrast to a centrally managed economy, production planning and the steering of goods are not regulated centrally but are carried out by the individual market participants themselves. The state does not actively intervene in market activities, but merely establishes framework conditions to ensure smooth competition. These are regulatory instruments, such as competition policy and environmental policy (Sauerland 2018). The possibilities for "direct intervention" by the state are anchored in so-called process policy. These include, for example, monetary, exchange rate, and fiscal policy. Process policy serves to stabilize the economic process or promote macroeconomic activities (Ramb 2018). Another task of the state in a market economy is the provision of public goods, such as legal order, and transport routes.

Adam Smith coined the term "the invisible hand of the market". It describes the essential elements of a market economy. Households and firms can make their production or consumption decisions. By matching supply and demand and striving to maximize profit utility, prices and the distribution of goods in the individual marketplaces regulate themselves independently. The scarcity of goods and service capacities on the market are opposed to the quantity demanded and the price demanded, thus creating a natural price formation that can be influenced by the market participants themselves. This permanent market mechanism leads to the fact that the interest of the individuals is brought together and relativized and thereby a "macroeconomically welfare-promoting" process develops. In this process, competition is a natural moderator. This is because the counter-reactions of market participants caused by selfish actions can ultimately lead to profit reduction. This ensures that every market participant always acts in a profit-oriented and utility-maximizing manner, but a natural moderation is created by competition (Sauerland 2018).

Since Deng Xiaoping and his opening-up policy, the Chinese market has been liberalized and brought into line with the market economy principle. However, in the eyes of America and Europe, China has not yet implemented enough market liberalization to be considered a market economy. These two actors are the reason why the WTO has not yet granted China the status of a market economy until the present, 3 years after the deadline. Over 80 countries, including Australia, Singapore, and Russia, recognize China as a market economy.

15.1.2.1 Importance of Market Economy Status in the WTO

WTO member states with market economy status are more difficult to accuse in an anti-dumping or anti-subsidy case, as market economy status means that the burden of proof lies with the complainant. Thus, the complainant must show that price dumping is taking place. Furthermore, an analogue country can no longer be used to determine duties (Schwenke 2016, p. 2). When talking about market economy status in relation to China, the discussion often centers on how to take action against China in terms of price dumping and which method may be used. The WTO understands price dumping as a scenario in which the export price is lower than the price paid for it domestically. In the case of suspected dumping in countries with a market economy, the price paid for domestic trade serves as the basis for calculating duties and penalties (Bellora and Jean 2016, p. 3). However, for non-market economy countries, there is the possibility to use the price of an analogue market economy country as the basis for calculating the dumping margin. The EU and other members of the WTO benefit from China being considered a non-market economy country. Since it is assumed that prices in China are determined by government influence and not by the principle of supply and demand, WTO members can choose a product price from other market economy countries as a basis for calculating import duties on Chinese goods. These comparative tariffs result in the possibility of relatively high safeguard tariffs on Chinese goods. Article 15 of China's accession treaty allows each country to decide independently whether to accept China as a market economy and which calculation method to use to determine the dumping margin. However, due to the expiry of the transitional period, the addition of Art. 15 (a) (ii) has now been dropped. This omission has led to a heated discussion, as it has not been clearly defined which calculation method must be used and how China is to be treated in the future (Schwenke 2016, p. 1).

15.1.2.2 European Perspective

The European Union sees itself as a fair trading partner because it is the "only member of the WTO that systematically takes into account the interests of all stakeholders" (European Parliament 2012). While the EU already imposes only lower import duties in market sectors where market economy pricing has been demonstrated, it is not prepared to grant China market economy status. From the EU's perspective, however, the primary issue is not that China does not meet certain conditions. When the transition period expired in 2016, the EU faced the following problem: By granting China market economy status, the EU would lose one of its most important trade policy procedural arrangements, the anti-dumping regime, which uses the prices of an analogue market economy country as the basis for calculating duties for imports from China to protect itself from cheap imports from China. Without this regulation, the EU fears being flooded with cheap steel from China, among other things. This would threaten thousands of jobs in the EU, as European steel companies would no longer be able to compete (Godement 2016, p. 2). The anti-dumping regime and the possible methods to fight it are regulated by EU Regulation 2016/103. Article 2(7)(b), (c) states that for imports from China and other non-market economy status countries, individual producers can apply to be treated under the market economy principle. To do so, they must demonstrate that they operate under market economy conditions.

- Firms make their decisions on prices, costs, and inputs, including, for example, raw materials, the cost of technology and labour, production, sales, and investment, based on market signals reflecting supply and demand, and without significant government intervention in this respect; in this respect, the costs of major inputs must be based essentially on market values;

- Companies have one clear set of basic accounting records which are independently audited in line with international accounting standards and are applied for all purposes;
- Production costs and the financial situation of the companies are no longer subject to significant distortions carried over from the former non-market economy system, in particular as regards depreciation of fixed assets, other write-offs, barter trade, and payment by way of debt compensation;
- Companies are subject to ownership and insolvency rules that ensure legal certainty and stability for the management of the company; and
- Currency translation is carried out at market rates (European Parliament 2016b, p. 26).

However, these five criteria do not only apply if individual producers or industries want to be treated according to the market economy principle. If countries such as China want to be accepted by the EU as market economies, they must fulfill these five criteria. In the EU's opinion, China has not met any of these criteria to date and therefore cannot be granted market economy status. Therefore, on 5 December 2016, China's application for market economy status was rejected by the EU Parliament with 546 votes in favour, 24 against, and 77 abstentions (European Parliament 2016a).

Although the EU is not prepared to treat China as a market economy, it nevertheless had to amend its anti-dumping and anti-subsidy legislation because it could no longer be applied to Chinese imports. As it was no longer compatible with the WTO regime after the end of the transition period. At the same time, the EU skirted the issue of whether China was a market economy or not. In the new version of the Anti-Dumping and Anti-Subsidy Act, no distinction will be made in the future as to whether the country is a market economy or not. Fulfillment meant of the above-mentioned five criteria will be verified. The method of calculating the dumping margin is the same for all countries and sectors. In addition, the EU has to prove price dumping or subsidization. In this way, the EU ensures that the changes conform with WTO guidelines. The EU Commission must prove that there is a "significant market distortion". In other words, the relationship between the sales price and production costs is significantly distorted. In addition, the EU Commission draws up reports on all countries or economic sectors in which price distortions can be identified. However, it is still up to companies in the European Union to lodge complaints regarding price dumping. However, they can refer to the Commission's reports when making their complaints. The timing of the amendment suggests that its main purpose is to protect the market from cheap Chinese imports. However, the EU Commission points out that the new regulation is formulated in a "country-neutral" way and is absolutely compliant with the current WTO rules (European Council 2019b). By amending the Anti-Dumping and Anti-Subsidy Act in this way, the EU has ensured that even if China is granted market economy status, the European market can still be protected.

15.1.2.3 American Perspective

The US continues to deny China's market economy status. They argue that the WTO Guidance does not require that China be automatically granted the status at the end of the transition period. In their view, China does not meet the necessary criteria. The US refers to China as a "non-market economy" (NME), a country without a market economy principle as understood in economics. China wants to be recognized as a market economy. For this, the government of the PRC had to submit a formal application to the American anti-dumping authority. This then examined whether China can demonstrate a market economy structure from the American point of view. The following points were examined:

1. To what extent is the respective currency of the applicant country convertible into a foreign currency?
2. To what extent does free wage bargaining between employees and employers contribute to the formation of the wage level?
3. To what extent is it possible for foreign companies to form joint ventures or make other investments in the applicant country?
4. The extent of state control over inputs and the share of state ownership.
5. The extent of government intervention in the distribution of commodities, price formation, and production decisions of companies.
6. Other factors which the selection board considers appropriate. In this respect, it may well be established that only certain areas are subject to market economy treatment, while other areas may continue to be considered as NMEs.

In October 2017, a new review of the Chinese market was conducted. The subsequent report by the US Department of Commerce indicates that the PRC should still be considered an NME because the Chinese state intervenes in the private sector. However, there is criticism that the points are very broadly worded and do not allow for a measurable outcome (Morrison 2019).

15.1.2.4 Chinese Perspective

If the Chinese had their way, market economy status should have been automatic after 15 years. China is not only interested in reaping all the benefits from an economic point of view. Recognition by the EU and the US would also have symbolic significance, showing that they regard China as an equal trading partner (Bellora and Jean 2016, p. 2). From China's perspective, there is a lack of a clear definition of the term market economy. Russia was granted market economy status in 2002, but China continues to be denied it. In an interview, Shi Mingde, the Chinese ambassador to Germany, said:

> We have been practicing a market economy for more than 30 years. We are convinced that we have been a market economy for a long time. The big discussion is actually about the validity of Article 15 of the WTO documents and not about the term market economy. There should be no confusion between the two. (Wei 2016)

China is less concerned with the notion of a market economy. Rather, China is concerned with being treated fairly. It is about Article 15 of China's accession agreement to the WTO. China demands that all WTO members abide by the treaty. Shi continues:

> We are WTO members and we have established WTO rules between China and Europe. And now, suddenly, nothing is to apply because Europe's steel industry is no longer competitive? For me, this is an expression of protectionism on the part of the Europeans. (Wei 2016)

The Chinese believe that the US and the EU are violating existing WTO law by denying China's market economy status or failing to comply with Article 15 accordingly (Wei 2016). More specifically, the issue is that Article 15 (a) (ii) expired in 2016. This article legitimized the use of special anti-dumping measures against China. The legal basis for treating China as a non-market economy country was Article 15 (a) (ii), which authorized WTO members to decide for themselves which method of calculation to use when there is price dumping or subsidization of the product by the government. This thus allowed members to resort to third-country prices. China considers that the removal of Article 15(a)(ii) automatically means that the right to freely choose methods no longer applies. China would, therefore, by implication, have to be treated as a market economy and domestic prices would have to be used as a basis for calculation in anti-dumping proceedings (Martinek 2017, pp. 204–205).

15.1.2.5 Consequences

A study conducted by the Ifo Institute (Detlof and Fridh 2007) found that the European Union's calculation method, which uses third-country prices as a benchmark, leads to a higher average dumping margin than when domestic prices are used as a basis for calculation. The average duty for companies in an industry without a market economy principle is 39 percent. By comparison: For companies in a market economy industry, the average anti-dumping duty is just 11 percent.

If, therefore, the calculation of anti-dumping duties by means of the third-country principle results in a price distortion, then, conversely, by abolishing this method, the dumping margin can be reduced. Thus, if the EU were to grant market economy status to China, this would have the effect of significantly reducing the level of the constructed dumping margins. Some industries or companies have already been granted market economy treatment because they have been able to demonstrate that their pricing is based on the market economy principle. Thus, the reduction of the anti-dumping duties would not decrease by the entire 28 percentage points as might be initially assumed. However, it can be seen that for countries that have already granted market economy status to China, the anti-dumping duties have decreased significantly. The total number of anti-dumping cases against China has also decreased significantly for these countries (Sandkamp and Yalcin 2016, p. 53). If the EU were to recognize China as a market economy, the artificially induced distortion of the dumping margin would decrease significantly.

15.1.3 European-Chinese Trade Relations

Before 1995, economic relations between the PRC and the EU were almost non-existent. In less than 30 years, China has become the EU's most important trading partner. Especially from 2002 to 2003, after China's accession to the WTO, the bilateral trade relationship experienced an enormous upswing. The EU is China's second-largest trading partner after the US. However, problems in the trade relationship are increasing, even if negotiations and communication between the two parties continue, the EU needs to agree on a unified and clear China strategy. The EU's Directorate-General for External Policies already noted in its 2011 study that while European companies are largely coping with the situation in the Chinese market, they still urgently need better support through a unified European policy adapted to the Chinese market (Directorate-General for External Policies of the Union 2011, p. 4). The EU and China have developed some joint strategies and issued joint statements since the beginning of their trade relations. But there is still no legally binding agreement between the two parties. With various strategy papers and agreements, the common course is constantly being readjusted (Baudin 2006, pp. 53–54).

15.1.3.1 EU-China Partnership and Cooperation Agreement (PCA)

The legal basis of the EU-China relationship is the EU-China Trade and Economic Cooperation Agreement, which was signed in 1985. In 2006, in the joint statement of the annual EU-China Summit, both sides decided to negotiate a Partnership and Cooperation Agreement (PCA) that would "reflect the full breadth and depth of today's overall strategic partnership between the EU and China" (Council of the European Union 2006, p. 2). Despite optimistic sentiment in 2006, the revised Agreement could not be adopted in 2007 as planned; instead, negotiations were interrupted in 2009 and have not resumed to date. The reason is, among other things, that the PCA represents all-partnership cooperation, it includes agreements in areas such as trade, military, economic cooperation, etc. On the EU side, this means that it affects different areas of competence in which not only the European Union has decision-making power, but also areas in which, according to EU regulations, decision-making power lies with the individual states. This makes the negotiations on the PCA considerably more difficult, as the EU member states still have very different opinions on China. From the Chinese point of view, the problem lies more in the content of the PCA; while the EU is aiming for an overall agreement, China is interested in reaching a trade policy agreement. China wants the EU to recognize it as a market economy and to lift the arms embargo against China. However, as the EU is not willing to meet these demands, negotiations have stalled (Yan 2015).

15.1.3.2 EU-China 2020

Negotiations on the PCA have been suspended, but both sides have been working on new strategies for the partnership. Meanwhile, China has integrated strategies for the development of the two-way partnership in its 12th and 13th Five-Year Plans. The EU has adopted its "EU-China 2020" strategy paper. And both sides have agreed to promote cooperation and strategic partnership over the next 10 years. In addition to points such as security and peace as well as sustainable development, the strategy paper sets out the framework conditions for the trade relationship (EU Delegation to China 2013, p. 2).

Annual meetings such as the High-Level Economic and Trade Dialogue are intended to serve as an important planning and steering instrument to promote the further development of the bilateral trade relationship. Together, problems in the economic and trade relations are to be addressed and strategic solutions developed (EU Delegation to China 2013, p. 5). However, in 2017, the EU-China Summit ended without a joint final declaration; this was because no agreement could be reached on various trade issues. This made it all the more desirable to agree on a joint final declaration at the 2019 summit. In this final declaration, the EU and China express their desire to deepen their strategic cooperation in the coming years. China agrees to further open its market to foreign investors. In the run-up to the summit, the EU had denounced the legally binding technology transfers in China. In the joint final declaration, both parties agreed on the importance of protecting intellectual property and punishing violations. The joint final statement shows a clear concession by China on trade and economic issues. To better track the progress of individual developments and concessions in the future, reports are now to be prepared at the end of the year, which will be presented to the heads of state (European Council 2019a, pp. 2–4). The negotiations are a good starting point for the final negotiations on the EU's investment agreement with China, which is to be adopted in 2020. The investment agreement is intended to create a regulatory framework for investment and improve access to investment markets and investor protection. It is intended to contribute to the sustainable development of European-Chinese economic relations (Saarela 2018, p. 14).

15.1.3.3 European Perspective

After 18 years in the WTO, China has not yet fulfilled all the agreements listed in the accession treaties. The EU respects the fact that China has already taken a major step towards market liberalization. However, there are still several points that China has not yet sufficiently addressed and which represent a problem from a European point of view. Apart from the lack of transparency, these are above the regulations that still discriminate against foreign companies, the continued strong influence of the Chinese state and its subsidies, as well as the disregard for the rule on the protection of intellectual property. Another problem from the EU's perspective is the existing foreign trade deficit. In 2018, 25 out of 28 EU member states had a trade deficit. Only Germany, Finland, and Ireland had a trade surplus. Although in 2018 the EU recorded the highest exports to China in the last decade, this did not offset the trade deficit. Over the past decade, Chinese imports to the EU have been growing steadily with only a brief dip during the financial crisis in 2008, resulting in China's persistent trade surplus with the EU. In 2018, this amounted to 185 billion euros (European Union 2019).

In 2019, the Federation of German Industries published a policy paper, "Partners and systemic competitors – how do we deal with China's state-owned economy?" (Gätzner 2019). In it, it becomes clear that China remains the most important trading partner.

However, the originally hoped-for alignment of the political system fails to materialize; instead, the People's Republic is increasingly aggressively pursuing its model of a state-controlled economy, not only in its own country. China is increasingly becoming a systemic competitor. The EU must respond to this new challenge and strengthen the market economy principle of the West (Gätzner 2019). The EU is working at full speed on the negotiations for the investment agreement, which is to be signed in 2020. In the strategy paper on China published in 2016, a possible free trade agreement is envisaged. This is to be sought after successful negotiation of the investment agreement and implementation of the associated reforms. However, the trade defense instruments must first be further modernized (European Commission 2016).

15.1.3.4 Chinese Perspective

The PRC lacks an understanding of Europe's political model and does not see the EU as a single entity. Only in matters of trade does the PRC see the EU as a single major player. At the same time, China is aware that individual states disagree on many issues and knows how to take advantage of this (Schmidt and Heilmann 2012, p. 150). For China, however, Europe is an important trading partner and through good cooperation, a counterbalance to the dominant American economy can be formed. Shi Mingde clarifies in an interview that despite interim differences, both sides must share the overarching goal of promoting free trade and multilateralism. From China's point of view, the "Occident" (Europe, North America, and Australia) must understand and accept that the weaker emerging economies are increasingly gaining economic importance and that the supremacy of the Occidental states is no longer appropriate. Economic actors such as the EU and the US must accept that it is important to find common solutions with other states (Shi 2018). But China perceives concern that the fear of Chinese investment and cheap imports is leading Europe to adopt protectionist measures. The European Union's anti-dumping measures, which were amended to protect its market from cheap imports from China, and the EU's outbound investment review framework adopted in February 2019 are each not as strict as their counterparts in the United States, but China nevertheless sees them as a clear step toward protectionism. The new foreign investment review framework aims to more strictly monitor and, if necessary, prevent direct investments with state influence. This new regulation, therefore, has a direct impact on Chinese direct investments and has already resulted in the first investment intention being successfully blocked (Hanemann et al. 2019, p. 14). German carmakers benefit from China's market liberalization, they dominate the Chinese car market. On the other hand, the EU is afraid of Chinese firms expanding their market share in Europe. But economic cooperation can only work if both sides are on an equal footing and exchange can take place in both directions and is accepted.

The paradox can be well explained using the example of the steel industry. The EU accuses China of subsidizing the steel industry, which means that European steel producers cannot keep up with Chinese prices. Jobs in Europe are in danger as a result, according to the EU. China, on the other hand, sees the reason for its cheap production in the low wage costs. Chinese workers earn almost a tenth of what a European steel worker earns in a year. Added to this is quality. China is aware that it produces lower quality steel than European companies. So the cheap steel production is not due to subsidies from the Chinese state, but solely due to low production costs. China, therefore, believes that the punitive tariffs imposed by the EU are unjustified. Moreover, companies such as carmakers benefit from the cheap costs and thus have cheaper production costs in turn (Ziedler 2019).

In 2018, only 4 years after the last strategy paper, China published its latest EU strategy paper. With the new strategy paper, which contains the thinking of Chinese leader Xi Jinping, it is clear that China wants to continue to develop its relationship with the EU and resolve any discrepancies. China has a long-term goal of a free trade agreement with the EU. To work towards this goal, China is clear on what they expect from the EU. For example, in Part III on trade cooperation, investment, connectivity, and fiscal and financial matters, it is made clear that China wants the EU to relax high-tech export controls and comply with WTO rules. The EU must curb discrimination against WTO members. But China did not only make demands. With its new strategy paper, it holds out the prospect of further liberalization and opening of the market. And it wants to achieve a win-win situation for China and the EU through good cooperation (Yang 2018).

15.2 Trade Conflict with the USA

The US is taking a very different approach to the EU in its dealings with the PRC. The trade conflict between the PRC and the USA has the world in its grip. With the presidency of Donald Trump, the previous consensus of foreign diplomacy, in which the respective countries try to reach compromises through negotiations, has been abolished. Until Trump's presidency, the consensus was that the United States and China would benefit from joint trade and therefore avoid conflict (Wu 2013, pp. 377–378; Harding 2013, p. 406). Since January 2018, the U.S. government has begun to impose trade tariffs, contributing to an escalation. These tariff levies have an impact on all countries, especially exporting nations like Germany. The US under Donald Trump is not limiting itself to a trade conflict with China but is challenging the successes and gains of all exporting nations (Giesen et al. 2018). However, the trade conflict between the PRC and the US is particularly glaring because China is challenging US global dominance and wants to take that position itself. The bottom line is that the position of the US as a world power is at stake, which is being challenged from the Chinese side. If the economic figures are considered, PR China should have caught up with the USA in about 10 years and subsequently overtaken it. From the American point of view, therefore, the time has come to put the Chinese government in its place to remain a world power.

15.2.1 Causes and Measures

In 2016, Donald Trump promised during his election campaign to reduce the US trade deficit with China, which at the time amounted to USD 334 billion (Hulverscheidt 2018). Another point of criticism from the American side was the insufficient protection of intellectual property from the Chinese side. Moreover, the PRC has been accused of trade protectionism because many markets are not freely accessible to foreign companies. In addition, the USA accuses Chinese companies of taking advantage of technology transfers and industrial and technological espionage. Furthermore, when foreign companies operate in the PRC, they must disclose much ownership and use rights for technologies (Deuber and Rahmann 2019). With the social credit system, this situation will become even more acute. This is causing resentment not only in the US.

15.2.1.1 Customs Duties

By and large, the tariff increases originate from the USA; the PR China usually reacts to these announcements with its tariff increases, which, however, are smaller. In January 2018, tariffs were imposed on solar panels and washing machines by the US side. Aluminum and steel were hit with higher tariffs by the U.S. side starting in March 2018. The PRC criticized this as an arbitrary measure. For its part, the PRC responded with higher tariffs on soybeans, aircraft as well as cars and threatened to file a complaint with the WTO (Giesen et al. 2018). In 2018 and 2019, the mutual spiral of tariff increases continued, so that in the summer of 2019, tariffs with a total value of approximately USD 300 billion were imposed by the US side and tariffs with a total value of USD 60 billion were imposed on the Chinese side in return. According to various reports and calculations, the impact on the trade deficit is small, so this reason seems to be of a rather marginal nature (Hulverscheidt 2018; Reuters 2019). Moreover, in the summer of 2019, the US government accuses the Chinese leadership of extending the trade conflict to the currency because it has devalued the RMB (Zeit 2019). Much more serious on the American – but also, for example, on the German – side seems to be the fear of being left behind by Chinese companies at the technological level.

15.2.1.2 Technology and Industrial Espionage

In addition to tariffs, decrees and other actions by the U.S. government regarding Chinese tech companies must be included in the trade dispute, as the conflict is about future technology dominance. The US has been more aggressive in its dealings with the PRC than the EU.

In April 2018, the U.S. placed a ban on American companies from supplying products to ZTE. ZTE is a Chinese company that equips the telecommunications industry. The reason for this sanction was that ZTE did not comply with sanctions against Iran, was sued for this in a US court by the US, and apologized in court. ZTE promised a change in leadership management (Stecklow et al. 2018). Subsequently, the export ban was lifted in exchange for a penalty payment that totaled \$1.4 billion (U. S. Department of Commerce 2018).

In May 2019, a decree was issued by the US president declaring a national emergency for telecommunications in the US. Since then, it has been prohibited to sell Huawei products. It is also prohibited to enter into business relationships with companies that are considered hostile by the American government. As a result, Google had to abandon its business relations with Huawei. In addition, Huawei no longer receives a license for the Android operating system from the US company Alphabet Inc., so Huawei smartphones no longer have access to various Google services, such as the Play Store. While Huawei has announced plans to develop its operating system, this is likely to take time (Laaff 2019). As a consequence, this constellation would lead to a two-part market for smartphones, an American-Western one with Android and a Chinese-Eastern one with Huawei's operating system.

Huawei was spied on in 2013 by the U.S. intelligence community, which could not find any evidence of espionage on the part of Huawei (Kühl 2018). In addition, Canada extradited Huawei's CEO Meng Wanzhou to the US, which accused her of violating Iran sanctions. This action is unprecedented in world politics and has caused irritation around the world as to the motives for this action (Hua 2019). In September 2019, Huawei accused the US government of obstructing its business of Huawei by threatening employees and using unfair methods (Chan 2019). The trade dispute is seemingly being waged by any means available, with no overarching or mediating body, such as the WTO, being sought by the disputing parties. However, the dispute over Huawei also suggests that it is basically about 5G technology, an innovation by a Chinese company that is not plagiarism, but which American companies cannot counter with their development.

15.2.2 Consequences and Possible Future Scenarios

The consequences of this trade conflict are devastating for the global economy, but especially for exporting nations (Röder 2018). Whether the trade conflict is more severe for the Chinese or the American economy can only be guessed. However, the overall economic impact on the respective countries seems small (Deuber and Rahmann 2019). Both sides will have an interest in presenting their situation as positive. On the American side, however, it has already been announced that the tariffs are most likely to hit American consumers (BBC 2019). It can be assumed that the trade conflict will hit President Xi at the worst possible moment, as the Chinese economy is weakening completely independently of the trade conflict, while the American economy has recovered under Trump (Yang 2019).

Some economists perceive Trump's approach as positive because it shows that the West does not take everything from the PRC. Especially in the context of the Silk Road, fears are arising that the PRC will take over world domination. On the other hand, it can be assumed that the trade conflict is not just about trade and technology, but about world domination per se. At least the Chinese leadership leaves no doubt that they seek world domination (Huang 2017, p. 159). When a hegemon is challenged or replaced, three possible scenarios exist. A confrontation occurs in which one party wins. In the case of the PRC and the US, this would be a military confrontation. Another possibility is that both parties are willing to compromise. Under the current leadership of Presidents Trump and Xi, this option is very unlikely. While it cannot be assumed that Donald Trump wants a war, he has not yet shown his diplomatic skills, which would be necessary for a compromise, to the world public. To make matters worse, the US has not invested in its foreign diplomatic service for some time (Farrow 2018). What conflict resolution options remain if a country more or less ceases its diplomatic service? The last scenario would be a stalemate situation in which there would be two worlds sealed off from each other – a kind of second Cold War (Nass 2019).

In contrast to the predominantly protectionist position of the USA, however, China itself is pursuing a foreign policy strategy antithetical to that of the USA. In fact, the People's Republic is developing into a global player in the world economy – within the world economic system built up primarily by the United States in the last century. So how is the People's Republic attempting to conquer the American-dominated world economic system?

15.3 China as a Global Player: The New Silk Road

Since US President Donald Trump took office, the US has increasingly focused on its economy. At the same time, it can be observed that the PRC is gaining importance. This is particularly evident in China's desire to restore the former Silk Road after just over a decade of membership in the WTO, which can be understood as a desire for the return of China as a former world power (Hartmann et al. 2018, pp. 13–15, 35).

But what exactly is this about? While the English historian Frankopan (2019, pp. 9–11) actually speaks of several "Silk Roads", the term "Belt and Road Initiative" is now mostly used, with a direct translation from Chinese previously circulating as "One Belt One Road" (Ngeow 2019, p. 75). In the following, the term "New Silk Road" is predominantly used for simplicity. The basic assumption is that economic development requires the political opening of a country, whereas protectionism and closure towards other countries lead to economic stagnation.

To this end, we will first take a look at the historical background of the so-called "Silk Road". Thereafter, it will be defined what the term means in the contemporary context, the extent of the project, and the actors involved. Then the political background of the project and the types of funding are presented.

15.3.1 Historical Background

The term "Silk Road" can be traced back to the German geographer Ferdinand von Richthofen in the 1870s. He used it for those trade routes through Central Asia that connected Europe with the South and East Asian countries. Even von Richthofen himself noted at the time that there was no single Silk Road per se and that the "Silk Road was a network of transcontinental trade routes" (Barisitz 2017, p. 10). This pluralistic expression is still used today. For example, the English historian Peter Frankopan (2019, pp. 10–11) speaks of the "Silk Roads" as a historical link between different peoples. Both cultural and economic aspects of different peoples came into contact on these roads. The trade routes existed for a very long period and enabled various great empires to prosper economically through contact with other peoples. In the form of a historical hypothesis, it can be claimed: When a country opens up, its economy flourishes; when it isolates itself, economic stagnation threatens (Freeman 2003, p. 3).

How did these trade routes develop their legendary reputation? Due to a lack of written sources, the earliest trade deceptions in the areas understood as the "Silk Road" can only be traced back to Chinese myths (Egel 2018, pp. 167–168), so the following argument is based on more recent sources, but it can be assumed that trade in the Central Asian areas occurred much earlier.

Historically, three periods of intense trade can be identified in the Silk Road areas: During the Han Dynasty (202 BC–220 AD), the Tang Dynasty (618–907 AD), and finally around the time of the Yuan Dynasty (1271–1368 AD) (Hoshmand 2019, pp. 95–97). Although Chinese exploration of Central Asian territories dates back to the imperial envoy Zhang Qian (195–114 BC) of the Han dynasty (Liu 2010, pp. 6–8), it was not until the Tang dynasty period that the boom occurred, during which the Chinese empire was able to trade a wide variety of goods, including silk, spices, and jade, with the Roman and later the Byzantine empires via Central European land routes (McBride 2015, pp. 1–2). Therefore, the actual peak period of the domestic silk routes can be placed in the Tang dynasty.

Legitimation based on history for one's policy by the ruling social class was already used in China in the Chinese imperial dynasties. This goes back to the Confucian wisdom "Wēn Gù Zhī Xīn", in English "recall the past to understand the future". This can be considered as the application of knowledge and facts from history. Currently, a Chinese renaissance is taking place; in the course of China's politically desired return to its formerly occupied position of world power in the early and medieval periods, the re-establishment of the ancient Silk Roads can be understood as a central component of Chinese planning (Hoering 2018, pp. 30–31, 52).

15.3.2 Significance and Scope

In the following, we will discuss how the New Silk Road differs from the historical trade routes and then explain both the communist government's intentions and Western perspectives on the initiative.

15.3.2.1 The New Silk Road

Since when was the term Silk Road revived in modern times? The first time the Chinese term for the New Silk Road "Yī Dài Yī Lù" was mentioned by President Xi Jinping in 2013, however, confusion arose among international audiences due to a lack of understanding. The original English direct translation "One Belt, One Road" was changed to "Belt and Road Initiative" in 2016 because non-Chinese people misunderstood the term. "Belt" refers to trade routes on land routes, while "road" refers to shipping routes from East Asia to Middle Eastern ports. However, this was only the first mention of the term used today. The underlying policy of increased overseas investment began as early as 1999 with the so-called "Go Out" policy, as the Chinese leadership identified dependence on domestic markets as a risk even then. Chinese overseas investment increased from 3 billion USD in 1991 to 35 billion USD in 2003 (Van 2019, p. 31). While some two decades ago the so-called "Go West" policy promoted the more remote areas within China through targeted infrastructure investments, the new overland Silk Road serves the PRC to open up new sales and commodity markets beyond its borders. Partly due to cost, however, sea freight remains the most important logistical aspect of trade; here, along the sea route, "ports are emerging as logistical hubs to service ships, transport goods to the hinterland, and conversely, thread products into global trade" (Hoering 2018, pp. 17–18). China has increasingly invested abroad over the past few years to diversify its markets; by 2018, Chinese outward investment already amounted to USD 96 billion (OECD 2019).

What is understood by the New Silk Road? As mentioned at the outset, no single trade route can be defined as "The Silk Road" (Frankopan 2019, pp. 9–11). Rather, the New Silk Road represents the entire connectivity of a land belt and a sea route; ergo, it refers to the totality of a network of inter-trading nations. More precisely, there is a connection and reconstruction of the traditional Silk Road of the Han Dynasty, as well as the Silk Road Economic Belt of the Yuan Dynasty, as well as a synthesis of the ancient sea route of the navigator, Zheng He, a kind of Columbus of China, with today's sea routes (Van 2019, p. 32).

The New Silk Road itself is not clearly defined beyond the theoretical concept, but it is an essential part of China's rise along global value chains. While new markets are being integrated into a larger infrastructural network, at the end of the day, China is the core of the Silk Road. While opportunities emerge further down the production chain, the New Silk Road serves China to gain prominence in the global system. Each country's position in the global economy is determined by patterns in specialization and competitive advantage. According to Chinese understanding, this system is not a fair competition, but certain participants – namely the former colonial states – dictate the rules, while developing and emerging countries try to gain importance. In the sense of a "feedback mechanism", other countries can be made dependent through increased investment, infrastructure, and trade. The first effects of this mechanism can be seen as, for example, Greece or Hungary taking Chinese positions in Brussels (Maçães 2018, pp. 29–30). The People's Republic expects capital returns for its investments, as happened in the West, for example, in the reconstruction of the Federal Republic of Germany with the help of American investments.

At least six overland economic corridors are emerging under the New Silk Road: the China-Mongolia-Russia Economic Corridor, the New-Eurasia Land Bridge, the Bangladesh-China-India-Myanmar Economic Corridor, the China-Pakistan Economic Corridor, the Central China Asia Economic Corridor, and the China-Indochina Peninsula Economic Corridor. By sea, the so-called Maritime Economic Corridor is emerging under the name of the "21st Century Maritime Silk Road" (Malik 2018, pp. 69–70).

The People's Republic is addressing existing global infrastructure needs through the Silk Road Initiative. While actual investments are estimated at between USD 2.2 and 4.4 trillion annually, investment needs exceed these figures; the required infrastructure investments are estimated at between USD 3.3 and 6.9 trillion annually. Regions not covered by the planned economic corridors of the New Silk Road are particularly problematic, as conflicts could potentially arise here due to regional disparities. The New Silk Road is currently the most extensive effort to address infrastructural needs, but China cannot do it alone and can only address it together with the support of other nations (OECD 2018, pp. 5–7).

Xi Jinping uttered the following words at the 2017 Belt and Road Forum in Beijing:

> Infrastructure connectivity is the foundation of development through cooperation. We should promote land, sea, air and cyberspace connectivity, focus our efforts on major passages, cities and projects, and connect networks of highways, railways and seaports. The goal of establishing six major economic corridors under the Belt and Road Initiative has been set, and we should strive to achieve it. We must seize the opportunities arising from the new round of energy mix and the revolution of energy technologies to develop the global energy interconnection structure and achieve environmentally friendly and low-carbon development. We should improve the cross-regional logistics network and promote the connectivity of policies, rules and standards to provide institutional guarantees to improve connectivity. (Xinhua 2017)

In the course of the New Silk Road, the question arises to what extent the gigantic investments are sustainable for the target countries, or whether they will fall into a debt trap and become financially dependent on China. In a study by the OECD (Bandiera and Tsiropoulos 2019, pp. 33–35), it was found that 37 percent of the 30 economies studied could expect an increase in their debt ratio as a result of the New Silk Road. The main problem here is the opacity of the investments made in the course of the New Silk Road, which makes it difficult to examine the financial sustainability of the overall initiative.

In the three planned economic corridors in Central Asia, the World Bank (Bird et al. 2019, pp. 12–14) predicts, among other things, transport cost reductions in various countries ranging from no reduction to 19.5 percent of trade value. In addition, real income increases of 1.4–1.9 percent on average are expected due to travel cost savings, although there is an uneven distribution here. For example, while no to minimal income increase is predicted in Turkmenistan, almost 5 percent is expected in Kyrgyzstan. The economic development of the various sub-regions is thus expected to be uneven.

Due to the various individual projects with different countries, the entire New Silk Road initiative becomes confusing and thus non-transparent. Moreover, the initiative includes a previously unseen set of countries or regions, defined by strong cultural and geographical differences, that are supposed to cooperate (Hartmann et al. 2018, pp. 46–47). Furthermore, it is questionable whether the different countries of the Central Asian region can effectively adapt to change (Bird et al. 2019, p. 28). This lack of transparency and clarity may be intentional on the part of the Chinese government to prevent other states from intervening. First, the various projects started by the PRC are mostly treaties between two independent states, in which no third country will interfere. And secondly, the overview is lost simply because of the number of sub-projects. To put it casually, it could be a kind of guerrilla tactic by the Chinese leadership to gain supremacy in the world.

In addition to the land route, projects are being initiated on the sea route. The maritime Silk Road of the twenty-first century connects Asia by the sea with destinations in Africa and Europe. One focus of the maritime route is Southeast Asia (Schüller and Nguyen 2015, p. 1). However, it is unclear what the scale of the new maritime Silk Road is: All maritime projects fall under the umbrella of the Belt and Road Initiative (Hartmann et al. 2018, p. 52).

The Chinese government describes the maritime route of the New Silk Road in the "Vision and Measures for the Joint Construction of the Silk Road Economic Belt and the Maritime Silk Road of the 21st Century" in the following words:

> The maritime Silk Road of the 21st century is envisioned to run from the Chinese coast to Europe via the South China Sea and Indian Ocean on one route, and from the Chinese coast to the South Pacific via the South China Sea on the other. (NDRC et al. 2015)

Similar to the regional differences in the various land routes, there are also sharp differences in the economic performance or development of the participating countries on the sea route. India, as the most populous nation in the region next to China, must also find its place in the newly emerging economic zones for successful multilateral development, otherwise, conflicts are predictable (Palit 2018, pp. 206–209).

15.3.2.2 Scope of the Investment and Investors Involved

As of July 2019, 136 nations and 30 international organizations have signed a total of 194 documents regarding cooperation in the course of the New Silk Road (Belt and Road Portal 2018). So far, no comparable investments have been made in infrastructure, neither in financial nor in geographical dimensions. In this context, China's focus is on connecting Asia more strongly within itself and with Europe (Hartmann et al. 2018, pp. 46–48).

The New Silk Road is primarily financially backed by Chinese institutions and banks, with four large state-owned commercial banks accounting for more than half of investments at the end of 2016 (Deloitte 2018). The OECD (2018, pp. 18–20) provides a very clear summary of the investors involved and the size of the respective investments, as can be seen in Table 15.1.

Although the majority of the financial needs of the New Silk Road are met by Chinese institutions and Chinese banks, other nations are also financially involved in the Asia Infrastructure Investment Bank (AIIB), for example, and are thus indirectly involved in the financing of the Silk Road projects. The AIIB gains credibility in the international community through the membership of non-Asian countries, such as Germany. Thus, it can raise money from various capital markets in an uncomplicated manner and act beyond its initial capitalization of USD 100 billion (Hoering 2018, p. 44).

15.3.3 Policy Objectives and Targets

One of the main problems for economic globalization is infrastructure gaps in developing countries. There are already approaches to improving infrastructure in large economic areas, such as the EU's comprehensive Trans-European Transport Network or the CAREC infrastructure program in Central Asia, which is supported by the Asian Development Bank. However, no single institution, multilateral or unilateral, can meet the financial needs alone. It is estimated that a total of up to USD 1 trillion will need to be invested to address this financing gap in the near future (Hoering 2018, pp. 35–36). It is questionable why the PRC is generating itself as a pioneer here. What exactly does the government of the PRC want to achieve with the New Silk Road?

The Chinese government's policy guidelines regarding the New Silk Road date back to March 28, 2015, when the National Development and Reform Commission, the Ministry

Table 15.1 Investors involved and size of the investment

Institution (excluding Ministry of Finance and Aid for Trade, etc.)	Estimated exposure (in USD billion)
China Development Bank	110
China Exim Bank	80
Agricultural Development Bank of China	–
Industrial and Commercial Bank of China	159
Bank of China	100
Silk Road Funds	40
China Construction Bank	10
New Development Bank	1261
China Export and Credit Insurance Corporation	570.56
Asia Infrastructure Investment Bank	2.33

Source: Own representation based on OECD (2018, pp. 18–19)

of Foreign Affairs, and the Ministry of Commerce jointly released a document, namely the "Vision and Proposed Measures for Jointly Building the 21st Century Silk Road Economic Belt and Maritime Silk Road." In this document, there is a list of five processing areas (Zhang 2017, p. 105). The five areas are: political communication, connectivity of facilities, unhindered trade, financial integration, and interpersonal ties. The following is a summary of the contents of each item (NDRC et al. 2015):

The goal of policy coordination means overall stronger cooperation between the nations involved in the New Silk Road. The intention is to "build a mechanism for multi-level exchange and communication of macro policies between states, develop common interests, strengthen mutual political trust, and build a new consensus on cooperation".

Under the aspect of connectivity of facilities, priorities are primarily assigned to "major thoroughfares, intersections and project[s]" for "linking unconnected road sections, removing traffic bottlenecks, improving road safety facilities and traffic management facilities and equipment, and improving road network connectivity". Under this heading, the need for sustainable implementation of projects is expressed, in addition to various aspects of infrastructure, such as maritime logistics, data, and energy infrastructure. This contrasts with the American trend of more lax treatment of environmental issues and is in line with the global trend towards sustainability.

Under the heading of unimpeded trade, more specific reference is made to the planned cooperation in various areas. In addition to reducing tariff barriers through improved bilateral and multilateral cooperation, greater cooperation is also planned in the area of investment and innovation. In addition to resource-enhancing industries, projects in more technically challenging areas are mentioned, with emphasis on the ecological nature of future projects and "cooperation on environmental conservation, biodiversity protection and climate change mitigation" to "jointly [make] the Silk Road an environmentally friendly road."

In the course of the goal of financial integration, it is intended, among other things, to establish through cooperation "a currency stabilization system, an investment and financing system, and a credit information system in Asia." In addition to financing institutions, some of which were still in the planning stage at the time, "[the] exchange and cooperation in dealing with cross-border risks and crises" are mentioned.

The fifth objective of people-to-people links addresses cooperation in cultural areas. In addition to increased exchanges of personnel and education, exchanges in the sciences are taken up to, among other things, "expand and advance practical cooperation between countries along the Belt and Road in the areas of youth employment, entrepreneurship education, vocational skills development, social security administration, public administration and management, and other areas of common interest." In addition to government collaboration, non-governmental organizations should collaborate and drive sustainability.

This document was analyzed by the GIGA Institute (Schüller and Nguyen 2015, p. 7). They conclude that China, as the central core of the Silk Road, would not conform to the outwardly propagated values of multilateralism due to conflicts of interest with the nations of ASEAN. The more precise course remains to be seen, they say, with resource sharing in the South China Sea showing "how serious China's claim to create a win-win situation for all countries in the region is."

In its foreign diplomacy, Chinese policy focuses on its strategic positioning within the global landscape. The main focus is on the development of the western border areas and security issues in the Indian Ocean sea lanes. As a result of the "Opening to the West" strategy and the "Belt and Road" initiative, the South Asia region is nowadays already a significant region in Chinese geostrategies (Feng 2017, p. 176). This is because China, on the one hand, covers raw materials via the sea routes running through South Asia, and on the other hand, sells its products all over the world via sea routes (Zhou 2018, p. 146).

In December 2016, at the 30th meeting of the Central Leadership Group on Comprehensive Reform Deepening, soft power was identified as a key aspect of the further development of the Belt and Road Initiative (Zhang 2017, pp. 105–106). This is consistent with the understanding of Western observers who assume Chinese supremacy as the goal. There are two aspects to this approach, one being the rise of China as a maritime world power and the other being the strategic realignment of all sectors of the Chinese economy (Hartmann et al. 2018, pp. 55–57, 66–71). In particular, a rival position to the US maritime supremacy position represents an extreme challenge with a conflict potential that should not be underestimated (Kirchberger 2015).

According to the military strategy published in 2015 by the Information Office of the State Council of the PRC, the communist government has two strategic goals for the near future. The first is to "build a moderately prosperous society in all respects by 2021" with the centennial of the CP and to build a modern socialist country that is "prosperous, strong, democratic, culturally advanced, and harmonious" by the PRC's 100th birthday (State Council of the PRC 2015; Liu 2017, pp. 139–140). The PRC's overarching goal is to use the New Silk Road as a means to achieve global supremacy. In doing so, China is promoting the initiative through both external institutions and internal instances, such as free trade zones, to establish the project globally through all channels (Wang 2018, pp. 4–5).

15.3.4 Types of Support

The New Silk Road is more than just a massive infrastructure project; according to the Chinese government, it is a commercial venture. Therefore, a variety of support measures are being taken. In addition, the economic conditions in the People's Republic are changing: the Chinese financial system has been changed, potential deals are scrutinized more strictly than in the past, and state-owned enterprises, as well as privately owned companies, can only invest in specifically defined areas (Deloitte 2018).

15.3.4.1 Loans

In the course of the New Silk Road, loans are largely granted to countries with relatively low creditworthiness. Thus, for a large part of the target countries, there is either only a poor rating or no rating at all. Since China alone, sometimes due to its own financial needs in poorer regions, cannot take over the financing of the New Silk Road, other large economies and international institutions must be involved in the New Silk Road. Likewise, more attention needs to be paid to the viability of projects, as in the past excessive financing had led to problem loans with lenders (OECD 2018, pp. 21–22). While the loans seem to represent a kind of new "debt colonialism" (Hornby 2018; Islam 2019, p. 14), many of the target countries, such as Sri Lanka, find it difficult or impossible to repay the borrowed capital. Sri Lanka had to sign over the Maritime Silk Road-financed Hanbanbota port and surrounding areas to China for 99 years in December 2017 (Abi-Habib 2018; Mathews 2019, p. 16). Thereby, there seems to be potential in the negotiations with the Chinese government, as, for example, the Malaysian government was able to negotiate much better conditions by renegotiating (Dieter 2019, p. 24).

The New Silk Road creates new opportunities in the target countries. The development of modern infrastructure can create framework conditions for the economic upswing, which was previously not possible due to a lack of financial resources.

15.3.4.2 Construction Work

Infrastructure projects include train or road connections in previously undeveloped areas, such as from China to Thailand or Laos (Jetin 2018, p. 143). This creates new structures in these areas, facilitating trade and creating opportunities for the local population.

The fact that much of the construction work is primarily undertaken by Chinese construction companies is due to cheaper prices and the practices of Chinese companies in the underlying tenders. For example, power plants and highways built by Chinese companies are indeed low-cost infrastructure solutions for target countries (Tong and Kong 2018, pp. 65–66). Moreover, Chinese contractors try to win tenders with low prices in the bidding process, which is common in the Chinese market where low material and labour costs prevail. However, according to some contracts, certain percentages of local labor must be hired, which requires Chinese companies to use local expert services in the respective project country to accurately determine the prospective project costs (Zhou 2018, p. 164). Thus, opportunities for local companies to generate profits arise both in the construction sector and beyond in the service sector.

15.3.4.3 Know-How

Chinese know-how finds new target markets in the course of the Silk Road in the target countries and there is – at least theoretically – a transfer of Chinese know-how to the respective countries. This allows countries such as Kazakhstan to benefit from new technologies and move up the global value chains. In this way, the Chinese economic model is in effect exported.

Although Chinese know-how in the service sector is still relatively weak (Wang 2018, p. 217), the New Silk Road could see a repetition of development patterns already observed in Japan's economic boom. The Japanese economist Kaname Amatsu (1935) drew up a model of "flying geese", in which Asian countries developed in a formation that followed the flight of geese according to their respective levels of development, with Japan taking the lead role as the top of the V grouping. Thus, one country leads the development and pulls the other countries along. In this model, in addition to capital, obsolete technologies are passed from more developed economies to developing countries.

Currently, the People's Republic is developing its know-how by concentrating research funds on core fields, such as smart manufacturing. These core fields can become competitive advantages in the future. Another strategy for developing know-how is 'Made in China 2025', which is seen as a direct competitor to the German model of 'Industrie 4.0', as both concepts refer to the third industrial revolution and the automation of industry (Wu and Duan 2018, pp. 596, 591–592). Indeed, foreign companies often complain about the knowledge drain caused by Chinese acquisitions or takeovers (Tu 2019, p. 9). But other examples also exist: According to a senior executive, knowledge and technology transfer of Chinese experts takes place on the Indonesian island of Sulawesi in the Indonesia Morowali Industrial Park, and after 5 years, few of the Chinese employees would stay (Damuri et al. 2019, p. 14).

While there is an increasing trend in the People's Republic towards a more sustainable energy industry, there are quite a few projects in other Asian countries in the course of which Chinese capital, as well as technology and know-how in the area of coal-fired power, are being used. Chinese overcapacities are finding buyers in Southeast Asia. Projects are being initiated in Vietnam, Malaysia or Indonesia and the trend is rising (Stiftung Asienhaus 2017, p. 7). In addition, there are gigantic Silk Road projects in the energy sector, in the course of which Chinese companies are selling ultra-high-voltage power lines or solar, wind and hydropower plants (OECD 2018, pp. 27–29).

15.3.5 Criticism

From a German perspective, many questions regarding the Silk Road remain unanswered. While new markets are being opened up for the overcapacities of Chinese state-owned enterprises and dependencies on individual nations are being weakened, there are doubts about the geostrategic goals of the People's Republic. On the Chinese side, the partners in the Silk Road Initiative are considered independent of their membership in international alliances of states, such as ASEAN, the African Union or even the EU (Röhr 2018, pp. 237–240).

The New Silk Road can be perceived both as a stimulus package for Chinese state-owned enterprises and as a debt trap for the target countries. However, it remains to be seen whether the planned progress can be achieved in the target countries. Furthermore, even the overall success of the project is doubted. "[An] Infrastructure development that is limited to expanding trade among industrialized countries and with their raw material suppliers [...] cannot bring lasting economic growth or even sustainable economic development for the countries involved" (Hoering 2018, pp. 39–41). Therefore, the question arises whether the New Silk Road is only beneficial for China or provides an economic boost to the participating nations. In any case, it can be assumed that there will be both winners and losers in the development of the New Silk Road. Winners will primarily be urban centers, where strong growth can be generated through wage increases; on the other hand, residents of rural areas will benefit only to a limited extent from the economic upswing. At the same time, risks will arise from regional inequalities, e.g. in Kazakhstan, which could lead to political instability and a slowdown in economic growth (Lall and Lebrand 2019, pp. 24, 26).

For investors in the countries of the New Silk Road, the main challenges exist within the confusing legal situation. The security of investments could be improved through the increased conclusion of comprehensive international investment agreements. However, effective law implementation is often lacking in these countries (Kher and Tran 2019, pp. 6–7, 37–38). International organizations need to reconsider their foreign policy positions; in some cases, international banks have lowered their standards by cooperating with the AIIB (Hoering 2018, pp. 45–47).

There is potential for conflict in the implementation of the New Silk Road. While on the surface there is mostly talk of 'win-win' scenarios, it can be seen that cooperation is only possible in line with the Chinese vision. While the Chinese government does effectively engage in social engineering in China through targeted media coordination, the 'Chinese dream' is no longer solely in the hands of the Chinese government, and there are certain conflicts of interest in areas such as maritime territorial claims that China can only indirectly influence (Ferdinand 2016, pp. 956–957).

With Italy officially joining the New Silk Road as the first G7 nation in March 2019, there is a debate in the EU about China's interference in European politics. With the increasing foreign investments of Chinese actors in Europe by 2017, as well as the potential endowment of Chinese technology in the field of 5G internet, Europe is facing certain security risks (Ewerts 2019).

As part of the New Silk Road, the People's Republic may be exporting debt. The Chinese financial sector is strongly characterized by extreme credit growth, which has always resulted in a financial crisis on a comparable scale. However, the Chinese government has so far controlled all problems in the economy through regulatory measures at home (Dieter 2019, pp. 7, 12–14).

15.4 Conclusion

In recent years, China has become one of the world's most important trading partners and is, therefore, an integral part of the global world order. The USA and the EU are critical of China's advance. Nevertheless, China's integration into the WTO and thus into global trade has so far had quite positive consequences for industrialized nations such as the EU and the USA, Japan, South Korea, etc. These industrialized countries benefit from a liberalized Chinese sales market. China imports many high-quality products. In return, consumers in industrialized nations benefit from low-cost imports. However, China is pursuing a new policy to export its high-quality innovations. As a result, Chinese companies are increasingly competing with Western manufacturers.

For Europe and the Western world, China's rise means a loss of power. China's integration into the WTO, with the aim of better guiding China's actions and guaranteeing its controlled integration into the world economy, has worked to a limited extent. Western public opinion mainly sees the negative consequences of China's rise. For them, the rise goes hand in hand with job losses, product piracy, and poor product quality at home. By implementing measures such as the new anti-dumping law, the European Union is trying to protect its domestic market from these negative consequences. By doing so, the EU runs the risk of falling into protectionism and abandoning the goal of free trade. To date, there is no legally binding trade agreement between the EU and China. But with the investment agreement, which is due to be signed in 2020, the two parties are taking a further step toward free trade.

The US has taken a different path. Since January 2018, the US government has been using trade tariffs to try to persuade China to open up its economy more, protect intellectual property and defend itself against perceived technological espionage. The outcome of this dispute is uncertain. The PRC is strategically repositioning itself through the Silk Road to gain alternative sales markets.

The term "Silk Road" does not describe a specific trade route or corridor, but has always been synonymous with trade routes between China and the West, or more precisely Europe. Trade has been conducted along these routes for millennia, both via land routes and sea routes. The Belt and Road Initiative, or the New Silk Road, represents a central aspect of China's policy for a return to the historic world power position once occupied by the Chinese imperial dynasties. In this way, Beijing is essentially addressing the infrastructural needs of developing countries that exist anyway and, unlike the West, is offering what appears at first glance to be an attractive proposition, as no comparable strategies have yet been implemented by the West. In addition to exporting foreign capital, China provides target countries with both infrastructure and know-how. China is accused of forcing other nations along the New Silk Roads into political dependence through loans. Although there are disadvantages, especially for target countries that are in great need of development, there are nevertheless economic opportunities. In the future, it remains to be seen to what extent the New Silk Road will develop and whether Chinese problems will spill over to other economies.

References

Abi-Habib, Maria. 2018. How China got Sri Lanka to cough up a port. *The New York Times*. https://www.nytimes.com/2018/06/25/world/asia/china-sri-lanka-port.html. Accessed 01 Oct 2019.

Bandiera, Luca, and Vasileios Tsiropoulos 2019. *A framework to assess debt sustainability and fiscal risks under the Belt and Road Initiative*. http://documents.worldbank.org/curated/

en/723671560782662349/pdf/A-Framework-to-Assess-Debt-Sustainability-and-Fiscal-Risks-under-the-Belt-and-Road-Initiative.pdf. Accessed 01 Oct 2019.

Barisitz, Stephan. 2017. *Central Asia and the silk road: Economic rise and decline over several millennia*. Singapore: Springer.

Baudin, Pierre. 2006. *China und die Europäische Union: Zwiespältige Beziehungen*. https://www.kas.de/c/document_library/get_file?uuid=478d0213-2256-3b86-07f1-cc134fb29e26&groupId=252038. Accessed 12 June 2019.

BBC. 2019. *Trump warns China against new tariffs*. https://www.bbc.com/news/business-48246821. Accessed 01 Oct 2019.

Beijing Review. 2011. *President Hu Jintao Reviews China's Experience in WTO*. http://www.bjreview.com/special/2011-12/13/content_411486.htm. Accessed 26 May 2019.

Bellora, Cecilia, and Sébastien Jean. 2016. *Granting market economy status to China in the EU: An economic impact assessment*. www.cepii.fr/PDF_PUB/pb/2016/pb2016-11.pdf. Accessed 07 June 2019.

Belt and Road Portal. 2018. *Yī tóng zhōngguó jiàndìng gòng jiàn yīdài yīlù hézuò wénjiàn guójiā yì*. https://www.yidaiyilu.gov.cn/info/iList.jsp?tm_id=126&cat_id=10122&info_id=77298. Accessed 01 Oct 2019.

Bird, Julia Helen, Mathilde Sylvie Lebrand, and Anthony J. Venables. 2019. *The Belt and Road Initiative: Reshaping economic geography in Central Asia?* https://elibrary.worldbank.org/doi/abs/10.1596/1813-9450-8807. Accessed 10 Oct 2019.

Chan, Edwin. 2019. *Huawei Accuses U.S. of harassing workers, attacking network*. Bloomberg. https://www.bloomberg.com/news/articles/2019-09-03/huawei-accuses-u-s-of-harassing-workers-attacking-its-systems. Accessed 01 Oct 2019.

Council of the European Union. 2006. *Nineth EU-China Summit Joint Statement*. https://www.consilium.europa.eu/ueDocs/cms_Data/docs/pressData/en/er/90951.pdf. Accessed on 08.06.2019.

Damuri, Yose Rizal, Vidhyanika Perkasa, Raymond Atje, and Fajar Hirawan. 2019. *Perceptions and readines of Indonesia towards the Belt and Road Initiative*. https://www.csis.or.id/uploads/attachments/post/2019/05/23/CSIS_BRI_Indonesia_r.pdf. Accessed 01 Oct 2019.

Deloitte. 2018. *Embracing the BRI ecosystem in 2018. Navigating Pitfalls and seizing opportunities*. https://www2.deloitte.com/us/en/insights/economy/asia-pacific/china-belt-and-road-initiative.html. Accessed 01 Oct 2019.

Detlof, Helena, and Hilda Fridh. 2007. EU treatment of non-market economy countries in anti-dumping proceedings. *The Global Trade and Customs Journal* 2(7): 265–281. https://heinonline.org/HOL/LandingPage?handle=hein.kluwer/glotcuj0002&div=47&id=&page=&t=1559981102. Accessed 08 June 2019.

Deuber, Lea, and Tim Rahmann. 2019. Was hinter den China-Zöllen von Trump steckt. *Wirtschaftswoche*. https://www.wiwo.de/politik/ausland/handelskrieg-was-hinter-den-china-zoellen-von-trump-steckt/21104972.html. Accessed 01 Oct 2019.

Dieter, Heribert. 2019. *Chinas Verschuldung und seine Außenwirtschaftsbeziehungen. Peking exportiert ein gefährliches Modell*. https://www.swp-berlin.org/fileadmin/contents/products/studien/2019S18_dtr_Website.pdf. Accessed 01 Oct 2019.

Directorate-General for External Policies of the Union. 2011. *Handelsbeziehungen EU-China*. EXPO/B/INTA/FWC 2009–01/Lot7–14. Europäischen Parlament. http://www.europarl.europa.eu/activities/committees/studies.do?language=DE. Accessed 30 May 2019.

Dreger, Christian. 2016. Ist China eine Marktwirtschaft? *DIW Wochenbericht* 21: 488. https://www.diw.de/documents/publikationen/73/diw_01.c.534441.de/16-21-4.pdf. Accessed 02 June 2019.

Egel, Nikolaus A. 2018. Forschungsstand zur historischen Seidenstraße. In *Chinas neue Seidenstraße: Kooperation statt Isolation – der Rollentausch im Welthandel*, 163–185. Frankfurt: Frankfurter Allgemeine Buch.

EU-Delegation to China. 2013. *EU-China 2020 Strategic Agenda of Cooperation.* https://eeas.europa.eu/sites/eeas/files/20131123.pdf. Accessed 08 June 2019.

European Commission. 2016. *Die Europäische Union setzt sich ehrgeizige Ziele für ihre Beziehungen zu China.* IP-16-2259_DE. Brüssel: Kocijancic, Maja; Kaznowski, Adam.

European Council. 2019a. *EU-China Summit Joint statement.* https://www.consilium.europa.eu/de/press/press-releases/2019/04/09/joint-statement-of-the-21st-eu-china-summit/. Accessed 09 June 2019.

———. 2019b. *Antidumping: EU einigt sich auf neue Vorschriften für den Schutz ihrer Erzeuger vor unlauteren Handelspraktiken.* https://www.consilium.europa.eu/de/press/press-releases/2017/10/11/anti-dumping-unfair-practices/. Accessed 07 June 2019.

European Parliament. 2012. *Trade defence instruments.* http://ec.europa.eu/trade/policy/accessing-markets/trade-defence/index_en.htm. Accessed 07 June 2019.

———. 2016a. *Resolution on China's market economy status.* 2016/2667(RSP). http://www.europarl.europa.eu/doceo/document/TA-8-2016-0223_EN.html. Accessed 07 Sept 2019.

———. 2016b. *Verordnung (EU) 2016/1036 des europäischen Parlaments und des Rates – über den Schutz gegen gedumpte Einfuhren aus nicht zur Europäischen Union gehörenden Ländern*, 21–54. https://eur-lex.europa.eu/legal-content/DE/TXT/PDF/?uri=CELEX:32016R1036&from=de. Accessed 07 June 2019.

European Union, Ed. 2019. *China-EU – International trade in goods statistics – Statistics explained.* https://ec.europa.eu/eurostat/statistics-explained/index.php/China-EU_-_international_trade_in_goods_statistics#EU_and_China_in_world_trade_in_goods. Accessed 09 June 2019.

Ewerts, Insa. 2019. *China as dividing force in Europe. Young China watchers.* http://www.youngchinawatchers.com/china-as-a-dividing-force-in-europe/. Accessed 29 Aug 2019.

Farrow, Ronan. 2018. *Das Ende der Diplomatie: Warum der Wandel der amerikanischen Außenpolitik für die Welt so gefährlich ist.* Reinbek: Rowohlt Buchverlag.

Feng, Chuanlu. 2017. An analysis on the geopolitical pattern and regional situation in South Asia. In *Annual report on the development of the Indian Ocean region*, ed. Wang Rong and Cuiping Zhu, 175–209. Singapore: Springer.

Ferdinand, Peter. 2016. Westward ho – The China dream and 'one belt, one road': Chinese foreign policy under xi Jinping. *International Affairs* 92 (4): 941–957.

Frankopan, Peter. 2019. *Die neue Seidenstraße: Gegenwart und Zukunft unserer Welt*, 3. Aufl. Berlin: Rowohlt.

Freeman, Alan. 2003. *Globalisation: Economic stagnation and divergence.* https://mpra.ub.uni-muenchen.de/6745/1/MPRA_paper_6745.pdf. Accessed 01 Oct 2019.

Gätzner, Stefan. 2019. *China – Partner und Wettbewerber.* https://bdi.eu/themenfelder/internationalemaerkte/china/#/artikel/news/china-partner-und-wettbewerber/. Accessed 11 June 2019.

Giesen, Christoph, Alexander Hagelüken, and Alexander Mühlauer. 2018. Aus dem Handelsstreit wird ein Handelskrieg. In *Süddeutsche Zeitung*. https://www.sueddeutsche.de/wirtschaft/us-strafzoelle-gegen-china-aus-dem-handelsstreit-wird-ein-handelskrieg-1.4017328. Accessed 01 Oct 2019.

Godement, François. 2016. *China's market economy status and the European interest.* European Council on Foreign Relations. https://www.ecfr.eu/page/-/ECFR_180_-_CHINA_MARKET_ECONOMY_STATUS_AND_THE_EUROPEAN_INTEREST_(002).pdf. Accessed 07 June 2019.

Hanemann, Thilo, Mikko Huotari, and Aatha Kratz. 2019. *Chinese FDI in Europe: 2018 trends and impact of new screening policies.* http://www.europarl.europa.eu/RegData/etudes/STUD/2018/570493/EXPO_STU(2018)570493_DE.pdf. Accessed 10 June 2019.

Harding, Harry. 2013. American visions of the future of U. S.-China relations. In *Tangled titans. The United States and China*, ed. David Shambaugh, 389–409. Lanham: Rowman & Littlefield Publishers.

Hartmann, Wolf D., Wolfgang Maennig, and Run Wang. 2018. *Chinas neue Seidenstraße: Kooperation statt Isolation – der Rollentausch im Welthandel*. Frankfurt: Frankfurter Allgemeine Buch.

Hilpert, Hanns Günther. 2014. *Chinas Handelspolitik. Dominanz ohne Führungswillen*. https://www.swp-berlin.org/fileadmin/contents/products/studien/2013_S22_hlp.pdf. Accessed 05 June 2019.

Hoering, Uwe. 2018. *Der lange Marsch 2.0: Chinas neue Seidenstraßen als Entwicklungsmodell*. Hamburg: VSA.

Hornby, Lucy. 2018. Mahatir Mohamad Warns against 'New Colonialism' During China Visit. *Financial Times*. https://www.ft.com/content/7566599e-a443-11e8-8ecf-a7ae1beff35b. Accessed 01 Oct 2019.

Hoshmand, A. Reza. 2019. Eurasian connection via the silk road: The spread of Islam. In *Silk road to belt road*, ed. Narzul Islam, 95–104. Singapore: Springer.

Hua, Sha. 2019. Huawei-Finanzchefin Meng Wanzhou – eine Kronprinzessin in Haft. *Handelsblatt*. https://www.handelsblatt.com/unternehmen/it-medien/tochter-des-konzernchefs-huawei-finanzchefin-meng-wanzhou-eine-kronprinzessin-in-haft/23733454.html?ticket=ST-20021456-Xt5EsUp4l3P2k93gefni-ap1. Accessed on 01.10.2019.

Huang, Yukon. 2017. *Cracking the China conundrum. Why conventional economic wisdom is wrong*. Oxford: Oxford University Press.

Hulverscheidt, Claus. 2018. US-Handelsdefizit mit China ist viel kleiner als gedacht. *Süddeutsche Zeitung* vom 01.08.2018. https://www.sueddeutsche.de/wirtschaft/usa-china-handelsdefizit-1.4076488. Accessed 30 Sept 2019.

Islam, M.N. 2019. *Silk road to belt road*. Singapore: Springer.

Jetin, Bruno. 2018. 'One belt-one road initiative' and ASEAN connectivity: Synergy issues and potentialities. In *China's global rebalancing and the new silk road*, ed. Bali Deepak, 139–150. Singapore: Springer.

Kher, Priyanka, and Trang Tran 2019. *Investment protection along the Belt and Road*. https://openknowledge.worldbank.org/bitstream/handle/10986/31247/134017-WP-MTI-Discussion-Paper-12-Final.pdf?sequence=1. Accessed 01 Oct 2019.

Kirchberger, Sarah. 2015. *Assessing China's naval power. Technological innovation, economic constraints, and strategic implications*. Berlin: Springer.

Kühl, Eike. 2018. Wer hat Angst vor Huawei? *Zeit online*. https://www.zeit.de/digital/mobil/2018-02/smartphones-china-huawei-zte-mate-10-spionage-risiken. Accessed 01 Oct 2019.

Laaff, Meike. 2019. Für dieses Smartphone ist kein Android-Update verfügbar. *Zeit online*. https://www.zeit.de/digital/mobil/2019-05/google-huawei-android-lizenz-sperre-zusammenarbeit-smartphones-folgen. Accessed 01 Oct 2019.

Lall, Somik, and Mathilde Lebrand. 2019. *Who wins, who loses? Understanding the spatially differentiated effects of the Belt and Road Initiative*. https://openknowledge.worldbank.org/bitstream/handle/10986/31535/WPS8806.pdf?sequence=4. Accessed 01 Oct 2019.

Liu, Xinru. 2010. *The silk road in world history*. New York: Oxford University Press.

Liu, Peng. 2017. The security structure in South Asia and its impacts on belt and road initiative. In *Annual report on the development of the Indian Ocean region*, ed. Wang Rong and Cuipng Zhu, 139–174. Singapore: Springer.

Maçães, Bruno. 2018. *Belt and road: A Chinese world order*. New York: Oxford University Press.

Malik, Ahmad Rashid. 2018. The China-Pakistan economic corridor (CPEC): A game changer for Pakistan's economy. In *China's global rebalancing and the new silk road*, ed. Bali Deepak, 69–83. Singapore: Springer.

Martinek, Madeleine. 2017. Zum Marktwirtschaftsstatus Chinas. Eine summarische Bestandsaufnahme des Diskussionsstands. *Zeitschrift für Chinesisches Recht* 24(3): 203–207. https://www.zchinr.org/index.php/zchinr/article/view/1844/1866. Accessed 07 June 2019.

Mathews, John A. 2019. *China's long term trade and currency goals: The Belt & Road Initiative.* https://apjjf.org/-John-A%2D%2DMathews/5233/article.pdf. Accessed 01 Oct 2019.

McBride, James. 2015. Building the new silk road. *Council on Foreign Relations* 22: 1–2.

Morrison, Wayne M. 2019. *Chinas status as a Nonmarket Economy (NME).* IF10385. https://fas.org/sgp/crs/row/IF10385.pdf. Zuletzt aktualisiert am 10.01.2019. Accessed 02 June 2019.

Nass, Matthias. 2019. Man sieht sich in Osaka. *Zeit online.* https://www.zeit.de/2019/27/g20-gipfel-osaka-donald-trump-xi-jinping. Accessed 01 Oct 2019.

National Development and Reform Commission, Ministry of Foreign Affairs, and Ministry of Commerce of the People's Republic of China. 2015. *Vision and actions on jointly building Silk Road Economic Belt and 21st-century Maritime Silk Road.* http://en.ndrc.gov.cn/newsrelease/201503/t20150330_669367.html. Accessed 01 Oct 2019.

Ngeow, Chow Bing. 2019. Religion in China's public diplomacy towards the belt and road countries in Asia. In *Silk road to belt road*, ed. Nazrul Islam, 75–93. Singapore: Springer.

Noesselt, Nele. 2018. *Chinesische Politik. Nationale und globale Dimensionen*, 2. Aufl. Baden-Baden: Nomos (Studienkurs Politikwissenschaft).

OECD. 2018. *The Belt and Road Initiative in the global trade, investment and finance landscape.* OECD Publishing. https://www.oecd.org/finance/Chinas-Belt-and-Road-Initiative-in-the-global-trade-investment-and-finance-landscape.pdf. Accessed 01 Oct 2019.

———. 2019. *FDI Flows.* https://data.oecd.org/fdi/fdi-flows.htm. Accessed 01 Oct 2019.

Palit, Amitendu. 2018. The MSRI, China, and India: Economic perspectives and political impressions. In *China's maritime silk road initiative and South Asia*, ed. Jean-Marc Blanchard, 203–228. Singapore: Palgrave.

Ramb, Bernd-Thomas. 2018. *Definition: Prozesspolitik.* https://wirtschaftslexikon.gabler.de/definition/prozesspolitik-43995/version-267316. Accessed 02 June 2019.

Reuters. 2019. Handelsdefizit der USA trotz Importzöllen gewachsen. In *Zeitonline.* https://www.zeit.de/wirtschaft/2019-09/handelsstreit-usa-china-us-wirtschaft-handelsdefizit. Accessed 30 Sept 2019.

Röder, Jürgen. 2018. Handelsstreit belastet die Börsen – "Potenzial für eine weltweite Rezession". *Handelsblatt.* https://www.handelsblatt.com/finanzen/maerkte/marktberichte/us-zoelle-gegen-china-handelsstreit-belastet-die-boersen-potenzial-fuer-eine-weltweite-rezession/22708962.html. Accessed 01 Oct 2019.

Röhr, Wolfgang. 2018. Berlin looking eastward: German views of and expectations from the new silk road. In *Rethinking the silk road*, ed. Maximilian Mayer, 227–246. Singapore: Palgrave Macmillan.

Saarela, Anna. 2018. *Eine neue Ära in den Beziehungen zwischen der EU und China: umfassendere strategische Zusammenarbeit?* Generaldirektion externe Politikbereiche der Union. http://www.europarl.europa.eu/RegData/etudes/STUD/2018/570493/EXPO_STU(2018)570493_DE.pdf. Accessed 09 June 2019.

Sandkamp, Alexander, and Erdal Yalcin. 2016. Chinas Marktwirtschaftsstatus und die Anti-Dumping-Gesetzgebung der EU. *ifo Schnelldienst* 69: 50–59. https://www.ifo.de/DocDL/sd-2016-04-sandkamp-yalcin-china-antidumping-2016-02-25.pdf. Accessed 08 June 2019.

Sauerland, Dirk. 2018. *Definition: Marktwirtschaft.* https://wirtschaftslexikon.gabler.de/definition/marktwirtschaft-38124/version-261550. Accessed 02 June 2019.

Schmidt, Dirk, and Sebastian Heilmann. 2012. *Außenpolitik und Außenwirtschaft der Volksrepublik China.* Wiesbaden: Springer VS.

Schüller, Margot, and Tam Nguyen. 2015. Vision einer maritimen Seidenstraße: China und Südostasien. *GIGA Focus Global* (07). https://www.giga-hamburg.de/de/system/files/publications/gf_global_1507_0.pdf. Accessed 01 Oct 2019.

Schwenke, Marcus. 2016. *Marktwirtschaftsstatus (MWS) für China.* Bundesverband Großhandel, Außenhandel, Dienstleistungen e.V. https://www.bga.de/fileadmin/user_upload/Publikationen/Geschaeftsbericht/Positionspapiere_Aussenwirtschaft/PosPap_MWS_China_27052016.pdf. Accessed 05 June 2019.

Shi, Mingde. 2018. Wirtschaft und Handel. Interview mit Shi Mingde. *Neue Osnabrücker Zeitung* (05.11.2018). http://de.china-embassy.org/det/sgyw/t1610421.htm. Accessed 01 Oct 2019.

State Council of the PRC. 2015, May 26. *The White Paper of China's Military Strategy, the Information Office of the State Council.* https://jamestown.org/wp-content/uploads/2016/07/China%E2%80%99s-Military-Strategy-2015.pdf. Accessed 01 Oct 2019.

Stecklow, Steve, Karen Freifeld, and Sijia Jiang. 2018. U.S. ban on sales to China's ZTE opens fresh front as tensions escalate. *Reuters.* https://www.reuters.com/article/us-china-zte-idUSKBN1HN1P1. Accessed 30 Sept 2019.

Stiftung Asienhaus. 2017. *Silk road bottom-up: Regional perspectives on the 'Belt and Road Initiative'.* https://www.eu-china.net/uploads/tx_news/Broschuere_Silk_Road_Bottom-Up_2017_02.pdf. Accessed 10 Oct 2019.

Tong, Sarah Y., and Tuan Yuen Kong. 2018. Singapore's role in the belt and road initiative. In *Securing the belt and road initiative*, ed. Alessandro Arduino and Gong Xue, 63–80. Singapore: Palgrave.

Tu, Changfeng. 2019. Chinese outbound investments. In *Chinese FDI in the EU and the US*, ed. Tim Wenniges and Walter Lohman, 1–13. Singapore: Palgrave Macmillan.

U. S. Department of Commerce. 2018. *Commerce Department Lifts Ban After ZTE Deposits Final Tranche of $1.4 Billion Penalty.* https://www.commerce.gov/news/press-releases/2018/07/commerce-department-lifts-ban-after-zte-deposits-final-tranche-14. Accessed 30 Sept 2019.

United States Trade Representative (USTR). 2018. *2017 Report to Congress on China's WTO Compliance.* https://ustr.gov/sites/default/files/files/Press/Reports/China%202017%20WTO%20Report.pdf. Accessed 04 June 2019.

Van, Dinh Trinh. 2019. The rise of China's past in the "belt and road initiative" (from historical perspectives). In *Silk road to belt road*, ed. Nazrul Islam, 25–38. Singapore: Springer.

von Unger, Eckart, Ed. 2019. China in der Welthandelsorganisation. *BDI.* https://bdi.eu/artikel/news/china-in-der-wto/. Accessed 23 May 2019.

Wang, Yiwei. 2018. Dealing with the risks of the belt and road initiative. In *China's global rebalancing and the new silk road*, ed. Bali Deepak, 207–225. Singapore: Springer.

Wei, Hongchen. 2016. Botschafter Shi Mingde im Handelsblatt-Interview. *Beijing Rundschau.* http://www.bjrundschau.com/International/201605/t20160526_800057759.html. Accessed 07 June 2019.

World Trade Organization. 2001a. *Protocol on the Accession of the People's Republic of China.* WT/L/432. https://www.wto.org/english/thewto_e/acc_e/a1_chine_e.htm. Accessed 24 May 2019.

———. 2001b. *WTO Ministerial Conference approves China's accession.* Press 252. https://www.wto.org/english/news_e/pres01_e/pr252_e.htm. Accessed 24 May 2019.

———. 2018. *Trade Policy Review Reported by the Secretariat China.* WT/TPR/S/375. Genf. https://www.wto.org/english/tratop_e/tpr_e/s375_e.pdf. Accessed 01 Oct 2019.

Wu, Xinbo. 2013. Chinese visions of the future of U. S.-China relations. In *Tangled titans. The United States and China*, ed. David Shambaugh, 371–381. Lanham: Rowman & Littlefield Publishers.

Wu, Yang, and Yichun Duan. 2018. "Made in China": Building Chinese smart manufacturing image. *Journal of Service Science and Management* 11 (6): 590–608. https://doi.org/10.4236/jssm.2018.116040. Accessed 01 Oct 2019.

Xinhua. 2017. *Full text of President Xi's speech at opening of Belt and Road forum.* Xinhua. http://www.xinhuanet.com/english/2017-05/14/c_136282982.htm. Accessed 01 Oct 2019.

Yan, Shaohua. 2015. *The EU-China partnership and cooperation agreement negotiation deadlock.* https://www.e-ir.info/2015/04/23/the-eu-china-partnership-and-cooperation-agreement-negotiation-deadlock/. Accessed 09 June 2019.

Yang, Yi. 2018. *China's Policy Paper on the European Union.* http://www.xinhuanet.com/english/2018-12/18/c_137681829.htm. Accessed 10 June 2019.

Yang, Xifan. 2019. Die Angst der Händler. Der Konflikt mit den USA kommt im chinesischen Alltag an – und setzt die Regierung unter Druck. *Zeit online*. https://www.zeit.de/2019/25/china-mittelschicht-handelsstreit-usa-donald-trump-xi-jinping. Accessed 01 Oct 2019.

Yuan, Yuan 2015. *Looking back 14 years after accession: Case of China. Intervention at session 2 of day 1: Transition from accession to membership – Maximizing the benefits of WTO membership and global economic*. MOFCOM. China Round Table on WTO Accessions. Dushanbe, 2015. https://www.wto.org/english/thewto_e/acc_e/Session2YuanYuanPostAccessionLookingback14yearafter.pdf. Accessed 26 May 2019.

Zeit. 2019. *China wertet eigene Währung ab.* https://www.zeit.de/wirtschaft/2019-08/handelsstreit-china-yuan-waehrung-abwerten. Accessed 01 Oct 2019.

Zhang, Jiadong. 2017. The current situation of the one belt and one road initiative and its development trend. In *Annual report on the development of the Indian Ocean region*, ed. Wang Rong and Cuiping Zhu, 103–137. Singapore: Springer.

Zhou, Haoming. 2018. China, securing "belt and road initiative": Risk management. In *Securing the belt and road initiative*, ed. Alessandro Arduino and Gong Xue, 163–177. Singapore: Palgrave.

Ziedler, Christopher. 2019, June 10. Interview mit Chinas Botschafter in Berlin: "Der Westen kann nicht mehr alles alleine bestimmen". *Interview mit Shi Mingde*. https://www.stuttgarter--nachrichten.de/inhalt.interview-mit-chinas-botschafter-in-berlin-der-westen-kann-nicht-mehr-alles-alleine-bestimmen.1c129fa6-d271-4bcf-93cf-f1858f2ac906.html. Accessed 01 Oct 2019.

Concluding Remarks

16

Barbara Darimont

There would be enough money, enough work, enough to eat if we would distribute the riches of the world properly instead of making ourselves slaves to rigid economic doctrines or traditions. (Einstein and Freud 1932/2005)

The discourse on the Chinese economic model is controversial both at home and abroad. Doubts exist as to whether it is possible to practice a market economy under an authoritarian state. Moreover, politicians and scholars from democratic states assume that democracy and the rule of law are indispensable for a flourishing market economy. This assumption is currently proving to be a fallacy. At the very least, the Chinese authoritarian economic system looks successful to some extent even to Europeans (Lu 2019). A change in Chinese conditions is not in sight; however, the question arises as to whether Western states are converging with Chinese values.

This discussion continues in the promotion of private enterprises or state-owned enterprises. The Chinese government meanders between promotion, which is associated with the granting of freedoms, and absolute control of the private sector. Whether these indifferent policy prescriptions lead to a thriving economy remains to be seen. To some extent, the CCP supervises private enterprises (Lu 2019). It can be assumed that the restrictive control does not promote the innovation capacity of the companies. In the start-up scene, it is evident that the government provides massive financial support to specific sectors of the economy; whether the principle of financial watering can bear fruit is doubtful. On the other hand, the PRC has a virulent start-up scene, the hurdles are low, and more is ventured

B. Darimont (✉)
East Asia Institute of Ludwigshafen University of Business and Society,
Ludwigshafen am Rhein, Germany
e-mail: darimont@oai.de

B. Darimont (ed.), *Economic Policy of the People's Republic of China*,
https://doi.org/10.1007/978-3-658-38467-8_16

than in Germany, for example. It is possible that the business-minded Chinese mentality, which is successful completely independently of any system, can be seen here.

It is part of Chinese culture to show off the wealth you have acquired. For foreign observers, this may seem like the airs and graces of the nouveau riche. But showing one's own identity among the masses of 1.4 billion people is satisfying for many Chinese consumers. Those who have achieved something are allowed to show it in the form of wealth. Getting rich is the goal of many Chinese (Lu 2019). A very questionable attitude for a communist country, but political ideologies have long been out of the question. As Deng Xiaoping replied to former German Chancellor Helmut Schmidt's question that he was no longer a communist after all: "So what?" (Schmidt 2012).

Consequently, the Chinese population is catching up with the rest of the world in terms of consumption. Luxury consumption may play a much greater role in the PRC than in Western nations because gifts are used to express respect for others. Therefore, an examination of Marcel Mauss' "gift economy" (1925) would be an asset for further academic research. Furthermore, the Chinese market produces its brand and fashions, and the global luxury consumption market is becoming Chinese. Contrary to the general view that the PRC almost only copies, innovative new trends are emerging in China, not only in luxury consumption but for example in the case of this particular area.

Until this decade, the Chinese labour market was characterized by a surplus of workers, which justified the strategy of being the world's cheap workbench. Since the cohorts of the one-child policy – i.e. significantly fewer workers – have been pouring into the labor market, the situation has changed. The PRC needs to educate the future generation very well to increase per capita income to finance the upkeep and care of the old generation. The importance of demographic change will be seen in the PRC labor market. The idea of the Chinese leadership is to flank this change with artificial intelligence and digitalization. The real pressure of having to solve the problems alone will give rise to new ideas of old-age provision and care in this area.

The Chinese e-commerce sector has developed rapidly in recent years. This has become possible because the e-commerce companies have not only limited themselves to the big cities but have also included the rural population in their trade. This included a development that was pushed in the areas of infrastructure, logistics, online payment systems, and digital networking. The big internet giants Baidu, Alibaba, and Tencent have shown in China what the networking of the future can look like and have created ecosystems that cover almost all consumer wishes. One thesis that has not yet been confirmed is that Chinese companies are particularly flexible networkers because they are used to putting personal animosities aside under an authoritarian system. This would be in contrast to European companies, which hardly ever cooperate with their competitors but are in constant competition.

From the Chinese perspective, consumer-to-business (C2B) seems to be the future, as the market from consumer to entrepreneur, that is, the consumer gives his wishes to the supplier, who implements them. This means that the market could become more individualistic. The Chinese population as the "shopping world champion" could become the leader in this area.

Financial policy in the PRC is subject to strict state restrictions. Although leeway is repeatedly allowed, such as in the case of shadow banks or crowdfunding, this leeway is immediately closed by the state through legal regulations if the risks are considered too high. As a result, there are hardly any investment opportunities besides the real estate market. Capital flight is strictly discouraged by the Chinese government. However, the FinTech market in China is flourishing and generating promising developments.

A difficult issue to assess is over-indebtedness in the PRC, which has increased more than in almost any other country in the last 10 years. By most estimates, it now stands at 300 percent of GDP (Lu 2019). The level of debt is less serious than the speed at which it has been built up. If the pace is maintained, a crisis is bound to occur. The risk of a financial crisis seems high at this point, as the Chinese government is trying to stabilize the economy with infrastructure projects, but this continues to drive up debt – a debt spiral. Moreover, although this method has worked in the past, generating economic growth in this way is not possible in the long run, as its effectiveness is exhausted (Joffe 2019). Many investments are no longer profitable because enough roads, airports, etc. have been built before. Currently, the Chinese government is trying to counteract this by encouraging local governments to only finance projects where the loans can be repaid (Liu 2019), but this development needs to be followed up to make concrete statements.

While in the 1970s it was still considered that the PRC might suffer from a food shortage due to the high population figures, the agricultural sector has grown in recent years so that the PRC has now become one of the largest exporters of pork and aquatic products in the world. However, with this comes the need for the PRC to import feed – especially soybeans and corn. At the very moment, African fever is posing immense challenges to Chinese pig farming. If the Chinese population were to eat less pork and more chicken, the problem of a possible food shortage would be obsolete. African fever may be a catalyst for this development, as pork prices in the PRC are rising at an above-average rate (Ng 2019).

The environmental problems of the PRC are well known: Air and water pollution and contaminated soil. In this area, the political leadership shows that it is addressing the problem. Many laws and policy directives have been issued. Implementation has improved in this area in recent years. In water supply, China seems to have abandoned the planned mega-projects, such as the South-North Water Transfer Project. The future will be in desalination plants. Nevertheless, water supply remains a problematic issue – especially for agriculture. Enormous progress has been made in the area of waste management, not least with the import ban on plastic from other countries. Innovations, such as cockroach farms for household waste, will come from China in the future, as the People's Republic needs to reduce its rising mountains of waste. Recycling is also being implemented in China, like these days in Shanghai, where the strictest recycling rules in the world were introduced in the summer of 2019 (Deuber 2019). The Germans may soon be replaced by the Chinese as "recycling world champions".

In its energy policy, the PRC relies on nuclear power. With the increasing energy demand, this is a solution to deal with the obvious pollution. Five different nuclear reactors are in use, which is unusual compared to, for example, France with one type of reactor. A problem could be that this also requires the training of considerable skilled personnel for the different reactor types, which already seems to be a problem. In the area of renewable energy, the PRC is implementing everything that is currently possible and innovative. This ranges from hydropower and wind power to solar technology. In the area of solar technology, the PRC has catapulted the German solar industry out of the market by dumping prices. Now China has overcapacities that it needs to reduce. This has been done to a large extent in its own country in recent years – to the benefit of its own and thus the global environment, as this means that the steadily rising demand for energy can increasingly be met from non-coal sources.

Many projects in the PRC are Potemkin villages. Huge solar and wind power plants are being built, but the power grid is not being maintained. The social credit system is being introduced, but the necessary databases have yet to be built. For this reason, it is impossible to predict how development will continue. However, in the context of the platform economy, digitalization, and artificial intelligence, the PRC is likely to be ahead of the rest of the world.

The PRC has become the world's largest exporting nation in the last decade, consequently, foreign trade is one of the most important economic areas. In 2001, the PRC joined the WTO. To date, the European Union and the USA do not recognize the PRC as a market economy, which means that in the case of suspected dumping, the price of analogue countries can be used. For the PRC, this is not least a loss of prestige. The European Union is trying to resolve the dispute with negotiations, the US has gone on a confrontational course by imposing tariffs since January 2018 and the trade conflict has escalated. The outcome is unclear.

With the New Silk Road project, the PRC is trying to build new value chains in Central Asia and throughout the world. This is intended as a strategy to break away from economic dependence on the USA and to gain new markets. The Central Asian regions in particular, but also Africa, have been neglected by the industrialized nations in recent decades, so the PRC at least offers economic opportunities here. To what extent the Silk Road also serves military purposes remains to be seen here.

The Silk Road Initiative shows that the Chinese government knows how to use Western concepts to its advantage. Many politicians in the West had hoped that trade would transform China into a democratic construct based on the rule of law. So far, this notion has not been borne out. Conversely, the new Silk Road seems to be working according to the principle of "change through trade" in Chinese. It remains questionable, however, in which direction these countries should transform themselves. To a kind of economic backyard of the PRC? Or are they equal partners?

So how does the global society – but also Germany – deal with these new challenges? In any case, a blockade would be fatal and not conducive to achieving our goals. Communicating and negotiating with each other is the only viable path. On the one hand, the Western community of values must grant the newcomer the place he deserves, on the other hand, PR China must accept that there are other value systems and respect them. Especially since it is unclear what values the PRC stands for. If it is free trade, it has to allow it in its markets. In part, it seems as if the PRC wants to buy the rest of the world, but then it must expect resistance – especially in the European world. The question of how venal the West is could be the real challenge for the Western community of values. How important is the car industry, which is so successful in the PRC, to the Germans? How important is German society's prosperity? Where would the "red" line be crossed? These limits will be tested by the Chinese rulers.

Possibly it is a worldwide distribution struggle? And the real question is: who will be poor and who will be rich in the future? This distribution struggle seems to have started within the nations since many political measures give the impression that it is primarily about preserving and increasing the sinecures of the respective rulers. Interest groups that use lobbying to push through their ideas are not unknown either in the PRC or in the rest of the world. It is questionable whether they are more assertive in a democratic or authoritarian system. In the PRC, many industries are run by political clans, and they have a hand in the creation of new achievements, whether in the military or the newly formed Shanghai Stock Exchange. At the same time, many are getting the impression that Trump is also led by power elites. It almost seems that both countries are ruled by a leader whose actual holders of power are more attached to the interest of maintaining power than to the common good of society. And this is completely independent of the political or economic system in place. In American society, the distributive struggle has arguably already begun (Schularik 2019). In Chinese society, it is suspected. So if the two largest economic powers in the world are struggling with internal distributional issues, it will spread to the rest of the world. It is therefore questionable whether a global distribution struggle is taking place.

China has a heterogeneous society whose differences have been intensified by the rapid economic development of recent decades. The PRC is a country of extremes, be it in education, economic growth or culture. The differences in income distribution, as indicated by the Gini coefficient, are particularly serious. Finding an order for this diversity and complexity is a challenge not only for China but for global society (Vogelsang 2012, p. 616).

For everyone involved, it is a matter of finding the right balance. This is well known in both Eastern and Western cultures: Confucius devoted an entire section in the Book of Rites to the measure and the middle (zhōngyōng). For Plato and Aristotle, it was prudence and moderation (σωφροσύνη). Making globalization sustainable for all is a challenge for the world economy.

References

Deuber, Lea. 2019, September 10. Untrennbar. In Shanghai wurden über Nacht die strengsten Recyclingregeln der Welt eingeführt. Was Chinas große Müllrevolution für die privaten Haushalte bedeutet. *Süddeutsche Zeitung*, S. 3.

Einstein, Albert, and Sigmund Freud. 1932/2005. *Warum Krieg? Ein Briefwechsel*, 21. Aufl. Zürich: Diogenes.

Joffe, Josef. 2019, June 25. Der China-Hype. Das Wachstum der Möchtegern-Weltmacht verfällt. *Die Zeit*, S. 9.

Liu, Xinning. 2019. Chinese local government funds run out of projects to back. *Financial Times*. https://www.ft.com/content/6aaa5bfe-efce-11e9-ad1e-4367d8281195?sharetype=blocked. Accessed 21 Oct 2019.

Lu, Franka. 2019. Das anpassungsfähigste autoritäre Regime der Welt. *Zeit online*. https://www.zeit.de/kultur/2019-09/70-jahre-volksrepublik-china-mythos-erfolg-autokratie-nationalismus. Accessed 05 Oct 2019.

Mauss, Marcel. 1925/2009. *Die Gabe. Form und Funktion des Austauschs in archaischen Gesellschaften*. Berlin: Suhrkamp (französisch: Essai sur le don. Forme et raison de l'échange dans les sociétés archaïques, 1925).

Ng, Eric. 2019. China's largest agricultural companies turn swine fever epidemic into growth opportunity while small farms perish. *South China Morning Post*. https://www.scmp.com/business/companies/article/3027165/chinas-largest-agricultural-companies-turn-swine-fever-epidemic. Accessed 05 Oct 2019.

Schmidt, Helmut. 2012. *Ich fürchtete um sein Leben*. Zeit online. https://www.zeit.de/zeitgeschichte/2012/01/Deng-und-Schmidt. Accessed 04 Oct 2019.

Schularik, Morizt. 2019, October 01. Chimerika im Handelskrieg. Trumps Politik zeigt die Verunsicherung eines Landes, das Zweifel am eigenen Erfolgsmodell hat. *Süddeutsche Zeitung*, S. 18.

Vogelsang, Kai. 2012. *Geschichte Chinas*. Ditzingen: Reclam.

Publisher Correction to: Economic Policy of the People's Republic of China

Publisher

Publisher Correction to:

B. Darimont (ed.), ***Economic Policy of the People's Republic of China,*** **https://doi.org/10.1007/978-3-658-38467-8**

Owing to an error on the part of the publisher, the chapter authors' names were missing in the initially published version of this translation. The chapter authors' names have now been added which has led to slight changes in the pagination, compared to the previous version. The content has not changed.

The updated version of this book can be found at
https://doi.org/10.1007/978-3-658-38467-8

B. Darimont (ed.), *Economic Policy of the People's Republic of China*,
https://doi.org/10.1007/978-3-658-38467-8_17

Printed in the United States
by Baker & Taylor Publisher Services